Lecture Notes in Computer Science 9553

Commenced Publication in 1973
Founding and Former Series Editors:
Gerhard Goos, Juris Hartmanis, and Jan van Leeuwen

Marco Nehmeier · Jürgen Wolff von Gudenberg
Warwick Tucker (Eds.)

Scientific Computing, Computer Arithmetic, and Validated Numerics

16th International Symposium, SCAN 2014
Würzburg, Germany, September 21–26, 2014
Revised Selected Papers

 Springer

Editors
Marco Nehmeier
Institute of Computer Science
University of Würzburg
Würzburg
Germany

Jürgen Wolff von Gudenberg
Universität Würzburg
Würzburg
Germany

Warwick Tucker
Department of Mathematics
Uppsala University
Uppsala
Sweden

ISSN 0302-9743   ISSN 1611-3349 (electronic)
Lecture Notes in Computer Science
ISBN 978-3-319-31768-7 ISBN 978-3-319-31769-4 (eBook)
DOI 10.1007/978-3-319-31769-4

Library of Congress Control Number: 2016934445

LNCS Sublibrary: SL1 – Theoretical Computer Science and General Issues

This Springer imprint is published by Springer Nature
The registered company is Springer International Publishing AG Switzerland

In Memory of Walter Krämer

This book is dedicated to Walter Krämer (1952 – 2014), the longtime chair of the SCAN Scientific Committee.

After having suffered from a long and heavy illness, Walter Krämer passed away Monday, October 27, shortly after this conference. His whole (scientific) life was dedicated to reliable computing. In the year 1979 he started his career in reliable floating point computations with a diploma (master) thesis on the evaluation of standard functions for the IBM/370 hexadecimal floating point architecture that became one of the cornerstones of the IBM ACRITH project. In 1987 he earned his PhD from the University of Karlsruhe. Six years later he advanced his topic further by finishing his habilitation on the calculation of functions and constants in computers. He worked on several research and management positions at the University of Karlsruhe until he became professor at the University of Wuppertal, holding the chair of scientific computing and software engineering. He provided a great service to the community by maintaining and further developing C-XSC, one of the most widely used interval arithmetic packages. He was also active in representing and promoting our field. He was the chair of the GAMM Activity Group on Computer Arithmetic and Scientific Computing. After the closing of this group he founded another one with similar content: Computer-Assisted Proofs and Symbolic Computations. He also was a well-accepted teacher at the University of Wuppertal and supervised more than 12 PhD theses. According to his mission to spread the idea of reliable computing, it is not surprising that several of his PhD students came from abroad, e.g., Brazil or Egypt.

All in all Walter Krämer was one of the most active promoters of our field. He was never afraid of difficult and time-consuming work when the idea of reliable computing was to be pushed. He traveled extensively and had contacts with researchers all around the world. Even in the advanced state of his disease, he tried to keep in touch with his research group.

The whole community and I in particular will always remember our colleague and friend Walter Krämer.

March 2016   Jürgen Wolff von Gudenberg

A bibliography of Walter Krämer's work can be found at the end of this book.
Picture: © Angelika Krämer-Maiss, consent for publication is granted.

Preface

SCAN2014, the 16th GAMM-IMACS International Symposium on Scientific Computing, Computer Arithmetic, and Validated Numerics, was held in Würzburg, Germany, September 21–26, 2014. This conference continued the series of international SCAN symposia initiated in the late 1980s by the University of Karlsruhe, Germany, and organized under the joint auspices of GAMM and IMACS. SCAN symposia have been held in Germany, Europe, the USA, Russia, and Japan. The next two SCAN conferences are scheduled to be held in Sweden and Japan.

The main concerns of research addressed by SCAN conferences are validation, verification, or reliable assertions of numerical computations. Interval arithmetic and other treatments of uncertainty are developed as appropriate tools. In computer science validation usually means finding the right model, whereas verification means implementing the model correctly. In our context, indeed, we use both validation and verification when solving real-life problems. It makes sense to consider the mathematical modeling technology in more detail and to understand what is coupled with this or that concept. We have:

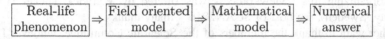

"Validation" applies to the first and second arrows in the scheme, while "verification" characterizes the third one. Numerical methods, algorithms, and computations are responsible for the third arrow, too; they are not present in the first and second stages of the modeling process. This is why in connection with numerical methods "verification" is more appropriate. Worrying about whether the thickness of the input intervals correctly captures uncertainty in the physical problem, however, is validation.[1]

One of the duties of the SCAN conferences is the awarding of the Moore Prize for the application of interval arithmetic. This year Prof. Kenta Kobayashi was selected for his research on the wave equation.

The symposium attracted more than 80 participants from 18 countries. The following seven invited lectures were given: Sylvie Boldo (Inria, France) gave a lecture about formal verification of tricky numerical computations. Ekaterina Auer (University of Duisburg-Essen, Germany) talked about result verification and uncertainty management in engineering applications. Bartlomiej Jacek Kubica (Warsaw University of Technology, Poland) presented interval methods for solving of quantified nonlinear problems. Interval arithmetic, one of the main subjects of the SCAN conferences, has been standardized as IEEE Std 1788. Many participants of the symposium were active in this working group. The technical editor John Price (Cardiff University, UK) introduced the architecture of this standard in a plenary talk. Andrej Bauer (University

---

[1] Discussion initiated by S. Shary.

of Ljubljana, Slovenia) reported on programming techniques for exact real arithmetic. John Gustafson (Ceranovo Inc., USA) proposed an energy-efficient and massively parallel approach to valid numerics, which may replace floating point arithmetic in the future. Algorithmic and software challenges at extreme scales were the topic of Jack Dongarra's (University of Tennessee and ORNL, USA) lecture.

Country	Participants	Country	Participants
Germany	18	UK	2
Japan	17	Austria	1
France	16	Brazil	1
USA	7	China	1
Russia	6	India	1
Czech Republic	3	The Netherlands	1
Bulgaria	2	Poland	1
Egypt	2	Slovenia	1
Hungary	2	Sweden	1

More than 60 contributed talks were given. A strict and careful reviewing process resulted in the 22 contributions that are collected in this volume. We take the opportunity to thank all the reviewers for their detailed comments presented in time. A great help in organizing this review process was offered by Warwick Tucker (Uppsala University, Sweden), the host of SCAN2016—many thanks.

I further want to thank the members of the Scientific Committee, whoever I asked gave me immediate feedback, as well as the organizers of the last three meetings, Vladik Kreinovich (El Paso), Nathalie Revol (Lyon), and Sergey Shary (Novosibirsk) in particular.

I express a special thanks to all our sponsors, their donations made the conference possible. Organizing such a conference involves a lot of work, but organizing this conference was a real pleasure for me, thanks to the tremendous assistance I got from the Organizing Committee comprising Alexander Dallmann, Fritz Kleemann, Marco Nehmeier, Anika Schwind, and Susanne Stenglin.

December 2015   Jürgen Wolff von Gudenberg

Organization

Program Committee

Jürgen Wolff von Gudenberg	Würzburg, Germany [Chair]
Vladik Kreinovich	El Paso, USA
Marco Nehmeier	Würzburg, Germany
Nathalie Revol	Lyon, France
Sergey P. Shary	Novosibirsk, Russia
Warwick Tucker	Uppsala, Sweden

Organizing Committee

Jürgen Wolff von Gudenberg	Würzburg, Germany [Chair]
Alexander Dallmann	Würzburg, Germany
Fritz Kleemann	Würzburg, Germany
Marco Nehmeier	Würzburg, Germany
Anika Schwind	Würzburg, Germany
Susanne Stenglin	Würzburg, Germany

Proceedings Coordinators

Marco Nehmeier	Würzburg, Germany
Anika Schwind	Würzburg, Germany

Scientific Committee

G. Alefeld	Karlsruhe, Germany
J.-M. Chesneaux	Paris, France
G.F. Corliss	Milwaukee, USA
T. Csendes	Szeged, Hungary
A. Frommer	Wuppertal, Germany
R.B. Kearfott	Lafayette, USA
W. Kraemer	Wuppertal, Germany
V. Kreinovich	El Paso, USA
U. Kulisch	Karlsruhe, Germany
W. Luther	Duisburg, Germany
G. Mayer	Rostock, Germany
S. Markov	Sofia, Bulgaria
J.-M. Muller	Lyon, France
M. Nakao	Fukuoka, Japan
M. Plum	Karlsruhe, Germany

Contents

Global Optimisation

Dynamical Systems

PDE

Interval Arithmetic and Interval Functions

Hausdorff Continuous Interval Functions and Approximations

Roumen Anguelov[1,2] and Svetoslav Markov[2(✉)]

[1] Deptartment Mathematics and Applied Mathematics,
University of Pretoria, Pretoria, South Africa
roumen.anguelov@up.ac.za
[2] Institute of Mathematics and Informatics,
Bulgarian Academy of Sciences, Sofia, Bulgaria
smarkov@bio.bas.bg

Abstract. The set of interval Hausdorff continuous functions constitutes
the largest space preserving basic algebraic and topological structural
properties of continuous functions, such as linearity, ring structure,
Dedekind order completeness, etc. Spaces of interval functions have important applications not only in the construction of numerical methods and
algorithms, but to problems in abstract areas such as real analysis, set-
valued analysis, approximation theory and the analysis of PDEs. In this
work, we summarize some basic results about the family of interval Hausdorff continuous functions that make interval analysis a bridge between
numerical and real analysis. We focus on some approximation issues formulating a new result on the Hausdorff approximation of Hausdorff continuous functions by interval step functions. The Hausdorff approximation
of the Heaviside interval step function by sigmoid functions arising from
biological applications is also considered, and an estimate for the Hausdorff
distance is obtained.

Keywords: Interval functions · Baire semi-continuous functions · Hausdorff continuous functions · Dilworth continuous functions · Sigmoid
functions

1   Introduction

Functions having discontinuities are encountered in many situations. A widely
used class of discontinuous functions is the class of Baire upper (lower) semi-
continuous functions [13]. Dilworth restricted the class of Baire semi-continuous
functions up to normal semi-continuous functions [15]. Both classes are conveniently reformulated in terms of interval-valued functions using graph completion operators [3]. For example, when a graph is completed, the Dilworth
normal semi-continuous functions are the Hausdorff continuous interval-valued
functions. Such a reformulation leads to interesting original results oriented to
practical applications.

© Springer International Publishing Switzerland 2016
M. Nehmeier et al. (Eds.): SCAN 2014, LNCS 9553, pp. 3–13, 2016.
DOI: 10.1007/978-3-319-31769-4_1

It has been shown that the space of Hausdorff continuous functions is the largest linear space of interval functions [3–5,22]. This space has important applications in the theory of PDEs and real analysis [1,6–9,12]. Moreover, the space of Hausdorff continuous functions has a special place in *Interval Analysis* as well, more specifically in the analysis of interval-valued functions [2]. It has been also shown that the practically relevant set, in terms of providing tight enclosures of sets of real continuous functions, is the set of Dilworth continuous interval-valued functions [5].

The relation between Baire semi-continuous functions and interval-valued functions establishes a new paradigm for *interval analysis* as part of *analysis* rather than (or in addition to) the field of *numerical methods*, where it is currently classified, see e.g. [23]. This provides a new research direction of applying interval analysis to abstract mathematical problems. It has been shown that the spaces of interval-valued functions[1] have important applications not only in the construction of numerical methods but to problems in more abstract areas like real analysis, set-valued analysis, approximation theory and analysis of PDEs. Some of the more interesting results are: (i) A generalization of the order convergence structure on the space of Hausdorff continuous functions to a convergence structure on the space of minimal upper semi-continuous compact set-valued (shortly: usco) maps [10]; (ii) All rational extensions and their metric completions of $C(X)$ are subspaces of the space of Hausdorff continuous functions [11]; (iii) The solutions of large classes of nonlinear systems of PDEs can be presented with Hausdorff continuous interval functions [6,7]; (iv) The theory of continuous viscosity solutions of Hamilton-Jacobi equations can be recast in the setting of Hausdorff continuous functions, where the discontinuous solutions are accommodated in a natural way [8].

In the next Sect. 2 we summarize some basic results concerning the class of interval Hausdorff continuous functions and the related classes of interval functions. Section 3 contains a new result on the Hausdoff approximation of Hausdorff continuous functions by interval step functions. Section 4 is devoted to a new result on the approximation of interval step functions by a class of sigmoid functions arising from biological applications.

2 Classes of Interval Functions: Basic Results

The concept of Hausdorff continuity (H-continuity) generalizes the familiar concept of continuity in such a way that many essential properties of the usual continuous real functions are preserved. The set $C(\Omega)$ of all continuous real functions defined on a subset $\Omega \subset \mathbb{R}^n$ is a commutative ring with respect to the point-wise defined addition and multiplication of functions and a linear space with respect to addition and multiplication by a scalar. Is it possible to extend the algebraic operations on $C(\Omega)$ to the set $\mathbb{H}(\Omega)$ of H-continuous functions in a way that preserves these two basic algebraic structures, that is, the set $\mathbb{H}(\Omega)$

---

[1] For brevity we shall further write "interval function" instead of "interval-valued function".

to become a commutative ring and linear space with respect to the extended operations? It turns out that the answer is affirmative as briefly shown in the sequel.

2.1   Basic Notation and Definitions: Baire Continuous Functions

Definition 1. *[13] A real-valued function f is* upper (lower) semi-*continuous at a point x_0 if the function values for arguments near x_0 are either* close *to $f(x_0)$ or less (greater) than $f(x_0)$.*

Intervals on the real line $\mathbb{R}$ are denoted as $a = [\underline{a}, \overline{a}] = \{x : \underline{a} \le x \le \overline{a}\}$, and the set of all intervals is denoted $\mathbb{IR} = \{[\underline{a}, \overline{a}] : \underline{a}, \overline{a} \in \mathbb{R}, \ \underline{a} \le \overline{a}\}$; denote also $w(a) = \overline{a} - \underline{a}$, and $|a| = \max\{|\underline{a}|, |\overline{a}|\}$.

Let $\Omega \subseteq \mathbb{R}^n$ be an open set. A real or interval function f on Ω is locally bounded if for every $x \in \Omega$ there exist $\delta > 0$ and $M \in \mathbb{R}$ such that $|f(y)| < M$, $y \in B_\delta(x)$, $B_\delta(x) = \{y \in \Omega : ||x - y|| < \delta\}$. Denote

$$\mathbb{A}(\Omega) = \{f : \Omega \to \mathbb{IR}, \ f \text{ locally bounded}\},$$
$$\mathcal{A}(\Omega) = \{f : \Omega \to \mathbb{R}, \ f \text{ locally bounded}\} \subseteq \mathbb{A}(\Omega).$$

Definition 2. *D is a dense subset of Ω. The lower/upper Baire operators $I(D, \Omega, \cdot)$, $S(D, \Omega, \cdot) : \ \mathbb{A}(D) \to \mathcal{A}(\Omega)$ are defined for $f = [\underline{f}, \ \overline{f}] \in \mathbb{A}(D)$ and $x \in \Omega$ by*

$$I(D, \Omega, f)(x) = \sup_{\delta > 0} \inf \ \{\underline{f}(y) : \ y \in B_\delta(x) \cap D\},$$

$$S(D, \Omega, f)(x) = \inf_{\delta > 0} \sup \ \{\overline{f}(y) : \ y \in B_\delta(x) \cap D\}.$$

Definition 3. *The graph completion operator $F : \mathbb{A}(D) \to \mathbb{A}(\Omega)$ for $f \in \mathbb{A}(D)$ is defined as*

$$F(D, \Omega, f)(x) = [I(D, \Omega, f)(x), \ S(D, \Omega, f)(x)], \quad x \in \Omega, \ \ f \in \mathbb{A}(D).$$

For $D = \Omega$ we write

$$I(f) = I(\Omega, \Omega, f), \quad S(f) = S(\Omega, \Omega, f), \quad F(f) = F(\Omega, \Omega, f).$$

Using end-point presentation of functions: $f = [\underline{f}, \ \overline{f}] \in \mathbb{A}(\Omega)$ we can write

$$I(D, \Omega, f) = I(D, \Omega, \underline{f}), \quad S(D, \Omega, f) = S(D, \Omega, \overline{f}),$$

$$F(D, \Omega, f) = [I(D, \Omega, \underline{f}), \ S(D, \Omega, \overline{f})].$$

Definition 4. *A function $f \in \mathbb{A}(\Omega)$ is* S-continuous, *if $F(f) = f$.*

Definition 5. *A function $f \in \mathbb{A}(\Omega)$ is* D-continuous, *if for every dense subset D of Ω, $F(D, \ \Omega, \ f) = f$.*

Definition 6. *A function* $f \in \mathbb{A}(\Omega)$ *is H-continuous, if for every function* $g \in \mathbb{A}(\Omega)$ *such that* $g(x) \subseteq f(x)$, $x \in \Omega$, $F(g)(x) = f(x)$, $x \in \Omega$.

Theorem 1. *For every* $f \in \mathbb{H}(\Omega)$ *the set* $W_f = \{x \in \Omega : w(f(x)) > 0\}$ *is of first Baire category (that is, H-continuous functions are "thin").*

H-continuous functions do not differ much from the usual real-valued continuous functions because they assume interval values only on a meagre set[2].

2.2   Arithmetic Operations in $\mathbb{H}(\mathbb{R})$

Interval arithmetic operations are denoted as usually: for $a = [\underline{a}, \overline{a}], b = [\underline{b}, \overline{b}] \in \mathbb{IR}$ we have $a + b = \{\alpha + \beta : \alpha \in a, \quad \beta \in b\}$, $a \times b = \{\alpha\beta : \alpha \in a, \quad \beta \in b\}$; or endpoint-wise: $[\underline{a}, \overline{a}] + [\underline{b}, \overline{b}] = [\underline{a} + \underline{b}, \ \overline{a} + \overline{b}]$, $[\underline{a}, \overline{a}] \times [\underline{b}, \overline{b}] = [\min\{\underline{ab}, \underline{a}\overline{b}, \overline{a}\underline{b}, \overline{ab}\}, \max\{\underline{ab}, \underline{a}\overline{b}, \overline{a}\underline{b}, \overline{ab}\}]$.

For functions $f, g \in \mathbb{A}(\Omega)$, $f = [\underline{f}, \overline{f}]$, $g = [\underline{g}, \overline{g}]$, $x \in \Omega$, we have [2]

$$(f + g)(x) = f(x) + g(x) = [\underline{f}(x) + \underline{g}(x), \overline{f}(x) + \overline{f}(x)],$$

$$(f \times g)(x) = f(x) \times g(x) = [\min M, \max M],$$

$$M = \{\underline{f}(x)\underline{g}(x), \underline{f}(x)\overline{g}(x), \overline{f}(x)\underline{g}(x), \overline{f}(x)\overline{g}(x)\}.$$

Example 1. Denote by $h \in \mathbb{H}(\mathbb{R})$ the (interval) Heaviside step function given by

$$h(x) = \begin{cases} 0, & \text{if } x < 0, \\ [0, 1], & \text{if } x = 0, \\ 1, & \text{if } x > 0, \end{cases} \tag{1}$$

and $g = (-1) \times h \in \mathbb{H}(\mathbb{R})$. For the sum $h + g$ we have

$$(h + g)(x) = h(x) + g(x) = \begin{cases} 0, & \text{if } x < 0 \text{ or } x > 0 \\ [-1, 1], & \text{if } x = 0, \end{cases}$$

showing that $h + g \notin \mathbb{H}(\mathbb{R})$.

Theorem 2. *(a) There exists a unique function* $p \in \mathbb{H}(\Omega)$ *such that* $p(x) \subseteq (f + g)(x)$, $x \in \Omega$; *(b) There exists a unique function* $q \in \mathbb{H}(\Omega)$ *such that* $q(x) \subseteq (f \times g)(x)$, $x \in \Omega$.

We define H-addition and H-multiplication of H-continuous functions $f, g \in \mathbb{H}(\Omega)$ via interval operations as follows.

Definition 7. *(a)* $f \oplus g$ *is the unique H-continuous function* $p(x)$ *as defined by Theorem 2 (a), that is satisfying* $(f \oplus g)(x) \subseteq (f + g)(x)$, $x \in \Omega$; *(b)* $f \otimes g$ *is the unique H-continuous function* $q(x)$ *as defined by Theorem 2 (b), that is satisfying* $(f \otimes g)(x) \subseteq (f \times g)(x)$, $x \in \Omega$.

---

[2] In topology, a *meagre set*, also called a *set of first Baire category*, is a set that, considered as a subset of a (usually larger) topological space, is *small or negligible*.

Example 2. For the H-sum of the Heaviside step function $h \in \mathbb{H}(\mathbb{R})$ given by (1) and $g = (-1) \times h \in \mathbb{H}(\mathbb{R})$ we have $(h \oplus g)(x) = 0$, $x \in \mathbb{R}$.

Theorem 3. *The set $\mathbb{H}(\Omega)$ is a commutative ring with identity with respect to the H-operations $\oplus$ and $\otimes$.*

Remark. Note that the H-operations $\oplus$ and $\otimes$ are not point-wise in general. At a point where both operands have interval values the value of the H-sum $\oplus$ or the H-product $\otimes$ are not determined by the values of the operands only at that point but rather by the values of the operands in a neighborhood of the point.

In the special case when one of the operands is a real (point) valued function the operations $\oplus$ and $\otimes$ coincide with the point-wise operations, namely we have:

$$(f \oplus g)(x) = (f + g)(x) \text{ if } w(f(x)) = 0 \text{ or } w(g(x)) = 0,$$
$$(f \otimes g)(x) = (f \times g)(x) \text{ if } w(f(x)) = 0 \text{ or } w(g(x)) = 0.$$

More properties of the H-operations can be found in [4].

2.3   The Set of H-Continuous Functions as a Linear Space

Multiplication by a scalar is defined as multiplication by a constant function. Since the value of this function is a real number this multiplication coincides with the point-wise multiplication

$$(\alpha * f)(x) = \alpha * f(x) = \begin{cases} [\alpha \, \underline{f}(x), \alpha \overline{f}(x)] \text{ if } \alpha \geq 0, \\ [\alpha \, \overline{f}(x), \alpha \underline{f}(x)] \text{ if } \alpha < 0. \end{cases}$$

The set $\mathbb{H}(\Omega)$ is a linear space with respect to "$\oplus$" and "$*$". Moreover, it is the largest space of interval functions as stated in the next theorem.

Theorem 4. *[5] Let $\mathbb{G}(\Omega)$ be the set of all D-continuous interval functions. Assume that the set $\mathcal{P} \subseteq \mathbb{G}(\Omega)$ is closed under inclusion in the sense that*

$$\left. \begin{array}{c} f \in \mathcal{P}, \ g \in \mathbb{G}(\Omega) \\ g(x) \subseteq f(x), \ x \in \Omega. \end{array} \right\} \implies g \in \mathcal{P}.$$

If $\mathcal{P} \subseteq \mathbb{G}(\Omega)$ is a linear space, then $\mathcal{P} \subseteq \mathbb{H}(\Omega)$.

Hence the H-operations "$\oplus$", "$\otimes$" cannot be extended further than $\mathbb{H}(\Omega)$ in a way preserving the algebraic structure of $C(\Omega)$.

3   Hausdorff Approximations Using Step Functions

3.1   Hausdorff Distance and Modulus of H-Continuity

Let us recall that the Hausdorff distance (H-distance) $\rho(f, g)$ between two functions $f, g \in \mathbb{A}(\Omega)$, $\Omega \subseteq \mathbb{R}^n$, is defined as the distance between their completed graphs $F(f)$ and $F(g)$ considered as closed subsets of $\Omega \times \mathbb{R}$ [17], [21]. More precisely,

$$\rho(f,g) = \max\{ \sup_{A\in F(f)} \inf_{B\in F(g)} ||A-B||, \sup_{B\in F(g)} \inf_{A\in F(f)} ||A-B||\}, \qquad (2)$$

wherein $||.||$ is a norm in $\mathbb{R}^{n+1}$. In technical proofs presented in the sequel we assume that the norm in $\mathbb{R}^{n+1}$ is the maximum norm, that is for $A = (a_1, ..., a_{n+1})$ we have $||A|| = \max\{|a_1|, ..., |a_{n+1}|\}$. However, all statements remain true for any norm due to the equivalence of the norms in $\mathbb{R}^{n+1}$.

In the space of S-continuous functions on Ω, the H-distance satisfies the axioms of a metric. There is a natural connection between the H-continuous functions and the H-distance. For example, one can easily see that an S-continuous function f is H-continuous if and only if $\rho(\underline{f}, \overline{f}) = 0$. Indeed, it follows from the definition that f is H-continuous if and only if $F(\underline{f}) = F(\overline{f})$ or, equivalently, $\rho(F(\underline{f}), F(\overline{f})) = 0$. The link between the two concepts is further discussed below in terms of the modulus of H-continuity.

For $\delta > 0$, the operators I_δ and S_δ are defined for $f \in \mathbb{A}(\Omega)$ as follows

$$I_\delta(f)(x) = \inf\{\underline{f}(y) : y \in B_\delta(x)\}, \ x \in \Omega, \qquad (3)$$

$$S_\delta(f)(x) = \sup\{\overline{f}(y) : y \in B_\delta(x)\}, \ x \in \Omega. \qquad (4)$$

It is easy to see that in terms of Definition 2 we have

$$I(f)(x) = \sup_{\delta>0} I_\delta(f)(x), \ \ S(f)(x) = \inf_{\delta>0} S_\delta(f)(x), \ x \in \Omega.$$

Definition 8. *The modulus of H-continuity $\tau(f;\delta)$ for given $f \in \mathbb{A}(\Omega)$ and $\delta > 0$ is the H-distance between $I_\delta(f)$ and $S_\delta(f)$, that is*

$$\tau(f;\delta) = \rho(I_\delta(f), S_\delta(f)) = \rho(F(I_\delta(f)), F(S_\delta(f))).$$

Theorem 5. *[21]   An S-continuous function f is H-continuous iff $\lim_{\delta\to 0} \tau(f;\delta) = 0$.*

3.2   Interval Step Functions as an Approximation Tool

The usual concept of step-functions of a real argument can be extended to $\mathbb{H}(\Omega)$ as follows.

Definition 9. *A function $f \in \mathbb{H}(\Omega)$ is called a step function if there exists a collection $\{U_1, U_2, ..., U_m\}$ of open subsets of Ω with the following properties*

(i) $U_i \cap U_j = \emptyset$ for $i \neq j$,

(ii) the set $V = \bigcup_{i=1}^{k} U_i$ is dense in Ω,

(iii) for every $i \in \{1, 2, ..., k\}$, f is a real constant on U_i.

It is easy to see that a step function is completely determined by its values on the open set V. In fact we have $f = F(V, \Omega, f|_V)$. Similarly to the real step functions, an interval step function f assumes finite number of values, namely the constant values on the sets U_i, $i = 1, 2, ..., k$, and some real intervals with end-points equal to the constant functional values. Further, we note that the set of step functions is a linear subspace of $\mathbb{H}(\Omega)$. Indeed, the sum of step functions is a step function and so is the product of a step function and a real number. In the next theorem we establish some approximation properties of the step functions.

Theorem 6. *Let $f \in \mathbb{H}(\Omega)$. For every $\varepsilon > 0$ there exists a step function φ such that $\rho(f, \varphi) < \varepsilon$.*

Proof. In view of Theorem 5, there exists $\delta > 0$ such that $\tau(f; \delta) < \varepsilon$. Then consider any collection of open sets $\{U_1, U_2, ..., U_m\}$ with the properties (i) and (ii) as given in Definition 9 and such that the diameter of each set is smaller than δ. These can be constructed for example by partitioning Ω via the planes $x_l = i\delta$, $i \in \mathbb{Z}$, $l = 1, 2, ..., n$. Since f assumes interval values only on a meagre subset of Ω, for every $i \in \{1, 2, ..., k\}$ there exists $x^{(i)} \in U_i$ such that $f(x^{(i)}) \in \mathbb{R}$. Define $\psi(x) = f(x^{(i)})$ for $x \in U_i$, $i = 1, 2, ..., k$, and $\varphi = F(V, \Omega, \psi)$.

We show that φ is the required function. First let us note that φ is a step function. Indeed, since ψ is continuous on any U_i, the operator F does not change the values of ψ on each of these sets, so that $\varphi(x) = \psi(x)$, $x \in V$. This implies both that φ is H-continuous and that it is a step function. Further, from the definition of φ it follows that

$$I_\delta(f) \leq \varphi \leq S_\delta(f).$$

Therefore, we have

$$\rho(f, \varphi) \leq \rho(I_\delta(f), S_\delta(f)) = \tau(f; \delta) < \varepsilon,$$

which proves the theorem.

In the special case of a real argument, the interval step-functions have a simple representation in terms of the Heaviside step function h given in (1). Indeed, when $\Omega = \mathbb{R}$, the sets U_i, $i = 1, ..., k$, associated with an interval step function f in terms of Definition 9 are open intervals of the form (d_{i-1}, d_i), where $d_0 = -\infty$, $d_k = +\infty$, and $d_1, d_2, ..., d_{k-1}$ is a finite increasing sequence of reals. Let $f(x) = c_i$ for $x \in (d_{i-1}, d_i)$. A familiar *rectangular pulse* on the interval $[d_{i-1}, d_i]$, $i = 1, ..., k-1$, is represented as

$$h(x - d_{i-1}) - h(x - d_i)$$

Then the step function f is given by

$$f(x) = c_1(1 - h(x - d_1)) \oplus c_2(h(x - d_1) - h(x - d_2)) \oplus ... \oplus c_{k-1}(h(x - d_{k-2}) - h(x - d_{k-1}))$$

$$\oplus c_k h(x - d_{k-1}) = c_1 + \sum_{i=1}^{k-1}(c_{i+1} - c_i)h(x - d_i).$$

Note that f is discontinuous only at the points $d_1, ..., d_{k-1}$ where it assumes interval values. More precisely, we have

$$f(d_i) = \begin{cases} [c_i, c_{i+1}] \text{ if } c_i < c_{i+1} \\ [c_{i+1}, c_i] \text{ if } c_i > c_{i+1} \end{cases}.$$

For other approximation properties using Hausdorff metric adapted to resolution analysis one may consult Sect. 6 of [3].

4   Approximation by Sigmoid Functions

Sigmoid functions find multiple applications to neural networks and cell growth population models [14,20]. A sigmoid function on $\mathbb{R}$ with a range $[a, b]$ is defined as a monotone function $s(t) : \mathbb{R} \to [a, b]$ such that $\lim_{t \to -\infty} s(t) = a$, and $\lim_{t \to \infty} s(t) = b$. One usually considers continuous (or even smooth) sigmoid functions. Within the class of H-continuous interval functions, the Heaviside step function is a particular case of sigmoid function.

4.1   Approximation by Sigmoid Logistic Functions

An important class of smooth sigmoid functions arises from population growth models. A classical example is the familiar Verhulst population growth model, also known as *logistic model*. One can arrive to this model starting from the reaction equation $U + X \xrightarrow{k} X + X$, where U is a nutrient substance, X is a particular population and k is the specific growth rate of the population. The biological interpretation ot this reaction equation is that the nutrient U is utilized by the population X leading to the reproduction of the population. Denoting the biomass of X by x and the mass (concentration) of U by u and applying the mass action law, one obtains the dynamical system

$$du/dt = -kxu,$$
$$dx/dt = kxu,$$
$$u(0) = u_0, \quad x(0) = x_0.$$

Noticing that $u' + x' = 0$, hence $u + x = x_0 + u_0 = \text{const} = a$, we can substitute $u = a - x$ in the differential equation for x to obtain the Verhulst differential equation $x' = kux = kx(a - x)$. The latter is usually written with a normalized rate constant $k := k/a$ as

$$\frac{dx}{dt} = \frac{k}{a}x(a - x) = kx\left(1 - \frac{x}{a}\right), \quad x(0) = x_0. \tag{5}$$

The solution x to Eq. (5) passing through the point $(0, x(0) = x_0 = a/2)$ is the (basic) logistic sigmoid function:

$$s_0(t) = \frac{a}{1 + be^{-kt}}; \quad b = \frac{a - x_0}{x_0} = 1. \tag{6}$$

In what follows we shall estimate the H-distance between a step function f and a logistic sigmoid function g. Without loss of generality we can consider the Heaviside step function $f = ah$ and the logistic sigmoid function (6): $g = s_0$. According to (2) the H-distance $\rho(f, g)$ between two functions $f, g \in \mathbb{A}(\Omega)$ for $\Omega \subset \mathbb{R}$

makes use of the maximum norm in R^2 so that the distance between the points $A = (t_A, x_A)$, $B = (t_B, x_B)$ in R^2 is $\|A - B\| = \max(|t_A - t_B|, |x_A - x_B|)$. In that case the H-distance $d = \rho(h, s_0)$ between the Heaviside step function ah and the sigmoid function (6) satisfies the relations $0 < d < a/2$ and $a - s_0(d) = d$, that is

$$(a - d)/d = e^{kd}, \quad (0 < d < a/2). \tag{7}$$

Obviously $d \to 0$ implies $k \to \infty$ (and vice versa). From (7), a straightforward expression for the rate parameter k in terms of d follows:

$$k = \frac{1}{d} \ln \frac{a - d}{d} = O(d^{-1} \ln(d^{-1})). \tag{8}$$

4.2   Estimate for the H-Distance in Terms of the Rate Parameter

Relation (8) gives an estimate of the rate k in terms of the H-distance d. The following theorem gives a relation for the H-distance d in terms of the rate parameter k. For simplicity we assume $a = 1$, denoting thus in the sequel $s_0(t) = (1 + e^{-kt})^{-1}$.

Theorem 7. *The Hausdorff distance $d = \rho(h, s)$ between the Heaviside step function h_0 and the sigmoid Verhulst function s_0 can be expressed in terms of the reaction rate k as follows:*

$$d = \frac{\ln(k + 1)}{k + 1} \left(1 + O\left(\frac{\ln \ln(k + 1)}{\ln(k + 1)} \right) \right). \tag{9}$$

Proof. Assuming $a = 1$ relation (7) becomes $(1 - d)/d = e^{kd}$ for $0 < d < 1/2$. This implies $kd = \ln(1/d) + \ln(1 - d)$. In order to express d in terms of k, let us examine the function

$$f(d) = kd - \ln(1/d) - \ln(1 - d), \ 0 < d < 1/2.$$

From $\lim f(d)_{d \to 0, d > 0} = -\infty$, $\lim f(d)_{d \to 1/2, d < 1/2} = k/2 > 0$ we conclude that $f(d) = 0$ possesses a solution in $(0, 1/2)$. From $f'(d) = k + 1/d + 1/(1 - d) > 0$ we conclude that function f is strictly monotonically increasing, hence $f(d) = 0$ has an unique solution $d(k)$ in $(0, 1/2)$. For $k \to \infty$ we have $d(k) \to 0$, hence $\ln(1 - d(k)) = -d(k) + O(d(k)^2)$. Consider then the function $g(d) = (k + 1)d - \ln(1/d)$ which approximates function f with $d \to 0$ as $O(d^2)$; in addition $g'(d) > 0$. So we can further denote by $d(k)$ the (unique) zero of g and study g instead of f. We look for two reals d_- and d_+ such that $g(d_-) < 0$ and $g(d_+) > 0$ (leading to $g(d_-) < g(d(k)) < g(d_+)$ and thus $d_- < d(k) < d_+$). Trying $d_- = 1/(k + 1)$ and $d_+ = \ln(k + 1)/(k + 1)$ we obtain $g(1/(k + 1)) = 1 - \ln(k + 1) < 0$ and $g(\ln(k + 1)/(k + 1)) = \ln \ln(k + 1) > 0$ proving the estimates $1/(k + 1) < d(k) < \ln(k + 1)/(k + 1)$. To find a better lower bound we compute

$$g\left(\frac{\ln(k + 1)}{k + 1} \left(1 - \frac{\ln \ln(k + 1)}{\ln(k + 1)} \right) \right) = \ln \left(1 - \frac{\ln \ln(k + 1)}{\ln(k + 1)} \right) < 0.$$

We thus obtain

$$\frac{\ln(k+1)}{k+1} - \frac{\ln\ln(k+1)}{k+1} < d(k) < \frac{\ln(k+1)}{k+1},$$

which implies (9).

Remark. In the general case $a \neq 1$ one should substitute in (9) $k+1$ by $k+a^{-1}$.

5   Conclusions

We briefly summarized some basic results (Theorems 1–5) about H-continuous functions and their application to problems in abstract areas such as real analysis, approximation theory, set-valued analysis and analysis of PDEs. We then formulated and proved a new result (Theorem 6) on the Hausdorff approximation of H-continuous functions by interval step functions defined on an open subset of $\mathbb{R}^n$. Finally we discussed some applications of H-continuous functions to biological dynamic processes, in particular, we considered the remarkable phenomenon that certain enzyme kinetic and population growth processes develop almost step-wise [16,20]. Such processes are usually described or approximated by smooth sigmoid functions (especially in the theory of artificial neural networks). However, it is possible that H-continuous step-wise functions can be also conveniently used. To substitute a sigmoid function by a step function, we need to know the approximation error (Theorem 7). Biological processes are often very sensitive and can be effectively studied within the framework of interval analysis [19]. In addition, the input data coming from biological experiments are usually rather uncertain and thus can be represented as interval data. If these interval data are guaranteed (that is they include the measurement errors), then numerical methods and programming tools with automatic result verification can be used [18].

The presented results suggest that interval analysis, apart from being currently associated with numerical analysis, can also be considered as belonging to the field of real analysis. We may thus consider interval analysis as a bridge between real and numerical analysis, a bridge that extends both subjects and unifies them into a common scientific area.

Acknowledgments. RA acknowledges partial support of the National Research Foundation of South Africa. RA and SM acknowledge partial support by the Institute of Mathematics and Informatics at the Bulgarian Academy of Sciences. The authors thank Prof. Kamen Ivanov for the analysis and derivation of formula (9). They are grateful to the anonimous reviewer for his careful reading and many remarks.

References

1. Anguelov, R.: Dedekind order completion of $C(X)$ by Hausdorff continuous functions. Quaestiones Mathematicae **27**, 153–170 (2004)

2. Anguelov, R., Markov, S.: Extended segment analysis, Freiburger Intervall-Berichte, Inst. Angew. Math, U. Freiburg i. Br. 10, pp. 1–63 (1981)
3. Anguelov, R., Markov, S., Sendov, B.: On the normed linear space of Hausdorff continuous functions. In: Lirkov, I., Margenov, S., Waśniewski, J. (eds.) LSSC 2005. LNCS, vol. 3743, pp. 281–288. Springer, Heidelberg (2006)
4. Anguelov, R., Markov, S., Sendov, B.: Algebraic operations on the space of Hausdorff continuous interval functions. In: Bojanov, B. (ed.) Constructive Theory of Functions, pp. 35–44. Marin Drinov Academic Publishing House, Sofia (2006)
5. Anguelov, R., Markov, S., Sendov, B.: The set of Hausdorff continuous functions–the largest linear space of interval functions. Reliable Comput. 12, 337–363 (2006). http://dx.doi.org/10.1007/s11155-006-9006-5
6. Anguelov, R., Rosinger, E.E.: Hausdorff continuous solutions of nonlinear PDEs through the order completion method. Quaest. Math. 28, 271–285 (2005)
7. Anguelov, R., Rosinger, E.E.: Solving large classes of nonlinear systems of PDEs. Comput. Math. Appl. 53, 491–507 (2007)
8. Anguelov, R., Markov, S., Minani, F.: Hausdorff continuous viscosity solutions of Hamilton-Jacobi equations. In: Lirkov, I., Margenov, S., Waśniewski, J. (eds.) LSSC 2009. LNCS, vol. 5910, pp. 231–238. Springer, Heidelberg (2010)
9. Anguelov, R., van der Walt, J.H.: Order Convergence on $C(X)$. Quaestiones Mathematicae 28(4), 425–457 (2005)
10. Anguelov, R., Kalenda, O.F.K.: The convergence space of minimal USCO mappings. CZECHOSLOVAK MATH. J. 59(1), 101–128 (2009)
11. Anguelov, R.: Rational extensions of $C(X)$ via Hausdorff continuous functions. Thai J. Math. 5(2), 261–272 (2007)
12. Anguelov, R., van der Walt, J.H.: Algebraic and topological structure of some spaces of set-valued maps. Comput. Math. Appl. 66, 1643–1654 (2013)
13. Baire, R.: Lecons sur les Fonctions Discontinues. Collection Borel, Paris (1905)
14. Costarelli, D., Spigler, R.: Approximation results for neural network operators activated by sigmoidal functions. Neural Netw. 44, 101–106 (2013)
15. Dilworth, R.P.: The normal completion of the lattice of continuous functions. Trans. Amer. Math. Soc. 68, 427–438 (1950)
16. Dimitrov, S., Markov, S.: Metabolic Rate Constants: some Computational Aspects, Mathematics and Computers in Simulation (2015). doi:10.1016/j.matcom.2015.11.003
17. Hausdorff, F.: Set theory, 2nd edn. Chelsea Publ., New York (1962) [1957], ISBN 978-0821838358 (Republished by AMS-Chelsea 2005)
18. Kraemer, W., Gudenberg, J.W.v (eds.): Scientific Computing, Validated Numerics, Interval Methods, Proc. SCAN-2000/Interval-2000, Kluwer/Plenum (2001)
19. Markov, S.: Biomathematics and interval analysis: A prosperous marriage. In: Christov, C., Todorov, M.D. (eds.) Proceedings of the 2nd International Conference on Application of Mathematics in Technical and Natural Sciences (AMiTaNS'10), AIP Conference Proceedings 1301, 26–36 (2010)
20. Markov, S.: Cell Growth Models Using Reaction Schemes: Batch Cultivation, Biomath 2/2, 1312301 (2013). http://dx.doi.org/10.11145/j.biomath.2013.12.301
21. Sendov, B.: Hausdorff Approximations. Kluwer, Boston (1990)
22. van der Walt, J. H.: The Linear Space of Hausdorff Continuous Interval Functions. Biomath 2, 1311261 (2013). http://dx.doi.org/10.11145/j.biomath.2013.11.261
23. Mathematics Subject Classification (MSC2010), AMS (2010). http://www.ams.org/mathscinet/msc/msc2010.html

Replacing Branches by Polynomials in Vectorizable Elementary Functions

Olga Kupriianova$^{(\boxtimes)}$ and Christoph Lauter

Sorbonne Universités, UPMC Univ Paris 06, UMR 7606, LIP6, 75005 Paris, France
{olga.kupriianova,christoph.lauter}@lip6.fr

Abstract. One of the goals for the mathematical function generator is to produce vectorizable codes. Therefore, in the generated code there should be no branching. As the most mathematical functions are implemented with domain splitting procedure and piecewise-polynomial approximation, there are several `if-else` statements in the final code to determine the corresponding polynomial coefficients. In this paper we propose a simple idea of replacing these `if-else` statements by the evaluation of a polynomial function. This is a novel approach that may not work for all the possible function implementation variants, and it needs to be improved with the use of some more sophisticated methods.

Keywords: Mathematical functions · Branching · Vectorizable code · Interpolation · Reconstruction

1 Introduction

Collection of codes to evaluate mathematical function in some programming language is called a `libm`. The standard `libms` provide a limited set of mathematical functions in single and double precisions. `Libms` contain code usually for elementary functions (sin, log, etc.) and several "special functions" like Gamma or Bessel functions [1]. Hardware producers spend a lot of manpower on maintenance and optimization of their proprietary `libms`. The existing Open source libraries are not flexible enough and do not provide user specific implementations of mathematical functions [2]. As one can define a huge number of implementation variants for each mathematical function (different domain, accuracy, degree of the approximation polynomial, etc.) it is not feasible to implement them all manually. Thus, we have been developing Metalibm, an automatic code generator that produces flexible implementations of mathematical functions [10]. Functions variants to be generated are defined by a set of parameters like function name (or algebraic expression), implementation domain, final accuracy and maximum degree of its polynomial approximation. The generated code evaluates the specified function the accuracy bounded by the final accuracy parameter.

Since the prevalence of SIMD instructions on modern processors, the code generation of vectorizable implementations is of big interest as well. To make

© Springer International Publishing Switzerland 2016
M. Nehmeier et al. (Eds.): SCAN 2014, LNCS 9553, pp. 14–22, 2016.
DOI: 10.1007/978-3-319-31769-4_2

Metalibm produce vectorizable codes, the algorithms without (or almost without) branching have to be used. For exponential and logarithmic functions vectorized loop calls reduce the computation time in 1.5–2 times.

Except for some rare cases, mathematical functions in Metalibm are approximated with piecewise-polynomial functions. Thus, in the produced code there is branching to determine the right polynomial coefficients for each subdomain. We propose here a method to avoid this branching. We are trying to find a mapping function (in mathematical sense of word function), that returns corresponding subdomain indexes for the input values from the implementation domain. This mapping function may be computed with one of the classical polynomial interpolation procedures. The proposed algorithm paves the way for research in generation of vectorizable functions implementations. We discuss here the weakness of the proposed method as well as the way to improve it. However, this improvement requires also research in interval arithmetic, basically in solving the linear systems with bounded variables.

In the next Section we give a short overview on the implementation process of mathematical functions, in Sect. 3 we explain in details the proposed method, provide a pseudocode and discuss its future evolution. In Sect. 4 there are the results of our method and the conclusion.

2 Mathematical Functions Implementation Workflow

The usual implementation workflow of a mathematical function is divided into three steps: argument reduction, approximation and reconstruction [3]. In Metalibm we use a modified Remez algorithm for polynomial approximation [4]. The larger is the approximation domain, the larger will be polynomial degree to approximate a function with the given accuracy. To save computation time and to avoid accumulation of errors, the degree of the polynomial has to be low. Therefore, the implementation domain for the function has to be reduced somehow. It may be done in two ways: property-based argument reduction [5–7] or domain splitting [8].

In the first case mathematical properties of the function allow to establish connection between the initial large domain $[a, b]$ and a small one $[\alpha, \beta]$. So, the approximation is computed for some $g(r)$, $r \in [\alpha, \beta]$ instead of $f(x)$, $x \in [a, b]$, where r is reduced argument. Then on reconstruction step the inverse argument transition has to be applied. As this approach is based only on function properties, it is limited and works only for several function families (e.g. exponential, logarithmic, symmetrical). When the function properties do not allow to reduce the domain or do not reduce it enough, piecewise-polynomial approximations are used. In this case, the reconstruction step contains several if-else statements to pick the right approximating polynomial for the function evaluation. To make the code vectorizable, branching during the computations of the finite function values has to be avoided.

Consider here that the domain splitting procedure returns a set of non-overlapping intervals $\{I_k\}_{k=0}^N$, such that $I_k = [a_k, a_{k+1}]$, so the adjacent intervals

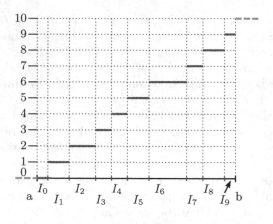

Fig. 1. Piecewise-constant mapping function $M(x)$

have the only point in the intersection $I_k \cap I_{k+1} = \{a_{k+1}\}$. The initial implementation domain is $[a, b] = I = \bigcup_{k=0}^{N} I_k$. Then, to compute $f(x)$ we execute if-else statements to determine subdomain's index k, where $x \in I_k$.

The code may be written without branches with the use of a mapping function that returns subdomain's index for each input value from I:

$$M(x) = k, \ x \in I_k, \ k = 0, 1, \ldots, N.$$

The function $M(x)$ is a piecewise-constant function as it is shown on Fig. 1.

For a naive domain splitting, when I is divided into N equal intervals we may use a linear function $m(x) = \frac{N}{b-a}(x - a)$ and then the mapping function is easily computed as $M(x) = \lfloor m(x) \rfloor$. However, this splitting is not optimal and a more sophisticated splitting algorithm is used instead [8].

3   Polynomial-Based Reconstruction Technique

3.1   How to Compute Polynomial Mapping

For a non-regular domain splitting as the one that is currently used in Metalibm the mapping function may be computed with an interpolation polynomial $p(x)$. This polynomial passes through the points (a_k, k), $k = 0, \ldots, N$, where $\{a_k\}$ are the splitpoints. Once the polynomial coefficients are computed, the mapping function can be computed as

$$M(x) = \lfloor p(x) \rfloor, \ x \in I.$$

Thereby, we obtain some conditions for this polynomial. Such a polynomial is shown on Fig. 2.

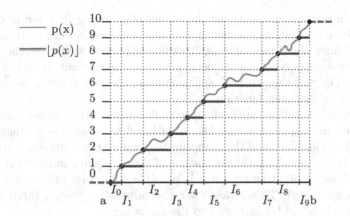

Fig. 2. Mapping function and a corresponding polynomial $p(x)$

3.2 Conditions for the Polynomial

As the mapping function stays constant on a subdomain I_k, the admissible range for the polynomial values on this subdomain is $[k, k+1)$. Thus, the task is to compute an interpolation polynomial p on the points (a_k, k), $k = 0, \ldots, N$ for which the following holds:

$$p(x) \in [k, k+1), \ x \in [a_k, a_{k+1}]. \tag{1}$$

A suitable polynomial p as well as the conditions (1) are shown on Fig. 3. As the classical interpolation procedures guarantees only that $p(a_k) = k$ by construction of the polynomial, conditions (1) have to be checked *a posteriori*. This can be done in Sollya [9] with the evaluation of this polynomial $p(x)$ over an interval $[a_k, a_{k+1}]$.

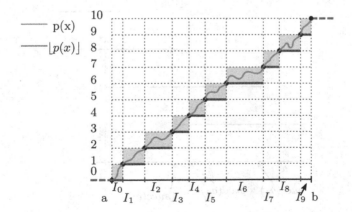

Fig. 3. Admissible ranges for the polynomial values.

There is a certain ambiguity for the values of mapping function in the split-points $\{a_k\}$. In splitpoints the two polynomials corresponding to the adjacent subdomains have the same value $p_{k-1}(a_k) = p_k(a_k) = k$. To get the index of the approximating polynomial at the point a_k we may admit $M(a_k) = k - 1$ or $M(a_k) = k$. Only in the "corner" splitpoints a_0 and a_N there is no ambiguity for the values of mapping function.

As all the computations are performed in floating-point numbers, the interpolation conditions $p(a_k) = k$ are no longer satisfied because of roundings. Taking into account the ambiguity of the mapping function in the splitpoints, conditions (1) have to be modified a little. As the set of floating-point numbers is discrete, for a given floating-point number a it is possible to find its predecessor $pred(a)$ and successor $succ(a)$. This means that the admissible ranges for polynomial values from (1) should be narrowed to the following:

$$p(x) \in [k, k+1), \text{ where } x \in [succ(a_k), pred(a_{k+1})] \subset I_k, 0 \leq k \leq N - 1. \quad (2)$$

The conditions for the splitpoints should be added then.

$$p(x) \in [k-1, k+1), \text{ where } x = a_k, k = 1, \ldots, N - 1 \quad (3)$$

The modified conditions for the polynomial ranges are shown on Fig. 4 with grey rectangulars, the range of polynomial values in split points is illustrated with a red line.

3.3   The Choice of the Interpolation Points

The interpolation points may be chosen in several different ways. With the set of splitpoints $\{a_k\}_{k=0}^{N}$ we compute four different polynomials. First, we may use "inner" polynomial with $N - 1$ points $\{a_k\}_{k=1}^{N-1}$. Then we can compute "left" and "right" polynomial with N points $\{a_k\}_{k=0}^{N-1}$ or $\{a_k\}_{k=1}^{N}$. And the last variant

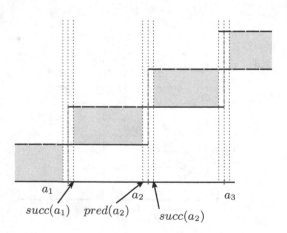

Fig. 4. Modified floating-point conditions for polynomial (Color figure online).

here is to compute a polynomial of degree N using all $N+1$ splitpoints. When *a posteriori* conditions are not verified for all the four polynomials (2–3), we may add some interpolation points. However, as the addition of new interpolation points raises the degree of the polynomial according to Runge's phenomenon it will oscillate in the ends, which means that the conditions (2–3) are rarely verified.

3.4   Towards *a Priori* Conditions

As the conditions for the polynomial values are checked only *a posteriori*, there is no guarantee that the polynomial for mapping function exists for arbitrary splitting. Contrariwise, our method finds it only for few splittings. When there are some points where the polynomial exceeds the admissible range, we can add them to the interpolation points, recompute the polynomial $p(x)$ and recheck the conditions (2)–(3). However, due to Runge's phenomenon the polynomial begins to oscillate [11] and the conditions are not verified. The choice of the interpolation points for this polynomial remains an open problem.

However, the ranges for the polynomial values are still checked *a posteriori* and there are some values out of the admissible range. These conditions may be taken into account if we operate intervals instead of points. The classical interpolation problem is a system of linear equations with Vandermonde's matrix:

$$\begin{pmatrix} 1 & x_0 & \cdots & x_1^N \\ 1 & x_1 & \cdots & x_1^N \\ \vdots & \vdots & \ddots & \vdots \\ 1 & x_N & \cdots & x_N^N \end{pmatrix} \begin{pmatrix} c_0 \\ c_1 \\ \vdots \\ c_N \end{pmatrix} = \begin{pmatrix} y_0 \\ y_1 \\ \vdots \\ y_N \end{pmatrix}, \tag{4}$$

where $(x_i, y_i), i = 0, \ldots, N$ are the interpolation points and $c_0, \ldots, c_N$ are the unknown polynomial coefficients. Computing the unknown coefficients means solving the system (4). When we use intervals instead of points to compute the interpolation polynomial, we take subdomains on abscissas and intervals $[k, pred(k+1)]$ on ordinates. Then, the task is almost the same: system of linear equations with unknown coefficients $c_0, \ldots, c_N$. Except of the numbers x_i, y_i we operate intervals in system (5).

$$\begin{pmatrix} 1 & \mathbf{x_0} & \cdots & \mathbf{x_1^N} \\ 1 & \mathbf{x_1} & \cdots & \mathbf{x_1^N} \\ \vdots & \vdots & \ddots & \vdots \\ 1 & \mathbf{x_N} & \cdots & \mathbf{x_N^N} \end{pmatrix} \begin{pmatrix} c_0 \\ c_1 \\ \vdots \\ c_N \end{pmatrix} = \begin{pmatrix} \mathbf{y_0} \\ \mathbf{y_1} \\ \vdots \\ \mathbf{y_N} \end{pmatrix} \tag{5}$$

The difference from the classical solution of the interval system is that we need only one vector $(c_0, c_1, \ldots, c_N)$, but not the set of all suitable coefficients. We may find the tolerance solution set of the system (5) in polynomial time, but it can be empty. In this case the united solution set may be found, but this problem is NP-hard [12]. Anyway, we have connected coefficients in the system matrix, and the existing methods do not take into account this type of connection. We leave this transition to *a priori* conditions for the future work.

3.5   Algorithm

To summarize the upper-mentioned, here is the pseudocode to obtain a reconstruction polynomial. Procedure `buildInterpolationPoly` takes subdomains
from splitting and computes the interpolation with divided differences. We bound
the degree of the reconstruction polynomial with the parameter `maxdegree`.

```
Input: domain, maxdegree
N = length(domains);
if (N - 3 > maxdegree) {
    print("Even the poly built by inner points has too large degree");
    poly = -1;
} else {
    innerpoints = extractOnlyInnerPoints(domains);
    ypoints = [| 1, ..., N - 1 |];
    poly = buildInterpolationPoly(innerpoints, ypoints);
    if (!checkConditions(poly, domains)) {
        xpoints = merge(inf(domains[0]), innerpoints);
        ypoints = [| 0, ..., N - 1 |];
        poly = buildInterpolationPoly(xpoints, ypoints);
        if (!checkConditions(poly, domains)) {
            xpoints = merge(innerpoints, sup(domains[N - 1]));
            ypoints = [| 1, ..., N |];
            poly = buildInterpolationPoly(xpoints, ypoints);
            if (!checkConditions(poly, domains)) {
                xpoints = merge(inf(domains[0]), xpoints);
                ypoints = [| 0, ..., N |];
                poly = buildInterpolationPoly(xpoints, ypoints);
                if (!checkConditions(poly, domains)) {
                    print("It was not possible to build such a polynomial.
                            Perhaps we need to add an interpolation point");
                    poly = -1;
                }
            }
        }
    }
}
return poly;
```

4   Conclusion

For the variation of the arcsin function on the domain $[-0.6; 0.6]$ with the target accuracy $\bar{\varepsilon} = 2^{-54}$ and maximum degree of the approximation Metalibm
splits the domain into six smaller subdomains. In this case Metalibm achieves
to compute the coefficients for the mapping polynomial. It uses "left" interpolation polynomial computed on all the split points except the last one. The
proposed method allows to obtain performance gain in 1.5–3 times. Thus, this
algorithm paves a way for generation of vectorizable function implementations.

The main disadvantage of the proposed method is that it worked in $\sim 30\%$ of tested cases. As it was mentioned, the main reason of failures of the algorithm is the *a posteriori* condition checking.

We invented a general method that is not based on any specific instructions, so can be used on a wide range of the machines. The use of the interval arithmetic may allow to generate vectorizable code for larger range of function variations. As we mentioned, there are two solutions for the interval system of linear equations suitable for our task. When the tolerance solution set is empty, we may try to find the united solution. The second problem is NP-hard but does not have to be avoided: we are interested not in the set of all the possible polynomial coefficients, but only in one combination of all the coefficients that gives the mapping function. Thus, this combination should be easier to find than the whole set. Furthermore, as this mapping function is computed during the function generation, there are no strict requirements on complexity and performance of this algorithm.

The proposed reconstruction technique with polynomial-based mapping function depends a lot on domain splitting. We tried to build an optimal domain split in terms of the quantity of subdomains and the polynomial degrees of the approximation on each of the subdomains. Mapping function for the regular splitting is easy to find but our splitting produces non-regular subdomains and computing the suitable polynomial for mapping gets impossible in some cases. Besides that, our splitting algorithm does not control the size (and the corresponding polynomial degree) of the last subdomain. Though practically we have not noticed such phenomenon, theoretically nothing prevents our splitting algorithm to get small last subdomain with low corresponding polynomial degree (one or two). This may be avoided with improving the splitting algorithm: instead of splitpoints we should get intervals for the possible splitpoint. Then, on the reconstruction step we fix splitpoints from these "tolerance" intervals so, that our polynomial for mapping function exists. Thus, the future research on this problem includes also establishing the compromise between splitting optimality and existence of polynomial for mapping.

References

1. Loosemore, R.M.S.e.a.S.: The GNU C Library Reference Manual. Free Software Foundation Inc
2. Kupriianova, O., Lauter, C.: Metalibm: a mathematical functions codegenerator. In: Proceedings of the 2014 Mathematical Software - ICMS 2014 - 4th International Congress, Seoul, South Korea, 5–9 August 2014, pp.713–717. http://dx.doi.org/10.1007/978-3-662-44199-2_106
3. Muller, J.-M.: Elementary Functions: Algorithms and Implementation. Birkhauser Boston Inc., Secaucus (1997)
4. Brisebarre, N., Chevillard, S.: Efficient polynomial l-approximations. In: 18th IEEE Symposium on Computer Arithmetic (ARITH-18 2007), 25–27 June 2007, Montpellier, France, pp. 169–176 (2007). http://doi.ieeecomputersociety.org/10.1109/ARITH.2007.17

5. Tang, P.T.P.: Table-driven implementation of the exponential function in IEEE floating-point arithmetic. ACM Trans. Math. Softw. **15**(2), 144–157 (1989)
6. Tang, P.T.P.: Table-driven implementation of the logarithm function in IEEE floating-point arithmetic. ACM Trans. Math. Softw. **16**(4), 378–400 (1990). http://acm.org/10.1145/98267.98294
7. Tang, P.T.P.: Table-driven implementation of the Expm1 function in IEEE floating-point arithmetic. ACM Trans. Math. Softw. **18**(2), 211–222 (1992). http://doi.acm.org/10.1145/146847.146928
8. Kupriianova, O., Lauter, C.: A domain splitting algorithm for the mathematical functions code generator. In: Asilomar Conference on Signals, Systems and Computers, Pacific Grove, CA, USA, 2–5 November 2014 (2014, to appear)
9. Chevillard, S., Joldeş, M., Lauter, C.: Sollya: an environment for the development of numerical codes. In: Fukuda, K., Hoeven, J., Joswig, M., Takayama, N. (eds.) ICMS 2010. LNCS, vol. 6327, pp. 28–31. Springer, Heidelberg (2010)
10. Brunie, N., de Dinechin, F., Kupriianova, O., Lauter, C.: Code generators for mathematical functions (2014). <hal-01084726v2>
11. Cheney, E.W.: Introduction to Approximation Theory. AMS Chelsea Publishing, New York (1982)
12. Shary, S.P.: Solving the linear interval tolerance problem. Math. Comput. Simul. **39**, 53–85 (1995)

The Forthcoming IEEE Standard 1788 for Interval Arithmetic

John Pryce[(✉)]

School of Mathematics, Cardiff University, Cardiff, UK
smajdp1@cardiff.ac.uk

Abstract. This is a slightly expanded form of the author's talk of the same title at SCAN 2014, Würzburg. Angled towards people who use interval numerical methods little or not at all, it briefly describes how interval arithmetic works, the mindset required to use it effectively, why an interval arithmetic standard was needed, the setting up of IEEE Working Group P1788 for the purpose, the structure of the standard it has produced, some difficulties we encountered, and the current state of the P1788 project. During production of these Proceedings the 1788 standard has been published, but the talk's original title has been kept.

This article, slightly expanded from the author's talk of the same title at SCAN 2014, briefly describes how interval arithmetic works, the mindset required to use it effectively, why an interval arithmetic standard was needed, the setting up of IEEE Working Group P1788 for the purpose, the structure of the standard it has produced, some difficulties we encountered, and the current state of the P1788 project. During production of these Proceedings the 1788 standard has been published, but the talk's original title has been kept.

The references include a recent survey [14], a recent textbook [17], and a current web site [8], that testify to the liveliness of this area.

1 What Intervals Are and Do

1.1 Basic Ideas

Interval Arithmetic (IA) implements "validated", also called "verified", numerical calculation. That is, it can *enclose* solution components x of a problem in an interval, i.e. between lower and upper bounds

$$x \in \boldsymbol{x} = [\underline{x}, \overline{x}] = \{\, t \in \mathbb{R} \mid \underline{x} \leq t \leq \overline{x} \,\}.$$

It does this even in finite-precision arithmetic, with roundoff errors present.

E.g. it makes Brouwer's fixed point theorem:

If $K \subset \mathbb{R}^n$ is compact convex, and function f is everywhere defined and continuous on K, and $f(K) \subseteq K$, then f has a fixpoint in K

© Springer International Publishing Switzerland 2016
M. Nehmeier et al. (Eds.): SCAN 2014, LNCS 9553, pp. 23–39, 2016.
DOI: 10.1007/978-3-319-31769-4_3

verifiable when K is a box (product of intervals) in the sense that during the evaluation of f, sufficient conditions for "everywhere defined and continuous" can be found—with ease in favourable cases, but maybe requiring both brute force and finesse in trickier cases..

The history of interval arithmetic might be traced back to Archimedes, in the sense that he rigorously proved the bounds

$$3\tfrac{10}{71} < \pi < 3\tfrac{1}{7},$$

see Thomas Heath [3]. As a systematic discipline it seems to have begun in the 20th century: Teruo Sunaga (Japan, 1958) [16]; Leonid Kantorovich (USSR, 1962) [4]. The most influential work of that time was the book by Ramon Moore (USA, 1966) [11], describing for instance the first implementation of a *validated ODE solver*.

Currently significant validated software exists for global optimisation, large sparse linear systems, particle beam design for the Large Hadron Collider, and many other applications. Rather than list extensive references we refer to those in Siegfried Rump's survey article [14] and in Vladik Kreinovich's web site [8], and also to the recent introductory book by Warwick Tucker [17].

1.2   Definition of Interval Operations

Interval operations take *all combinations* of points in the inputs, i.e.

$$\boldsymbol{x} \bullet \boldsymbol{y} = \{\, x \bullet y \mid x \in \boldsymbol{x} \text{ and } y \in \boldsymbol{y} \,\}, \quad \text{where } \bullet \text{ is one of } \{+ \ - \ \times \ \div\}$$

For $\div$, disallow $0 \in \boldsymbol{y}$ for now. In finite precision *round outward*. With these definitions one has the following fact, probably first stated by Moore:

Theorem 1 (Fundamental Theorem of Interval Arithmetic). *If a function $f(x_1, \ldots, x_n)$, defined by an expression, is evaluated with interval operations on interval inputs to get $\boldsymbol{y} = \boldsymbol{f}(\boldsymbol{x}_1, \ldots, \boldsymbol{x}_n)$ then*

$$\boldsymbol{y} \text{ contains the range of } f \text{ over box } \boldsymbol{x}_1 \times \cdots \times \boldsymbol{x}_n \text{ in } \mathbb{R}^n.$$

Example 1. Let $f(x_1, x_2) = x_1 + x_2/x_1$, suppose 2-digit decimal arithmetic is used, and let the input intervals be $\boldsymbol{x}_1 = [3, 4]$, $\boldsymbol{x}_2 = [3, 5]$. We compute

$$\boldsymbol{x}_1 + \frac{\boldsymbol{x}_2}{\boldsymbol{x}_1} = [3, 4] + \frac{[3, 5]}{[3, 4]} = [3, 4] + \left[\frac{3}{4}, \frac{5}{3}\right] \overset{\text{round}}{\longrightarrow} [3, 4] + [.75, 1.7]$$

$$= [3.75, 5.7] \overset{\text{round}}{\longrightarrow} [3.7, 5.7] = \boldsymbol{f}(\boldsymbol{x}_1, \boldsymbol{x}_2) = \boldsymbol{y}.$$

$\boldsymbol{y}$ *does* contain the range of f over $\boldsymbol{x}_1 \times \boldsymbol{x}_2 = [3, 4] \times [3, 5]$, which with a bit of calculus is found to be $[4, 5.25]$.   □

2   Why Do Intervals Need New Algorithms?

2.1   Example: Interval Version of Newton's Iteration

Consider *Newton's method* for solving a 1-D nonlinear equation $f(x) = 0$.

A Wrong Approach. The usual formula is:

$$x_{k+1} = x_k - \frac{f(x_k)}{f'(x_k)}$$

A direct interval transcription of this would be

$$\boldsymbol{x}_{k+1} = \boldsymbol{x}_k - \frac{\boldsymbol{f}(\boldsymbol{x}_k)}{\boldsymbol{f}'(\boldsymbol{x}_k)}$$

where $\boldsymbol{f}$ and $\boldsymbol{f}'$ are interval versions of (the computer code for) f and f'.

Unfortunately addition and subtraction of intervals—in infinite precision—just *adds* their widths. In symbols, $w(\boldsymbol{a} \pm \boldsymbol{b}) = w(\boldsymbol{a}) + w(\boldsymbol{b})$. In finite precision the result is even a little wider owing to roundoff.

So the width of $\boldsymbol{x}_{k+1}$ equals the width of $\boldsymbol{x}_k$ plus that of $\boldsymbol{f}(\boldsymbol{x}_k)/\boldsymbol{f}'(\boldsymbol{x}_k)$. The latter width is usually strictly > 0, so each interval cannot be narrower than the last, and usually is wider. Convergence of an interval algorithm to a root must involve the interval becoming smaller. The above simple transcription of Newton to intervals cannot possibly do that, and is bound to diverge!

A Right Approach. For a sensible solution to this problem, go back to basic theory. Let f be a C^1 function on a real interval I. Then by the Mean Value Theorem MVT, for any root z and any x, both in the interval, there is some ξ in the interval such that

$$f(x) = f(x) - f(z) = (x - z)f'(\xi) \tag{1}$$

so provided $f'(\xi) \neq 0$, see later,

$$z = x - \frac{f(x)}{f'(\xi)}. \tag{2}$$

A pointer to the right algorithm is in the quantifiers: $\boxed{\forall}$ root z, $\boxed{\forall}$ x, $\boxed{\exists}$ ξ. These give a geometric interpretation to Eq. (2), shown in Fig. 1:

> *For any $x \in I$, a searchlight shone from the point $(x, f(x))$ on the curve, its rays bounded by the lowest and highest slopes of f on I, is certain to illuminate any root z (identified with $(z, 0)$ in the plane) in I.*

To convert this to something computable note that in (2):

- $f(x)$ must be computed as an *interval*, since f is program code, hence liable to roundoff.
- $f'(\xi)$ must also be an interval, for two reasons: (a) f' is program code; (b) ξ is only known to *exist*—its exact position is unknown ($\boxed{\exists}$).
- However x can be a *point*. It is an arbitrary programmer-selected point in I ($\boxed{\forall}$)—typically the midpoint is used in practice.

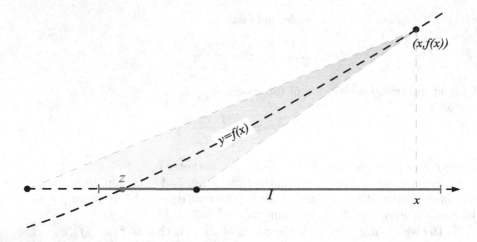

Fig. 1. Geometric view of Interval Newton: one-sided searchlight.

Hence, ($\forall$) for any root $z \in I$ we *also* have

$$z \in \left(x - \frac{[f(x)]}{[f'(\xi)]} \right)$$

using classical interval notation that $[\ldots]$ means "some interval containing". In more current notation, renaming the interval as $\boldsymbol{x}$

$$z \in \left(x - \frac{\boldsymbol{f}([x])}{\boldsymbol{f'}(\boldsymbol{x})} \right) \qquad = \left(point - \frac{interval \text{ function of } point}{interval \text{ function of } interval} \right) \qquad (3)$$

where $[x]$ is 1-point interval $\{x\}$ and $\boldsymbol{f}, \boldsymbol{f'}$ are interval versions of f, f'. This is the start of a satisfactory algorithm.

More General Picture. Actually the searchlight shines in both directions, crucial when the range of slopes includes both positive and negative slopes, see Fig. 2:

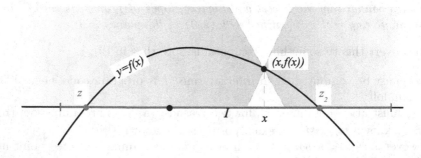

Fig. 2. Geometric view of Interval Newton: two-sided searchlight.

The same argument works as before, provided one interprets division as *reverse multiplication*, write this as $/\!/$:

$$c /\!/ b = \{ \text{ all solutions of } bx = c\} \qquad \text{1788's } \texttt{mulRev}(b,c).$$

So 0/0 is "whole real line" instead of "undefined". That is, (3) is replaced by

$$z \in \Big(x - \big(\boldsymbol{f}([x] /\!/ \boldsymbol{f}'(\boldsymbol{x}))\big)\Big). \tag{4}$$

This is just a restatement of formula (1), and its controlling quantifiers, two $\forall$ and one $\exists$. That is *always* valid even when (2) might divide by zero.

Now we enclose all roots even when many exist! However, the searchlight can "split" I into two pieces, as Fig. 2 shows. 1788 provides a "two-output reverse multiplication" operation $\texttt{mulRevToPair}$, adapted to this situation.

Interval Newton Iteration. We assumed the root z is in the interval $\boldsymbol{x}$ so it is safe to intersect the interval given by (4) with $\boldsymbol{x}$. This gives the 1983 method of Hansen and Greenberg [2], later refined by R.B. Kearfott [6], G. Mayer [10], P. van Hentenryck *et al.* [18], and others.

We assume the function f is C^1 on the initial interval (Fig. 3).

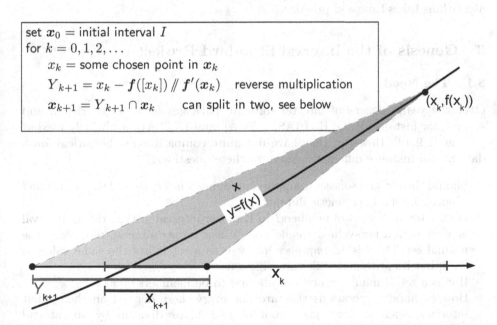

set $\boldsymbol{x}_0 = $ initial interval I
for $k = 0, 1, 2, \ldots$
    $x_k = $ some chosen point in $\boldsymbol{x}_k$
    $Y_{k+1} = x_k - \boldsymbol{f}([x_k]) /\!/ \boldsymbol{f}'(\boldsymbol{x}_k)$ reverse multiplication
    $\boldsymbol{x}_{k+1} = Y_{k+1} \cap \boldsymbol{x}_k$ can split in two, see below

Fig. 3. One step of Interval Newton method (one-sided searchlight case).

2.2   Lessons from the Example

Comments on the Algorithm. Here Y_{k+1}, hence $\boldsymbol{x}_{k+1}$, is potentially a union of two intervals that can be handled independently.

This structures the computation as a binary tree that progressively divides root-clusters into smaller sets, trying (but not always succeeding) to isolate each individual root. This can be done by various tree-traversal methods, exploiting parallelism if available.

By construction it is clear that all roots in $\boldsymbol{x}_k$ must be in Y_{k+1}, hence in (the possibly split in two) $\boldsymbol{x}_{k+1}$. However the algorithm has other remarkable and less obvious properties:

- if $\boldsymbol{x}_{k+1}$ is empty, then *no root exists* in $\boldsymbol{x}_k$.
- If $0 \notin \boldsymbol{f}'(\boldsymbol{x}_k)$, then *at most one root exists* in $\boldsymbol{x}_k$ (which must be in $\boldsymbol{x}_{k+1}$).
- If Y_{k+1} is nonempty, bounded, and contained in $\boldsymbol{x}_k$, then *exactly one root exists* in $\boldsymbol{x}_k$.

Comments on the "Interval Mindset". The analysis leading to the algorithm wasn't rocket science—just a careful look at how the quantifiers $\forall$, $\exists$ appeared in a use of the Mean Value Theorem.

In general however, to learn how to convert mathematics to effective interval algorithms takes time and practice.

3   Genesis of the Interval Standard Project

3.1   The Need

Over the years, dozens of interval software packages have been written, and several, for instance PROFIL/BIAS, Filib++ and INTLAB, are widely used at present [7,9,13]. However, they have not quite compatible mathematical foundations, for instance different answers to these questions:

- Should theory, and software, support unbounded intervals and the empty set? Moore's interval arithmetic did not.
- Is an interval $\boldsymbol{x}$ a set of numbers? In Kaucher interval arithmetic, an interval is a set with a two-valued "mode". [3, 4] is a *proper* interval, essentially the normal set. [4, 3] is an *improper* interval; as a set it has the same value as [3, 4] but its arithmetic rules are different.
- If $\boldsymbol{x}$ is a set of numbers, are $\pm\infty$ allowed to be members of $\boldsymbol{x}$?
- How to handle operations that are not everywhere defined on their input intervals, such as the square root of $[-2, 2]$, or division by an interval containing 0?

In addition, they have different software interfaces. Thus, currently one cannot write algorithms that are portable at a mathematical level, let alone write portable software.

3.2   Setting up a Working Group

In January 2008 at a conference in Dagstuhl, Germany, a project was started with the aim of producing an IA standard. In July that year it was approved by the IEEE as Working Group P1788 "A standard for interval arithmetic". In September, a conference in El Paso, USA, hosted its first face to face meeting at which the following officers were appointed.

Nathalie Revol, *Chair*
R. Baker Kearfott, *Vice Chair and Acting Chair*
William Edmonson, *Secretary*
Guillaume Melquiond, *Archivist*
J. Wolff von Gudenberg, *Web Master*
George Corliss, *Voting Tabulator*
John Pryce, *Senior Technical Editor*

Also during the project Christian Keil acted as *Deputy Technical Editor*, and Michel Hack, Vincent Lefèvre, Ian McIntosh, Dmitry Nadezhin, Ned Nedialkov and J. Wolff von Gudenberg were *Assistant Technical Editors*. About 140 people registered on the mailing list, of whom around 45 were regular voting members.

The group approved a final text in May 2014. Further revision, up to publication, then became the responsibility of the project's *sponsor ballot* group (whose membership overlaps with P1788's) and of IEEE editorial staff.

4   1788 Interval Principles

4.1   Definition of an Interval

There is a framework—so called *flavors*—to support *alternative mathematical foundations*. The standard currently has just one flavor called Set-Based, in which

- An interval x is a plain *set*, whose members are *real numbers*. This excludes $\pm\infty$ as members, so intervals are subsets of the real line $\mathbb{R}$.
- Open or half-open intervals are not allowed, but unbounded intervals are.
- The empty set is an interval.

This amounts to the mathematically simple definition:

Interval means *topologically closed and connected subset of* $\mathbb{R}$.

A potential alternative flavor is *Kaucher* (or very similar *modal*) interval arithmetic [5]: an interval is not a plain set, but an ordered pair $(\underline{x}, \overline{x})$ of reals:

$$(\underline{x}, \overline{x}) \text{ "means"} \begin{cases} \text{set } [\underline{x}, \overline{x}] \subset \mathbb{R} & \text{if } \underline{x} \le \overline{x} \text{ ("proper" interval)} \\ \text{something other if } \underline{x} > \overline{x} \text{ ("improper" interval).} \end{cases}$$

Another potentially important flavor is *Siegfried Rump's* interval arithmetic [15], which can support open or half-open intervals, and can handle finite precision under- and over-flow in a consistent and elegant way.

4.2   The Levels Structure

As in the IEEE floating-point standard 754, the 1788 standard manages complexity by distinguishing four specification levels:

Level 1. *Mathematical theory* of the set $\mathbb{IR}$ of **intervals** and their operations.
Level 2. *Finite precision* intervals—**datums**—and operations, independently of their representation.
Level 3. *Representation* of datums by **objects**, e.g. by a data structure comprising two floating-point numbers.
Level 4. *Encoding* of Level 3 objects as **bit strings**.

Inter-level Maps. Maps between levels are crucial—especially those between Level 1 and Level 2, or L1 $\longleftrightarrow$ L2 for short. The P1788 group made the following design decisions, which apply to all flavors:

– Each datum *is* a mathematical interval, so the map from L2 to L1 is just inclusion:

$$\text{L2 datums} \xrightarrow{\text{identity map}} \text{L1 intervals} \qquad (*)$$

(Not quite true: the datum also "knows what type it belongs to", i.e. is tagged with a unique name of its type. A programming language needs this information, since different types will be represented differently at L3.)

– Datums are organised into finite sets $\mathbb{T}$ called *interval types*. Thus each $\mathbb{T}$ may be regarded as a finite subset of $\mathbb{IR}$. The implementation has discretion on what types to provide.

– A L1 interval x maps to an interval of type $\mathbb{T}$—a $\mathbb{T}$-*interval* or $\mathbb{T}$-*datum*—by the $\mathbb{T}$-*hull* operation

$$\text{hull}_T(x) = \text{smallest } \mathbb{T}\text{-interval that contains } x,$$

where "contain" has a flavor-defined meaning (which for Set-Based intervals is the usual one). This defines the map back from L1 to L2:

$$\text{L1 intervals} \xrightarrow{\mathbb{T}\text{-hull}} \text{L2 datums of type} \mathbb{T} \qquad (**)$$

– To do an operation $x \bullet y$ at L2 on $\mathbb{T}$-datums, in any flavor:

map x, y to L1 by (*);
do the operation at L1;
map back to L2 by (**).

This specification of the relation between mathematics and finite precision looks trivial but is not: it defines the whole character of the standard. Not all IA theories are clear on this issue. Time will tell whether our choice was a wise one.

This choice affects implementations. For instance an arbitrary-precision interval package must be structured as a potentially infinite set of types, each containing finitely many intervals. It cannot comprise a single type containing potentially infinitely many intervals.

Maps for Levels 3 and 4. There are two fairly obvious rules:

  L2 ⟨ ⟩ L3: Each L2 datum is *represented by* at least one L3 object; each L3 object *represents* at most one L2 datum.
- L3 ⟷ L4: Each L3 object is *encoded by* at least one L4 bitstring; each L4 bitstring *encodes* at most one L3 object.

5   Exception Handling

5.1   A Hypothetical Scenario

Less than 10 years hence in the Old Bailey, London, ...

The case *Crown versus Google* concerns the Google Driverless Car, GDC. One of them badly injured a pedestrian who stepped into the road in front of it.

The GDC's emergency stop system is *designed* to act faster than a good human driver (undisputed) but is it badly *implemented* (disputed)?

The software uses an interval algorithm, built on a 1788-conforming library, which applies *Brouwer's fixed point theorem*. Could this have an error? E.g., it may have thought it had enclosed a root of an equation when it had not.

Depending on what software bugs are found (if any), liability might lie with the pedestrian's negligence. Or with GDC's software implementers. Or with the 1788 library implementers. Maybe even with the mathematicians who claimed to have proved the design of 1788 is correct?

A lot of money rides on whether 1788-based code might be wrong, when deciding that a function is defined and continuous on a box.

5.2   Theoretical Context

Basic Problem. How (at Level 1) to treat operations that are not everywhere defined and/or continuous on their input box? For example

$$\text{(real) square root } \sqrt{x} \qquad \frac{x}{y} \qquad \texttt{floor}(x)$$

$$\sqrt{[-2,2]} \qquad \frac{[2,3]}{[-1,1]} \qquad \texttt{floor}([2.5,4.5])$$

undefined on $-2 \le x < 0$ undefined if $y = 0$ discontinuous at $x = 3, 4$

We decided the default is "evaluate where defined, ignore where undefined", called *non-stop* or *loose* evaluation. For instance

$$\sqrt{[-2,2]} = \{\, \sqrt{x} \mid x \in [-2,2] \text{ and } x \ge 0 \,\} = [0, \sqrt{2}]$$

with no error reported. This is similar to IEEE 754 floating-point, which responds to an invalid operation such as $0/0$ or $\infty - \infty$ by returning the result NaN and continuing to compute.

This is a valid approach for, e.g., many *global optimisation* methods. It is not valid when applying *Brouwer's theorem*, which needs a guarantee that a function is everywhere defined and continuous on a box.

It also will not do for some *graphics rendering* algorithms, which need to know a function is everywhere defined on a box, but are not bothered about continuity.

Tracking Function Properties. One needs a mechanism to track whether a function has these desirable properties of definedness and/or continuity. This leads to a powerful extension of the Fundamental Theorem of interval arithmetic based on well-known theorems of set theory and analysis, which can be summarised:

> If for function f given by an expression, each individual library operation in f is everywhere defined on its input set, then the same holds for f.
> The same is true when *defined* is replaced by *defined and continuous*.

Example 2. Let $f(x) = 1/\sqrt{x}$, composed of library operations $\mathtt{sqrt}(t) = \sqrt{t}$ followed by $\mathtt{recip}(t) = 1/t$. Evaluate $z = f(x)$ in exact (Level 1) interval arithmetic. Abbreviate *defined and continuous* to DAC.

Let the input to f be $x = [1,4]$. We do $y = \mathtt{sqrt}(x) = [1,2]$, followed by $z = \mathtt{recip}([1,2]) = [\frac{1}{2},1]$. Each operation is (everywhere) DAC on its input: $\mathtt{sqrt}$ on $[1,4]$ and $\mathtt{recip}$ on $[1,2]$. We conclude f is DAC on this x.

If x is $[0,4]$ then $\mathtt{sqrt}$ is DAC on this x but $\mathtt{recip}$ is not DAC on the resulting $y = [0,2]$, so we cannot say f is DAC on x.

Similarly, if x is $[-2,4]$ then $\mathtt{sqrt}$ fails to be DAC on this x and again we cannot say f is DAC on x. □

In the two last cases of this example it is easy to prove f is definitely *not* DAC on x, but for complicated functions in the presence of roundoff, to prove such a negative—*definitely not everywhere DAC*—is nearly impossible. Therefore the 1788 system only provides for a definite positive, which is cheap to compute. It can also say *definitely nowhere defined* on the input, which is cheap too.

5.3 Decorations

To provide a mechanism to track such properties of functions, we rejected the IEEE 754 standard's method of *global flags*, as being obsolete for today's massively parallel platforms. Instead, 1788 provides for *decorated intervals*. Such an interval is a pair (y, dy), also written y_{dy} when convenient:

– an ordinary interval y,
– a tag dy called a decoration, giving data about definedness, continuity, etc.[1].

Formally, a decoration d is a label for an assertion (boolean-valued function) $p_d(f, x)$ about a function $f : \mathbb{R}^n \to \mathbb{R}$ and a box $x \subseteq \mathbb{R}^n$, for arbitrary n. Five decorations are defined in order of "goodness", $\mathtt{ill} < \mathtt{trv} < \mathtt{def} < \mathtt{dac} < \mathtt{com}$:

ill Label for ill-formed intervals, formally "f is nowhere defined".
trv (trivial) Always true = "no information".
def f is everywhere defined on x.
dac As **def**, plus f is everywhere continuous on x.
com As **dac**, plus f is bounded on x at Level 2, meaning that no overflow occurred while computing it[2].

---

[1] dy is just a mnemonic, "decoration for y". It has nothing to do with differentials.
[2] **com** means "common", see Sect. 6, but also that code can *verify* it is common.

Let $(\mathbf{y}, d\mathbf{y})$ result from evaluating an arithmetic expression $f(x_1, \ldots, x_n)$

- on correctly initialised decorated interval inputs $(\mathbf{x}_1, d\mathbf{x}_1), \ldots, (\mathbf{x}_n, d\mathbf{x}_n)$ (the programmer's responsibility),
- using correctly written decorated interval library operations (the implementation's responsibility).

Then Moore's Fundamental Theorem says

$$\mathbf{y} \text{ contains the range of } f \text{ over } \mathbf{x}_1 \times \cdots \times \mathbf{x}_n,$$

and in addition,

$$\text{the decoration } d\mathbf{y} \text{ makes a true assertion about } f \text{ over } \mathbf{x}.$$

For instance if $d\mathbf{y}$ is computed to be **def** then f has been proved to be everywhere defined on $\mathbf{x}$.

As with a computed range enclosure, a computed decoration is often *not sharp*. E.g. it may be **trv** (no information) or **def** (defined) when actually **dac** (defined and continuous) is true. Much of the craft of IA is knowing how to "sharpen" such information, e.g. by cutting an input box into smaller boxes handled separately.

Example 3. Consider the fixpoint problem, to solve $g(x) = x$ where

$$g(x) = 2\sqrt{x} - \tfrac{1}{2}.$$

Roots are $x = \tfrac{3}{2} \pm \sqrt{2} = 0.0858\ldots$ or $2.9142\ldots$
  We aim to use interval fixpoint iteration

$$\mathbf{x}_0 = \text{initial guess}; \quad \mathbf{x}_{n+1} = g(\mathbf{x}_n) \text{ for } n = 0, 1, \ldots$$

First, use ordinary undecorated interval arithmetic.
  Case A: $\mathbf{x}_0 = [2, 3]$. Then

$$\mathbf{x}_1 = \left[2\sqrt{2} - \tfrac{1}{2}, 2\sqrt{3} - \tfrac{1}{2}\right] = [2.3\ldots, 2.9\ldots] \subset \mathbf{x}_0.$$

This is genuine and (by Brouwer's Theorem) it proves a fixpoint exists in $\mathbf{x}_1$.
  Case B: $\mathbf{x}_0 = [-1, \tfrac{1}{16}]$. Then

$$\mathbf{x}_1 = 2\sqrt{[-1, \tfrac{1}{16}]} - \tfrac{1}{2} = 2\,[0, \tfrac{1}{4}] - \tfrac{1}{2} = [0, \tfrac{1}{2}] - \tfrac{1}{2} = [-\tfrac{1}{2}, 0], \text{ again } \subset \mathbf{x}_0 \,!$$

But there is no root in $\mathbf{x}_0$, let alone $\mathbf{x}_1$! This is spurious, due to 1788's (undecorated) square root function discarding the negative part of $\mathbf{x}_0$ without comment.
  Now use decorated interval arithmetic. The rule for propagating decorations is, roughly, that an operation outputs the worst decoration, in the "goodness" order defined on p. 10, out of the decorations on its operands and the decoration

generated while performing the operation. Showing decorations as subscripts, Case B gives

$$x_1 = [2]_{\text{dac}} \times \sqrt{[-1, \tfrac{1}{16}]_{\text{dac}}} - [\tfrac{1}{2}]_{\text{dac}}$$

$$= [2]_{\text{dac}} \times [0, \tfrac{1}{4}]_{\text{trv}} - [\tfrac{1}{2}]_{\text{dac}},$$

$$= [0, \tfrac{1}{2}]_{\text{trv}} - [\tfrac{1}{2}]_{\text{dac}} \qquad\qquad = [-\tfrac{1}{2}, 0]_{\text{trv}}.$$

Recall $\text{trv} = $ "no information", so the calculation is, correctly, unable to verify the conditions of Brouwer's Theorem. But Case A produces

$$x_1 = [2.3\ldots,\ 2.9\ldots]_{\text{dac}},$$

proving g is DAC on x_0, as well as mapping x_0 into itself—the conditions for applying Brouwer's Theorem have been verified.   □

The decoration system is the feature that most distinguishes 1788 from earlier IA systems. An annex in the Standard contains a rigorous proof of correctness: a *Fundamental Theorem of Decorated Interval Arithmetic*.

6   Difficulties the Group Encountered

Certain issues caused long and heated debate. We are grateful for the diplomatic skills the Chair and Vice-chair sometimes needed to deploy, and the good sense of IEEE procedural guidelines for "online democracy". Here are a few examples.

The Choice of Foundational Mathematical Model. Most users of IA are in the academic community, and most of these use some form of "interval is just a set of numbers" theory and software. But Kaucher/modal theory—with intervals like $[4, 3]$—has its proponents. One of them is Nate Hayes, whose company does high-quality graphics rendering for the movie industry. For its specialised interpolation algorithms, Kaucher methods are reported to give tighter enclosures and greater speed.

The resulting tension between "intervals for knowledge" and "intervals for profit" was fruitful. Faced with two related kinds of object, it is natural to look for a theory that supports both, in a tightly coupled sense that lets both exist in a computer program and inter-operate.

We tried this over a period of many months with set-based and Kaucher intervals, but failed. For instance, unbounded intervals within Kaucher theory needed arbitrary restrictions or led to logical contradictions—briefly, Kaucher cannot handle $[3, \infty)$ consistently, and set-based cannot handle $[4, 3]$.

This was the main motivation for the *flavor* concept. It allows different theories that are "recognisably 1788" in a loosely coupled sense. The main requirement is that each flavor's intervals must include Moore's original (closed, bounded, nonempty) real intervals, called *common intervals*, and a library of operations that at Level 1, when acting on common intervals, produce the same results in all flavors.

A Kaucher standard document was promised but has not materialised yet. However at least one other theory that holds promise for effective interval computation fits into the flavor mould—that of Rump [15], especially if coupled with the Gustafson *universal numbers* system [1] (and see elsewhere in this volume). So I feel the effort put into this part of the standard has not been wasted.

The Decoration Scheme. The group saw from the start that checking definedness and continuity of a function can in principle be automated, and early on rejected global flags in favour of decorating individual intervals. The chosen scheme took nearly two years, off and on, to decide. Initially we used separate boolean flags for "defined", "continuous", etc. Arnold Neumaier of Vienna first proposed that decorations should be a linear sequence. Of several such schemes, we nearly adopted one with 6 decorations, till Guillaume Melquiond pointed out that one of them gave no information not already available to a programmer, so we removed it to give the current 5-decoration system.

(We are also indebted to Arnold for his earlier document, the Vienna Proposal for Interval Standardization [12], from which many ideas in 1788 are drawn.)

Input/Output. I/O is important. A key reason why the Algol 60 language died and Fortran, its arguably inferior contemporary, thrived is that the latter had a language-defined I/O scheme and the former did not.

The working group debated at length on what I/O should be required and how prescriptive the standard should be. Eventually it agreed to specify an external text representation of intervals, so called *interval literals*. Examples are [empty], [1.23,4.56] and the uncertainty form 1.23?4 which means $1.23 \pm$ (4 units in the last place), i.e. $[1.19, 1.27]$.

An implementation shall provide functions to read such literals in free format, and write them in either free (e.g., for interactive work) or fixed (for tabulation) format. However we did not standardise the conversion specifiers (such as C's %8.3f or Fortran's F8.3 for output of floating-point numbers), leaving this implementation-defined, to be standardised at a future revision.

In addition to the above transformations, which generally incur roundoff, each finite-precision type $\mathbb{T}$ shall define an *exact text representation*, giving loss-free conversion between $\mathbb{T}$-intervals and text strings. For types based on IEEE 754 numbers, 1788 specifies an *interchange encoding*, giving loss-free conversion between $\mathbb{T}$-intervals and bit strings. This last is 1788's only Level 4 requirement.

What to say About Accuracy? The accuracy issue for intervals is different from that for floating-point. The result $\widetilde{y}$ of an interval library operation must enclose[3] the mathematical result y. If not, it is wrong, period. If it does—even if it is the useless result $[-\infty, +\infty]$—it is valid.

---

[3] This has been called the "Thou Shalt Not Lie" principle.

After much discussion we agreed that *tightness*—how close $\widetilde{y}$ is to the enclosed y—is a quality-of-implementation issue, barring a few cases where requiring optimal tightness is reasonable.

So the standard acts as a "regulatory authority" here—it does not specify an accuracy, but it requires a conforming implementation to state the accuracy of each of its library operations, in a verifiable way, using a format specified by the standard. There is typically a trade-off between accuracy and speed, and the aim is to make it possible for users to make a fair comparison of the merits of different implementations.

7   Current State

The main text has around 70 pages, of which roughly 60 % are Level 1, 35 % Level 2, 5 % Level 3, with a half-page of Level 4.

Following approval by a vote of the group in May 2014, the text was extensively reworked with help from IEEE editorial staff to fit their style guidelines.

It was signed off in November 2014 to enter the Sponsor Ballot phase, and examined by a selected group representative of academia, software developers, industry, etc., and of geographical regions. They approved it after various changes, both editorial and technical. Finally, IEEE Std 1788^{TM}-2015 was approved by the IEEE Standards Board in June 2015, and published at the end of that month.

In addition, a *Basic Standard for Interval Arithmetic* (BSIA) has been written by Ned Nedialkov. At around 20 pages it is a cut down version, simpler to implement and suitable for undergraduate teaching. A program that runs under an implementation of the BSIA should run and give identical results up to round-off under an implementation of the full standard. The IEEE have approved a project, P1788.1, for the BSIA to become a separate but related standard.

A   Proof of Interval Newton Properties

This appendix proves the properties stated in Sect. 2.2. It may be of interest because item (iv) of the Theorem does not seem to have appeared in the literature before. An *interval extension* of a real function f of real variables means an interval function f of corresponding interval variables such that $y = f(x_1, \ldots, x_n)$ is in $y = f(x_1, \ldots, x_n)$ whenever x_i is in x_i for each $i = 1, \ldots, n$.

Theorem 2. *Let* $f : \mathbb{R} \to \mathbb{R}$ *be* C^1 *on an interval* x, *which may be unbounded. Let* f *and* f' *be interval extensions of* f *and its derivative* f', *and let* x *be any point of* x. *Define the set*

$$Y = x - f([x]) \;/\!/\; f'(x)$$

where $/\!/$ *denotes division in the sense of reverse multiplication. (Thus* Y *may be empty, an interval, or the union of two disjoint unbounded intervals.)*

Then

(i) Y contains all zeros of f in $\boldsymbol{x}$.
(ii) If $Y \cap \boldsymbol{x} = \emptyset$, there are no zeros of f in $\boldsymbol{x}$.
(iii) If $0 \notin \boldsymbol{f'}(\boldsymbol{x})$, there is at most one zero of f in $\boldsymbol{x}$.
(iv) If Y is nonempty, bounded and $\subseteq \boldsymbol{x}$, there is exactly one zero of f in $\boldsymbol{x}$.

Proof. (i) Let $z \in \boldsymbol{x}$ with $f(z) = 0$. By the Mean Value Theorem

$$f(x) = f(x) - f(z) = (x - z)f'(\xi). \tag{5}$$

for some $\xi \in \boldsymbol{x}$. By definition of interval extension, $f'(\xi) \in \boldsymbol{f'}(\boldsymbol{x})$ and $f(x) \in \boldsymbol{f}([x])$. Hence by the definition of reverse multiplication

$$x - z \in \boldsymbol{f}([x]) \mathbin{/\!/} \boldsymbol{f'}(\boldsymbol{x}),$$

that is

$$z \in x - \boldsymbol{f}([x]) \mathbin{/\!/} \boldsymbol{f'}(\boldsymbol{x})$$

as required.

(ii) This is immediate from (i).

(iii) In (5) let both z and x be roots in $\boldsymbol{x}$. Then we have $0 = f(x) - f(z) = (x - z)f'(\xi)$. By hypothesis $0 \notin \boldsymbol{f'}(\boldsymbol{x})$ which implies $f'(\xi) \neq 0$. Hence $x - z = 0$, $x = z$, so there is at most one root.

(iv) Write $\boldsymbol{b} = \boldsymbol{f'}(\boldsymbol{x})$, $\boldsymbol{c} = \boldsymbol{f}([x])$, both being nonempty by the definition of interval extension. By hypothesis $Y = x - \boldsymbol{c} \mathbin{/\!/} \boldsymbol{b}$ is nonempty and bounded, so $Z = \boldsymbol{c} \mathbin{/\!/} \boldsymbol{b}$ is nonempty and bounded.

I claim $0 \notin \boldsymbol{b}$. For suppose $0 \in \boldsymbol{b}$. Then $0 \notin \boldsymbol{c}$, for if $0 \in \boldsymbol{c}$ then Z is the unbounded set $\mathbb{R}$, contrary to hypothesis. Now two subcases arise.
– Either $\boldsymbol{b}$ is singleton $[0]$, making Z empty, contrary to hypothesis.
– Or, $\boldsymbol{b}$ contains 0 and another point, in which case it contains points arbitrarily close to 0. Since $0 \notin \boldsymbol{c} \neq \emptyset$, $\boldsymbol{c}$ contains a nonzero point. Together these imply $\boldsymbol{c} \mathbin{/\!/} \boldsymbol{b}$ is unbounded, again contrary to hypothesis.

Thus all the cases of $0 \in \boldsymbol{b}$ give a contradiction, proving $0 \notin \boldsymbol{b}$. Hence by part (iii) there is *at most* one root in $\boldsymbol{x}$ and we must show there is *at least* one. If $f(x) = 0$ there is nothing more to prove, so assume $f(x) \neq 0$.

Let b^* be the bound of $\boldsymbol{b}$ nearest 0, so it is finite, $\neq 0$ and in $\boldsymbol{b}$. Let z^* be the intercept on the x-axis of the line through $(x, f(x))$ with slope b^*, so

$$z^* = x - f(x)/b^*, \tag{6}$$

equivalently

$$f(x) = (x - z^*)b^*. \tag{7}$$

Since $f(x) \in \boldsymbol{c}$ and $b^* \in \boldsymbol{b}$, (6) shows $z^* \in Y \subseteq \boldsymbol{x}$. Also $x \in \boldsymbol{x}$ so by the Mean Value Theorem there is $\xi \in \boldsymbol{x}$ with

$$f(x) - f(z^*) = (x - z^*)f'(\xi). \tag{8}$$

Subtracting this from (7) gives

$$f(z^*) = (x - z^*)(b^* - f'(\xi)) \tag{9}$$

Now $f'(\xi)$ is in b by the latter's definition, so by the definition of b^* it has the same sign as b^* and at least as large absolute value, i.e. $f'(\xi)/b^* \geq 1$. Dividing (9) by (7) (recalling $f(x) \neq 0$) now gives

$$f(z^*)/f(x) = 1 - f'(\xi)/b^* \leq 0,$$

so $f(z^*)$ has opposite (in the weak sense) sign to $f(x)$. By the Intermediate Value Theorem f has a zero z between x and z^*. Since both the latter are in x we have $z \in x$, and the result is proved. $\qquad\square$

References

1. Gustafson, J.L.: The End of Error: Unum Computing. Chapman and Hall/CRC. Taylor & Francis, Upper Saddle (2015)
2. Hansen, E.R., Greenberg, R.I.: An interval Newton method. J. Appl. Math. Comput. **12**, 89–98 (1983)
3. Heath, T.L.: A History of Greek Mathematics. Number v. 1 in A History of Greek Mathematics. Clarendon Press, Oxford (1921)
4. Kantorovich, L.V.: On some new approaches to computational methods and to processing of observations. Siberian Math. J. **3**(5), 701–709 (1962). In Russian
5. Kaucher, E.W.: Interval analysis in the extended interval space IR. Comput. Suppl. **2**, 33–49 (1980)
6. Baker Kearfott, R.: A Fortran 90 environment for research and prototyping of enclosure algorithms for nonlinear equations and global optimization. ACM TOMS **21**(1), 63–78 (1995)
7. Knüppel, O.: PROFIL/BIAS – a fast interval library. Computing **53**(3–4), 277–287 (1994)
8. Kreinovich, V.: Interval computations web site. http://cs.utep.edu/interval-comp, August 28, 2014
9. Lerch, M., Tischler, G., von Gudenberg, J.W., Hofschuster, W., Krämer, W.: FILIB++, a fast interval library supporting containment computations. ACM Trans. Math. Softw. **32**(2), 299–324 (2006)
10. Mayer, G.: Epsilon-inflation in verification algorithms. J. Comput. Appl. Math. **60**, 147–169 (1995)
11. Moore, R.E.: Interval Analysis. Prentice-Hall, Englewood Cliffs, N.J. (1966)
12. Neumaier, A.: Vienna proposal for interval standardization. August 28, 2014. December 2008. http://www.mat.univie.ac.at/~neum/ms/1788.pdf
13. Rump, S.M.: INTLAB - INTerval LABoratory. In: Csendes, T. (ed.), Developments in Reliable Computing, pp. 77–104. Kluwer Academic Publishers, Dordrecht (1999). http://www.ti3.tuhh.de/rump/
14. Rump, S.M.: Verification methods: Rigorous results using floating-point arithmetic. Acta Numer. **19**, 287–449 (2010). Cambridge University Press
15. Rump, S.M.: Interval arithmetic over finitely many endpoints. BIT Numer. Math. **52**(4), 1059–1075 (2012)

16. Sunaga, T.: Theory of an interval algebra and its application to numerical analysis. Res Assoc. Appl. Geom. Memoirs **2**, 29–46 (1958)
17. Tucker, W.: Validated Numerics: A Short Introduction to Rigorous Computations. Princeton University Press, Princeton (2011)
18. van Hentenryck, P., Macallester, D., Kapur, D.: Solving polynomial systems using a branch and prune approach. SIAM J. Numer. Anal. **34**(2), 797–827 (1997)

Uncertainty

Numerical Probabilistic Approach
for Optimization Problems

Boris Dobronets[(✉)] and Olga Popova

Siberian Federal University, Krasnoyarsk, Russia
{BDobronets,olgaarc}@yandex.ru

Abstract. In the paper a new approach to optimization problems with random input parameters, which is defined as random programming, is discussed. This approach uses a numerical probability analysis and allows us to construct the set of solutions of an optimization problem based on the joint probability density function.

1 Introduction

The studies of many practical problems, including the problem of decision-making, require the implementation of the optimization approach. The effectiveness of the solutions depends on several factors. Such factors primarily include the data necessary for the description and the solution of the problem. One of the important factors that should be considered when solving such problems is uncertainty of input data.

We can distinguish three basic models of uncertainty: stochastic, fuzzy and set-valued (in particular — interval-valued).

The nature of undefined data may be random errors associated with measurement, or the incompleteness of information. This is described as random, inaccurate, incomplete data.

The paper deals with the numerical probabilistic approach to solving optimization problems with random inputs. Using methods of mathematical programming for these problems, we obtain optimal solutions that depend on these parameters. In the cases where probability densities of input parameters are known, it is possible to construct a probability density function of the joint probability of the optimal solutions on the basis of numerical probability analysis. In contrast to the stochastic programming [6,8], where the optimal solution is a fixed solution, this approach allows us to obtain the whole set of solutions of an optimization problem defined by the constructed joint probability density function.

By *random programming*, we mean the methods for the construction of the solution set for an optimization problem with random input parameters based on the application of *numerical probabilistic analysis*.

It is important to note that after representation of the obtained uncertainties we deal with the problem of choosing a method that allows us to perform the subsequent calculations in such a way as to get real results without additional uncertainties [5].

© Springer International Publishing Switzerland 2016
M. Nehmeier et al. (Eds.): SCAN 2014, LNCS 9553, pp. 43–53, 2016.
DOI: 10.1007/978-3-319-31769-4_4

To this end, nowadays mathematical tools for uncertain programming are developed. Uncertain programming is the theoretical basis for solving optimization problems for various uncertainty conditions [6].

Since an interval number can be considered as a special case of an imprecise quantity, interval analysis, interval arithmetic, and interval programming fall into imprecise programming.

In most of stochastic programming algorithms, the operator of mathematical expectation is used and averaging procedures are performed.

In this paper, we develop a technique that uses Numerical Probabilistic Analysis to solve various problems with stochastic data uncertainty [2,4].

The basis of NPA consists of numerical operations on probability density functions of random variables. They involve the operations "+", "−", ".", "/", "↑", "max", "min", as well as binary relations "≤", "≥" and some others. The numerical operations of the histogram arithmetic constitute the major component of NPA [1].

Using the arithmetic of probability density functions and probabilistic extensions, we can construct numerical methods that enable us to solve systems of linear and nonlinear algebraic equations with stochastic parameters [2].

2   Formulation of the Problem and Background

Let us formulate the problem of *random programming* as follows:

$$\min_x f(x, \xi), \tag{1}$$

subject to (s.t.)

$$g_i(x, \xi) \leq 0, \quad i = 1, ..., m, \tag{2}$$

where x is the solution vector, ξ is the vector of parameters, $f(x, \xi)$ is the objective function, $g_i(x, \xi)$ are constraint functions.

The vector x^* is a solution of problem (1)–(2), if

$$f(x^*, \xi) = \inf_{x \in U} f(x, \xi),$$

where

$$U = \{x | g_i(x, \xi) \leq 0, \quad i = 1, ..., m\}.$$

We will suppose that ξ have random components. As x^* is a function of the vector ξ, then

$$x^* = x^*(\xi),$$

it is also a random vector and its joint probability distribution is what we are interested in.

The solution set of (1)–(2) is defined as follows

$$\mathcal{X} = \{x^* | f(x^*, \tilde{\xi}) = \inf_{x \in U} f(x, \tilde{\xi}), g_i(x, \tilde{\xi}) \leq 0, \quad i = 1, ..., m, \tilde{\xi} \in \text{supp}(\xi)\}.$$

So in contrast to the deterministic problem, for x^* it is necessary to determine the probability density function for each component of x_i^* as the joint probability density.

When both objective function and constraint functions are linear functions, the problem is called a problem of linear programming. Otherwise, the problem is called a problem of nonlinear programming.

For example the problem of linear programming with random data is formulated as follows:

$$\min_x c^T x, \tag{3}$$

s.t.

$$Ax = b, x \geq 0, \tag{4}$$

where A is a matrix, b, c are vectors.

The vector x^* is the solution of problem (3)–(4) provided that

$$c^T x^* = \inf_{x \in U} c^T x,$$

where

$$U = \{x | Ax = b, x \geq 0\}.$$

Let $\boldsymbol{A}$ be a random matrix, $\boldsymbol{b}$, $\boldsymbol{c}$ are random vectors. So x^* is a random function of the variables in $\boldsymbol{A}$, $\boldsymbol{b}$ and $\boldsymbol{c}$.

The solution set of (3)–(4) is

$$\mathcal{X} = \{x^* | c^T x^* = \inf_{x \in U} c^T x, Ax = b, x \geq 0, A \in \mathrm{supp}(\boldsymbol{A}), b \in \mathrm{supp}(\boldsymbol{b}), c \in \mathrm{supp}(\boldsymbol{c})\}.$$

3   Numerical Probabilistic Analysis

3.1   Operations on Probability Densities of Random Variables

Let us consider operations on histograms. Let $p(x, y)$ be a joint probability density function of two random variables $\boldsymbol{x}$ and $\boldsymbol{y}$. Let p_z be a histogram approximating the probability density of the operations between two random variables $\boldsymbol{x} * \boldsymbol{y}$, where $* \in \{+, -, \cdot, /, \uparrow\}$. Then the probability to find the value z within the interval $[z_i, z_{i+1}]$ is determined by the formula [1,2]

$$P(z_k < z < z_{k+1}) = \int_{\Omega_k} p(x, y) \, dx \, dy, \tag{5}$$

where $\Omega_k = \{(x, y) | z_k \leq x * y \leq z_{k+1}\}$ and the value P_k of the histogram on the interval $[z_k, z_{k+1}]$ is defined as

$$P_k = \int_{\Omega_k} p(x, y) \, dx \, dy / (z_{k+1} - z_k).$$

Then we extend the order relation $\succ \in \{<, \leq, \geq, >\}$ to random variables:

$$x \succ y \text{ if and only if } x \succ y \text{ for all } x \in \boldsymbol{x}, y \in \boldsymbol{y}.$$

If the support of $\boldsymbol{x}$, $\boldsymbol{y}$ are intersected, then we can talk about the probability of $\boldsymbol{x} \succ \boldsymbol{y}$

$$P(\boldsymbol{x} \succ \boldsymbol{y}) = \int_\Omega p(x, y) dx dy,$$

where $\Omega = \{(x, y) | x \succ y\}$ is the set of points $(x, y) \in R^2$ such that $x \succ y$, $p(x, y)$ is the joint probability density of $\boldsymbol{x}$, $\boldsymbol{y}$.

For example, consider the operation $\max(\boldsymbol{x}, \boldsymbol{y})$. The probability $P(\max(\boldsymbol{x}, \boldsymbol{y}) < z)$ is determined by the formula

$$P(z) = \int_{\Omega_z} p(x, y) dx dy,$$

where $\Omega_z = \{(x, y) | (x < z) \text{ and } (y < z)\}$ and the value P_i of the histogram on the interval $[z_i, z_{i+1}]$ is defined as

$$P_i = (P(z_{i+1}) - P(z_i))/(z_{i+1} - z_i).$$

3.2   Probabilistic Extensions

One of the most important problems that NPA deals with is to construct probability density functions of random variables. Let us start with the general case where $(x_1, \ldots, x_n)$ is a system of continuous random variables with the joint probability density function $p(x_1, \ldots, x_n)$ and the random variable z is a function $f(x_1, \ldots, x_n)$

$$z = f(x_1, \ldots, x_n). \tag{6}$$

By *probabilistic extension* of the function f, we mean the probability density function of the random variable z.

Let us construct the histogram F approximating the probability density function of the variable z. Suppose the histogram F is defined on a grid $\{z_i \mid i = 0, \ldots, n\}$. The domain is defined as $\Omega_i = \{(x_1, \ldots, x_n) | z_i < f(x_1, \ldots, x_n) < z_{i+1}\}$. Then the value F_i of the histogram on the interval $[z_i, z_{i+1}]$ is defined as

$$F_i = \int_{\Omega_i} p(x_1, x_2, \ldots, x_n) dx_1 dx_2 \ldots dx_n/(z_{i+1} - z_i). \tag{7}$$

By *histogram probabilistic extension* of the function f, we mean the histogram F constructed according to (7).

Let $f(x_1, \ldots, x_n)$ be a rational function. To construct the histogram of F, we replaced the arithmetic operation by the histogram operation, while the variables $x_1, x_2, \ldots, x_n$ are replaced by the histogram of their possible values. It makes sense to call the resulting histogram of F as *natural histogram extension* (similar to "natural interval extension").

Case 1 [3]. Let $x_1, \ldots, x_n$ be independent random variables. If $f(x_1, \ldots, x_n)$ is a rational expression where each variable x_i occurs no more than once, then the natural histogram extension approximates a probabilistic extension.

Case 2. Let the function $f(x_1, \ldots, x_n)$ admit a change of variables, so that $f(z_1, \ldots, z_k)$ is a rational function of the variables $z_1, \ldots, z_k$ satisfying the conditions of Case 1. The variable z_i is a function of x_i, $i \in Ind_i$ and the Ind_i are mutually disjoint. Suppose for each z_i it is possible to construct the probabilistic extension. Then the natural extension of $f(z_1, \ldots, z_k)$ is approximated by the probabilistic extension of $f(x_1, \ldots, x_n)$.

Case 3. We have to find the probabilistic extension for the function $f(x_1, x_2, \ldots, x_n)$, but the conditions of Case 2 are not fulfilled. Suppose for definiteness that only x_1 occurs a few times.

If, instead of the random variable x_1, we substitute a determinate value t, then it is possible to construct the natural probabilistic extension for the function $f(t, x_2, \ldots, x_n)$.

Suppose that t is discrete random value approximating x_1 as follows. Let t take the value t_i with probability P_i and for each function $f(t_i, x_2, \ldots, x_n)$ it is possible to construct the natural probabilistic extension.

Then the probabilistic extension f of the function $f(x_1, \ldots, x_n)$ can be approximated by the probability density φ as follows [3]

$$\varphi(\xi) = \sum_{i=1}^{n} P_i \varphi_i(\xi). \tag{8}$$

3.3   Systems of Linear Algebraic Equations

Consider a system of linear algebraic equations

$$\boldsymbol{A}\boldsymbol{x} = \boldsymbol{b}, i = 1 \ldots n, \tag{9}$$

where $\boldsymbol{x} \in \boldsymbol{R}^n$ is a random vector solution, $\boldsymbol{A} = (a_{ij})$, $\boldsymbol{b} = (b_i)$ are a random matrix and a right-hand side vector, respectively. Suppose that the random matrix a_{ij} and the vector b_i have independent components with probability densities pa_{ij}, pb_i respectively.

The support of the solution set can be represented as follows [2,7]

$$X = \{x | Ax = b, A \in \text{supp}(\boldsymbol{A}), b \in \text{supp}(\boldsymbol{b})\}.$$

With each $x \in X$ we can associate the following subset of coefficients $A_x \subset \text{supp}(\boldsymbol{A}), b_x \subset \text{supp}(\boldsymbol{b})$

$$\Omega_x = \{A, b | Ax = b, A \in \text{supp}(\boldsymbol{A}), b \in \text{supp}(\boldsymbol{b})\}.$$

Note that for fixed x, the coefficients of the matrix and the right-hand side vector are related by

$$\sum_{j=1}^{n} a_{ij} x_j - b_i = 0, i = 1, \ldots, n,$$

therefore

$$\Omega_x = \{A, b | \sum_{j=1}^{n} a_{ij}x_j - b_i = 0, i = 1, ..., n\}.$$

Suppose we want to find the probability $P(X_0)$ that the solutions x falls in a subset $X_0 \subset X$. With X_0 we associate the set $\Omega_0 = \{\Omega_x | x \in X_0\}$.
Then

$$P(X_0) = \int_{\Omega_0} \prod_{i=1}^{n} \prod_{j=1}^{n} pa_{ij} \prod_{i=1}^{n} pb_i d\Omega.$$

Since $P(X_0)$ in many cases is proportional to the volume of Ω_0, we can a priori determine the areas with the lowest and highest probability.

4   Random Linear Programming

It is known that for the problem (3)–(4) the optimal solution x^* is achieved at the corner of the set U.

Theorem 1 [9]. *Let the set U be defined the conditions (4). A point $x = (x_1, ..., x_n) \in U$ is a corner point if and only if there exist numbers $j_1, ...j_r$:*

$$A_{j_1}x_{j_1} + ... + A_{j_r}x_{j_r} = b; x_j = 0, j \neq j_l, l = 1, ..., r,$$

where the columns of the $A_{j_1}, ..., A_{j_r}$ are linearly independent.

Example 1. Let U is defined by a matrix A and vector b

$$A = \begin{pmatrix} 1 & 1 & 3 & 1 \\ 1 & -1 & 1 & 2 \end{pmatrix}, b = \begin{pmatrix} 3 \\ 1 \end{pmatrix},$$

then to the columns of the matrix A_1, A_2 there corresponds to a corner point with coordinates $(2, 1, 0, 0)$, to A_1, A_3 there corresponds $(0, 0, 1, 0)$, and to A_2, A_4 there corresponds $(0, 5/7, 0, 4/3)$.

Note that out of the n columns, we can choose r linearly independent columns in no more than C_n^r ways. Hence, the number of corner points of the set U is finite.

This means that we can try to solve the canonical problem (3)–(4) in the following way:

(1) find all corners points x of the set U,

(2) to calculate the value of the function (c, x) at each of the corner points and to determine the smallest ane.

However, this approach is not necessarily valid, as even in problems of small dimension the number of corner points can be very large.

Nevertheless, the idea of searching the corner points of a set is very fruitful and served as basis for a number of methods for solving linear programming problems. One of these methods is the so-called simplex method.

For the problem (3)–(4), construct the joint probability density of the vector x^*. For this purpose, we use a method for the solution of deterministic problems of linear programming, for example, the simplex method.

Consider the auxiliary problem.

$$\min_{x} c_t^T x, \tag{10}$$

s.t.

$$A_t x = b_t, x \geq 0 \tag{11}$$

and

$$A_t \in \text{supp}(A), b_t \in \text{supp}(b), c_t \in \text{supp}(c), \tag{12}$$

find a solution x_t^* and the corresponding corner point with numbers $j_1, ... j_r$.

We solve the random system of linear algebraic equations by numerical probabilistic analysis [4]

$$(A_{j_1} ... A_{j_r}) x = b.$$

The joint probability density of the obtained solution corresponds to x_t^*. If the supports of the input parameters are small enough, then due to continuity x_t^* coincides with x^*. In the case of arbitrary supports of the input parameters the search procedure for A_t, b_t, c_t, should be repeated, using the Monte Carlo method or genetic algorithms. In the case that different solutions x_t^* are obtained, they can be compared calculating the probabilistic extension $f_t = c^T x_t^*$.

4.1   Numerical Example

As a numerical example, consider the following problem

$$\min_{x} c^T x, \tag{13}$$

s.t.

$$Ax = b, x \geq 0 \tag{14}$$

and

$$A \in \text{supp}(A), b \in \text{supp}(b), c \in \text{supp}(c), \tag{15}$$

where $A = (a_{ij})$ is a uniform random matrix, each its element is a uniform random variable with support $[\underline{a}_{ij}, \overline{a}_{ij}]$, similarly, b, c are random vectors whose elements are uniform random variables.

The supports are defined as follows

$$A = \begin{pmatrix} [1-r, 1+r] & [1-r, 1+r] \\ [1-r, 1+r] & [-1-r, -1+r] \end{pmatrix}$$

$$\begin{pmatrix} [3-r, 3+r] & [1-r, 1+r] \\ [1-r, 1+r] & [2-r, 2+r] \end{pmatrix},$$

$$b = \begin{pmatrix} [3-r, 3+r] \\ [1-r, 1+r] \end{pmatrix},$$

$$c = (-1, -1, 0, 0).$$

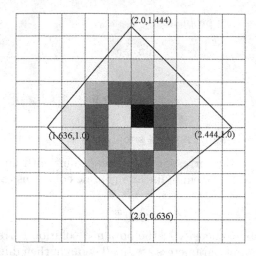

Fig. 1. Joint density of the vector (x_1, x_2)

For $r = 0$, which corresponds to the deterministic case, the solution is $x^* = (2, 1, 0, 0)$, the columns of the matrix A_1, A_2 correspond to a corner point.

In Fig. 1 the joint density of the vector x_1, x_2 for $r = 0.1$ with components $x_3 = 0, x_4 = 0$ is shown. The solid line is the boundary of the set of solutions on the (x_1, x_2) plane. The set $\mathcal{X}$ of solutions is the quadrangle with the vertices (2.0,0.636), (2.444,1.0), (2.0,1.444), (1.636,1.0). Value of the probability is represented by shades of gray. As can be seen from the Fig. 1 the probability density is non-uniformly distributed, the highest density is achieved at the center, near the point (2.0, 1.0).

The area of $\mathcal{X}$ strongly depends on the r, and it increases with increasing r and becomes infinite for $r = 1$. This is due to the fact that among the matrices

$$\begin{pmatrix} 0 & 0 \\ 0 & 0 \end{pmatrix} \in \begin{pmatrix} [0,2] & [0,2] \\ [0,2] & [-2,0] \end{pmatrix},$$

there are linearly dependent columns.

5   Applications

As an illustration, we consider optimization of hydroelectric power generation. Power generating electricity p can be expressed as

$$p = Chu,$$

where C is a constant; h is height of the water level, $h \in [h_{min}, h_{max}]$, u is water passing through the turbine, $u \in [u_{min}, u_{max}]$.

Height h depends on the amount V of water in the reservoir:

$$h = h(V)$$

and

$$V(t) = V_0 + \int_0^t q(\xi) - u(\xi) - u_r(\xi)d\xi.$$

where $q(t)$ is inflow; $u_x(t)$ is water passing through the spillway (is known and is determined by plant personnel); u is water passing through the turbine, $u \in [u_{min}, u_{max}]$.

Suppose we want to maximize the generation of electricity in the time interval $[0, T]$

$$\max_u \ P(u) = \int_0^T C h \left(V_0 + \int_0^T q(t) - u(t) - u_x(t)dt \right) u(t)dt.$$

Simplify the Problem

We represent volume of the reservoir as

$$V = V_0 + S(h - h_0),$$

and

$$h(t) = h_0 + (V(t) - V_0)/S = h_0 + (\int_0^t q(\xi) - u(\xi) - u_x(\xi)d\xi)/S.$$

It is now clear that the problem to be solved is to maximize the objective

$$P(u) = C \int_0^T \left(h_0 + (\int_0^t q(\xi) - u(\xi) - u_x(\xi)d\xi)/S \right) u(t)dt,$$

where $q(t)$ is inflow; $u_x(t)$ is water passing through the spillway; u is water passing through the turbine, $u \in [u_{min}, u_{max}]$.

Let us consider *a discrete model*. Let $\omega = \{t_0 < t_1 < \ldots < t_n\}$ be a grid, q_i be a random input value of water inflow for a time $[t_{i-1}, t_i]$, u_{xi} be water passing through the spillway for a time $[t_{i-1}, t_i]$, u_i be water passing through the turbine for a time $[t_{i-1}, t_i]$, $U = (u_i), i = 1, \ldots, n$,

$$\max_U \ P(U) = C \sum_{i=1}^n \left(h_0 + (\sum_{j=1}^i q_j - u_j - u_{xj})/S \right) u_i.$$

In some cases, the problem can be reduced to the solution of a random system of linear algebraic equations. Here, only the right-hand side of the system is random

$$2u_1 + u_2 + \ldots + u_n = Sh_0 + q_1,$$

$$u_1 + 2u_2 + \ldots + u_i + \ldots + u_n = Sh_0 + q_1 + q_2,$$

$$\ldots$$

$$u_1 + u_2 + \ldots + 2u_i + \ldots + u_n = Sh_0 + \sum_{j=1}^{i} q_j,$$

$$\ldots$$

$$u_1 + u_2 + \ldots + 2u_n = Sh_0 + \sum_{j=1}^{n} q_j.$$

Numerical Example. Note that u_i, $i = 1, \ldots, n$ can be expressed as a linear combination of q_i, $i = 1, \ldots, n$.

For $n = 3$ we get:

$$u_1 = \frac{-q_3 - 2q_2 + q_1 + Sh_0}{4},$$

$$u_2 = \frac{-q_3 + 2q_2 + q_1 + Sh_0}{4},$$

$$u_3 = \frac{3q_3 + 2q_2 + q_1 + Sh_0}{4}.$$

Let $q_i \in [\underline{q}_i, \overline{q}_i]$ be uniform random variables, $S = 1$, $q_1 = [0.1, 0.2]$, $q_2 = [0.2, 0.3]$, $q_3 = [0.3, 0.4]$ be supports and $h_0 = 0.9$.

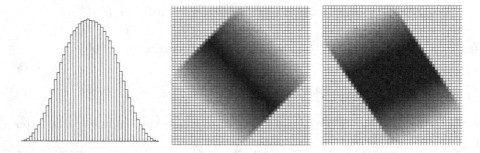

Fig. 2. Histogram u_1 and joint probability density (u_1, u_2), (u_2, u_3)

The supports are $u_1 = [0.0, 0.1]$, $u_2 = [0.25, 0.35]$, $u_3 = [0.575, 0.725]$. In Fig. 2 piecewise constant approximations of the joint probability density of the vectors (u_1, u_2), (u_2, u_3) are shown. The value of the probability is represented by shades of gray.

6   Conclusion

The considered methods for solving the above problems of linear optimization enable one to represent random programming as an effective method for solving optimization problems with uncertain input parameters. Methods of random programming allow one to build a joint probability density function on the set of optimal solutions. This approach helps the decision maker to choose the best solutions and enables risks assessment.

References

1. Dobronets, B.S., Krantsevich, A.M., Krantsevich, N.M.: Software implementation of numerical operations on random variables. J. Siberian Federal Univ. Math. Phys. **6**(2), 168–173 (2013)
2. Dobronets, B.S., Popova, O.A.: Numerical operations on random variables and their application. J. Siberian Federal Univ. Math. Phys. 4(2), 229–239 (2011). (Russian)
3. Dobronets, B.S., Popova, O.A.: Elements of numerical probability analysis. SibSAU Vestnik **42**(2), 19–23 (2012). (Russian)
4. Dobronets, B.S., Popova, O.A.: Numerical probabilistic analysis under aleatory and epistemic uncertainty. Reliable Comput. **19**, 274–289 (2014)
5. Schjaer-Jacobsen, H.: Representation and calculation of economic uncertainties: Intervals, fuzzy numbers, and probabilities. Int. J. Prod. Econ. **78**, 91–98 (2002)
6. Liu, B.: Theory and Practice of Uncertain Programming, 2nd edn. Springer, Heidelberg (2009)
7. Popova, O.A.: Optimization problems with random data. J. Siberian Federal Univ. Math. Phys. **6**(4), 506–515 (2013)
8. Shapiro, A., Dentcheva, D., Ruszczynski, A.: Lectures on Stochastic Programming: Modeling and Theory. SIAM, Philadelphia (2009)
9. Vasil'ev, F.P.: Numerical Methods for Solving Extremal Problems. Nauka, Moscow (1988). (Russian)

Towards the Possibility of Objective Interval Uncertainty

Luc Longpré, Olga Kosheleva, and Vladik Kreinovich[✉]

University of Texas at El Paso, El Paso, TX 79968, USA
{longpre,olgak,vladik}@utep.edu

Abstract. Applications of interval computations usually assume that
while we only know an interval containing the actual (unknown) value
of a physical quantity, there *is* the exact value of this quantity, and
that in principle, we can get more and more accurate estimates of this
value. Physicists know, however, that, due to the uncertainty principle,
there are limitations on how accurately we can measure the values of
physical quantities. One of the important principles of modern physics
is *operationalism* – that a physical theory should only use observable
properties. This principle is behind most successes of the 20th century
physics, starting with relativity theory (vs. un-observable aether) and
quantum mechanics. From this viewpoint, it is desirable to avoid using
un-measurable exact values and to modify the mathematical formalisms
behind physical theories so that they explicitly only take objective uncertainty into account. In this paper, we describe how this can be done for
objective interval uncertainty.

Keywords: Interval uncertainty · Algorithmic randomness · Physics

1 Formulation of the Problem

Is Interval Uncertainty Subjective? Applications of interval computations
usually assume that while we only know an interval $[\underline{x}, \overline{x}]$ containing the actual
(unknown) value of a physical quantity x, there *is* the exact value x of this
quantity, and that in principle, we can get more and more accurate estimates of
this value.

This assumption is in line with the usual formulations of physical theories – as partial differential equations relating exact values of different physical
quantities, fields, etc., at different space-time locations and moments of time;
see, e.g., [2]. Physicists know, however, that, due to the uncertainty principle,
there are limitations on how accurately we can measure the values of physical
quantities [2,8].

This is not just a theoretical concern: for example, the International Union of
Pure and Applied Chemistry (IUPAC) has recently officially recognized that the
atomic weight of a chemical element is not an exact number; depending on where
the sample came from, the atomic weight may differ within a certain interval;
see, e.g., [11].

© Springer International Publishing Switzerland 2016
M. Nehmeier et al. (Eds.): SCAN 2014, LNCS 9553, pp. 54–65, 2016.
DOI: 10.1007/978-3-319-31769-4_5

It is Desirable to Take Objective Uncertainty into Account. One of the important principles of modern physics is *operationalism* – that a physical theory should only use observable properties. This principle is behind most successes of 20th century physics, starting with relativity theory (vs. un-observable aether) and quantum mechanics. From this viewpoint, it is desirable to avoid using un-measurable exact values and to modify the mathematical formalisms behind physical theories so that they explicitly only take objective uncertainty into account.

Objective Uncertainty is About Probabilities. According to quantum physics, we can only predict probabilities of different events. Thus, uncertainty means that instead of exact values of these probabilities, we can only determine intervals; see, e.g., [3, 4].

Let us give a simple example. In the ideal world, atoms of the same element have exactly the same atomic weight, and – if their are radioactive – exactly the same probability p that they will decay within a given moment of time. In such an ideal situation, by taking larger and larger samples and measuring the decay frequency, we can get more and more accurate estimates of the desired decay probability p.

In practice, as we have mentioned, different objects made of the same element have, in general, slightly different average atomic weight – and similarly, they have slightly different average decay probabilities, probabilities that may take different values from the corresponding interval $[\underline{p}, \overline{p}]$. Depending on which objects we take, we may get frequencies close to the lower bound $\underline{p}$ and we may also get frequencies close to the upper bound $\overline{p}$. As we increase the sample size, we will get frequencies oscillating between $\underline{p}$ and $\overline{p}$, without ever converging to a single value.

Formulation of the Problem. What is the observational meaning of such interval-valued probabilities?

2   Analysis of the Problem

What is the Observational Meaning of Probability? Probability refers to repeated events: we repeat the same experiment (or perform many similar observations) and record the results as a binary sequence $\omega_1 \omega_2 \ldots$ For example, when we talk about the probability of a coin falling heads, we mean that we repeatedly flip the coin and record the resulting sequence: for example, we can take $\omega_i = 1$ if the coin falls heads in the i-th experiment, and $\omega_i = 0$ if this coin falls tails.

In there terms, the fact that the probability of heads is $1/2$ means that in the limit, when $n \to \infty$, the ratio of 1 s in a sequence $\omega_1 \ldots \omega_n$ tends to $1/2$. However, this is only the part of this meaning: for example, for a sequence $0101\ldots$, the ratio tends to $1/2$, but we would not call it a random sequence corresponding to probability $1/2$.

From the practical viewpoint, when we say that a sequence $\omega_1 \omega_2 \ldots$ is random, we assume that this sequence satisfies *all* the probability laws (such as the law of large numbers or the Central Limit Theorem); these probability laws are what practitioners use to check whether the sequence is random.

From this viewpoint, if a sequence satisfies all probability laws, then for all practical purposes we can consider it random. Thus, we can formally define a sequence to be random if it satisfies all probability laws. In precise terms, a probability law is a property ℓ which is true with probability 1: $P(\ell) = 1$. So, a sequence is random if it satisfies all the properties which are true with probability 1.

Properties are in 1-1 correspondence with sets – to each property, we can assign the set of all the sequences that satisfy this property and, vice versa, to every set, we can assign a property of belonging to this set. When we talk about probability laws, we mean only properties which can be described by finitely many symbols from a certain formal language; the corresponding sets are known as *definable sets*. Thus, we can say that a sequence is random if it belongs to all definable sets of probability measure 1.

A sequence belongs to a set of measure 1 if and only if it does not belong to its complement $C = -S$ with $P(C) = 0$. So, we can equivalently say that *a sequence is* random *if it does not belong to any definable set of measure 0*. This is, in effect, Kolmogorov-Martin-Löf's (KML) definition of a random sequence; see, e.g., [7].

Each definable set is determined by a finite sequence of symbols. There are no more than countably many finite sequences of symbols, thus, there are countably many definable sets. So, the union of all such sets has measure 0. Therefore, almost all sequences are KML-random.

Probability Interval: What Is Its Observational Meaning? We have recalled what is an observational meaning of an exact probability p. What is the observational meaning of a probability interval, when instead of a single probability measure we have several possible probability measures?

This is not an easy question: in [1,6], we have shown that in seemingly reasonable formalizations, every random sequence is actually random relative to one of the possible probability measures. In such a formalization, every random sequence – including the sequence of observations – corresponds to one specific probability measure. In other words, there is a probability, we just do not know it – this is exactly the subjective interval uncertainty that we are trying to avoid.

We Consider Independent Repeated Events. Probabilities have direct observational meaning only for repeating events. In mathematical terms, independent repeating events correspond to a *product measure*, when the probability of two events A and B happening in two consequent tests is equal to the product of the corresponding probabilities: $P(A \& B) = P(A) \cdot P(B)$.

Traditional Case. The traditional case is when we know the exact probability p. Then, observable sequences $\omega_1 \omega_2 \dots$ are KLM-random relative to a product of p-measures.

It may be that in practice, we do not know the exact value of this probability p, we only know the interval $[\underline{p}, \overline{p}]$ containing this probability. In this case, we have an interval uncertainty, but this interval uncertainty is *subjective* in the following sense: there is the actual exact value of the probability, the value which can be determined, e.g., by taking larger and larger samples; then the corresponding frequencies will be closer and close to the actual probability.

What We are Trying to Describe. What we are trying to describe is the case when there is no such objective probability: e.g., the case when the corresponding frequencies do not have an exact limit: limit frequencies oscillate between $\underline{p}$ and $\overline{p}$.

3   Objective Interval Uncertainty: Definitions and the First Result

Definition 1. We say that a sequence is $[\underline{p}, \overline{p}]$-*random* if it is random for some product measure with $p_i \in [\underline{p}, \overline{p}]$.

Definition 2. *We say that a sequence* $\omega_1\omega_2\ldots$ *is objectively* $[\underline{p}, \overline{p}]$-*random if this sequence is* $[\underline{p}, \overline{p}]$-*random, and it is not* $[\underline{q}, \overline{q}]$-*random for any proper subinterval* $[\underline{q}, \overline{q}] \subset [\underline{p}, \overline{p}]$.

Proposition 1. *For every interval* $[\underline{p}, \overline{p}]$, *there exist objectively* $[\underline{p}, \overline{p}]$-*random sequences.*

Proof. We will show that any sequence $\omega_1\omega_2\ldots$ corresponding to p_i for which $\liminf p_i = \underline{p}$ and $\limsup p_i = \overline{p}$ is objectively $[\underline{p}, \overline{p}]$-random.

Since $p_i \in [\underline{p}, \overline{p}]$, this sequence is $[\underline{p}, \overline{p}]$-random. Let us prove that this sequence $\omega_1\omega_2\ldots$ is not $[\underline{q}, \overline{q}]$-random for any proper subinterval $[\underline{q}, \overline{q}] \subset [\underline{p}, \overline{p}]$, i.e., that it is not random w.r.t. any sequence $q_i \in [\underline{q}, \overline{q}]$.

It is known that if two measures are mutually singular, then no sequence is random w.r.t. both measures. For product measures, singularity is equivalent to the following equality (see, e.g., [7,8]):

$$\sum_{i=1}^{\infty} \left[\left(\sqrt{p_i} - \sqrt{q_i}\right)^2 + \left(\sqrt{1 - p_i} - \sqrt{1 - q_i}\right)^2 \right] = +\infty.$$

For a proper subinterval, either $\underline{p} < \underline{q}$ or $\overline{q} < \overline{p}$. Without loss of generality, let us consider the case when $\underline{p} < \underline{q}$.

When $\liminf p_i = \underline{p}$ then, for every $\varepsilon > 0$, there are infinitely many i for which $\sqrt{p_i} \le \sqrt{\underline{p}} + \varepsilon$. For these i, we have $q_i \ge \underline{q}$, so $\sqrt{q_i} \ge \sqrt{\underline{q}}$. Thus, $\sqrt{q_i} - \sqrt{p_i} \ge \sqrt{\underline{q}} - \left(\sqrt{\underline{p}} + \varepsilon\right) = \left(\sqrt{\underline{q}} - \sqrt{\underline{p}}\right) - \varepsilon$. For $\varepsilon = (\sqrt{\underline{q}} - \sqrt{\underline{p}})/2$, we have $\sqrt{q_i} - \sqrt{p_i} > \varepsilon > 0$ and therefore, the above sum is infinite. So, a $\{p_i\}$-random sequence $\omega_1\omega_2\ldots$ cannot be $\{q_i\}$-random. The proposition is proven.

4   Objective Interval Uncertainty: A Stronger Definition and the Second Result

Discussion. We want to describe the idea that all we know is an interval $[\underline{p}, \overline{p}]$.

The above definition means, in effect, that all the values p_i from the sequence p_i are in between $\underline{p}$ and $\overline{p}$ and that, even if we dismiss finitely many probabilities, no narrower interval contains all the remaining values of p_i.

In general, however, such sequences may satisfy additional laws, in addition to $p_i \in [\underline{p}, \overline{p}]$. For example, if we have $p_{2i} = \underline{p}$ and $p_{2i+1} = \overline{p}$, then we satisfy the above condition – but we also satisfy the additional condition, that all even-placed probabilities are equal to $\underline{p}$ and all odd-placed probabilities are equal to $\overline{p}$.

Is it possible to have a sequence of probabilities p_i whose *only* meaningful property is that all these values are from the interval $[\underline{p}, \overline{p}]$? In other words, is it possible to find a sequence p_i which does not satisfy any other meaningful property?

What Is a "Meaningful Property". In order to answer the above question, we need to formalize what is meant by a meaningful property.

In foundations of mathematics, the main object is a set. Properties naturally correspond to sets: namely, to each property, we can put into correspondence the set of all the sequences that satisfy this property. In these terms, describing what we mean by a property is equivalent to describing the corresponding sets.

First, a meaningful property must be described by a finite sequence of symbols in an appropriate mathematical language. Corresponding sets are known as *definable* sets. It is important to realize that while every example of a set that we can give is definable, not all sets are definable: for example, there are more than countably many subsets of the set of all natural numbers, but since there are only countably many finite sequences, there are only countably many definable sets.

Second, it is reasonable to only consider *observable* properties, i.e., properties whose validity can be determined based on observations. Let us show that, because of this requirement, it is reasonable to require that for each observable property, the corresponding set of sequences is *closed* in the sense of component-wise convergence: if for all k, the sequences $p^{(k)} = \{p_i^{(k)}\}$ belong to this set, then their limit $p = \{p_i\}$, with $p_i \stackrel{\text{def}}{=} \lim_k p_i^{(k)}$, should also belong to the corresponding set.

Indeed, in practice, we do not observe probabilities, we only observe *frequencies* which are close to probabilities. If the actual probabilities are the limit values $(p_1, p_2, \ldots)$, this means that for every $\varepsilon > 0$ and for a sufficiently large sample, we will observe frequencies f_i which are ε-close to these limit value p_i. Since p is the limit of $p^{(k)}$, for sufficiently large k, the values p_i – and thus, the frequencies f_i – are close to the probabilities $p_i^{(k)}$, and therefore, consistent with the assumption that the actual probabilities are $p_i^{(k)}$ (and thus, with the assumption that the actual probabilities satisfy the given property).

So, if the actual probabilities are equal to the limit, then no matter how large a sample we take, the resulting observations will always be consistent with the given property. Thus, it is reasonable to add the limit p to the set of all probability sequences that satisfy the given property. Because of this argument, in the following text, we will assume that for each observable property, the corresponding set of sequences is closed.

The third property of the corresponding sets comes from the need to distinguish trivial unavoidable "properties" like $p_1 = \underline{p}$ – properties that do not really restrict any values beyond a few first ones – from non-trivial properties that we are trying to avoid. In other words, we need to formulate the idea that if we only know approximate values of the first n probabilities, then we cannot guarantee that the corresponding property will be satisfied.

This requirement can be described in precise terms, if on the set of all the sequences p we introduce a topology in which the basis is formed by "boxes" $(\underline{p}_1, \overline{p}_1) \times \ldots \times (\underline{p}_n, \overline{p}_n)$ corresponding to different n and different bounds $\underline{p}_i$ and $\overline{p}_i$. (Convergence in this topology corresponds to the above point-wise convergence.) In terms of this topology, the above requirement means that within every element from the basis – and thus, within every open set – there should be a sequence that does not belong to the corresponding set S. For closed sets, this requirement means that the set S is *nowhere dense*.

Summarizing, we can formalize our requirements by saying that by a meaningful property, we mean a closed nowhere dense definable set, and that the actual sequence of probabilities p should not belong to any of such sets.

Definition 3. *Let $\mathcal{S}$ be a set of all sequences $p_i \in [\underline{p}, \overline{p}]$, with topology whose basis is formed by the boxes $(\underline{p}_1, \overline{p}_1) \times \ldots \times (\underline{p}_n, \overline{p}_n)$.*

- *By a* meaningful property, *we mean a definable closed nowhere dense set $S \subseteq \mathcal{S}$.*
- *We say that a sequence p satisfies a meaningful property S if $p \in S$.*
- *We say that a sequence $p \in \mathcal{S}$ has no other properties if p does not satisfy any meaningful property S.*

Proposition 2. *For every two definable values $\underline{p} < \overline{p}$, there exists a sequence p_i for which $p_i \in [\underline{p}, \overline{p}]$ for all i and which has no other properties.*

Mathematical Comment. Not only there *exist* such sequences, but there are *many* such sequences: as we can see from the proof, "almost all" sequences $p \in \mathcal{S}$ (almost all in some reasonable sense) have no other properties.

Practical Comment. The above description is, at this stage, very theoretical. We do not have a full understanding of how to check whether an experimentally observed sequence has no other properties. One consequence that we *can* check is that the limit frequencies should fill the whole interval $[\underline{p}, \overline{p}]$ and not be bound by any narrower subinterval. In other words, as we increase the sample size, the frequencies should always fill this whole interval.

However, we may recall that in statistics, in addition to observing frequencies, there are other criteria that describe p-random sequences – e.g., according to the Central Limit Theorem, deviations of frequencies from the probability must be, asymptotically, normally distributed. It is therefore desirable to come up with similar criteria for the case of sequences which are $[\underline{p}, \overline{p}]$-random for some non-degenerate interval $\underline{p} < \overline{p}$.

Proof. By definition, a sequence p has no other properties if its does belong to any property-related set S (in the sense of Definition 3), i.e., equivalently, if it does not belong to the union U of all such sets S.

Each property-related set S is, by definition, a definable closed nowhere dense set. As we have mentioned, there are no more than countably many definable objects, so U is a union of countably many closed nowhere dense sets. Such unions are known as *meager* sets, or sets of *first Baire category*. It is known that the set of all sequences is *not* meager; this is the main gist of the corresponding *Baire's theorem*; see, e.g., [9]. Thus, there are sequences p which do not belong to U, i.e., which have no other properties. Moreover, "almost all" sequences p – in the sense of all sequences except for a meager set – do not belong to U, i.e., have no other properties. The proposition is proven.

5   Why This Is Interesting: Objective Interval Uncertainty Can Potentially Help in Solving NP-Hard Problems Faster

What We Do in This Section. Objective interval uncertainty means that the corresponding series of repeated experiments, the sequence of observations $\omega_1 \omega_2 \ldots$ is random with respect to some sequence of probabilities p_i for which $p_i \in [\underline{p}, \overline{p}]$ and which has no other property.

In this section, we prove that by using such sequences ω, it is, in principle, possible to drastically speed up the solution of NP-complete problems.

Practical Comment. It should be emphasized that our result only says that it is *theoretically* possible to speed up the solution of NP-hard problem. At this point, we do not know how to actually achieve such a speed-up – but we hope that the proof of theoretically possibility of the speed-up will eventually lead to practical algorithms.

Mathematical Comment. From the mathematical viewpoint, the result from this section is a modification of a similar result from [5].

What Is an NP Problem? Brief Reminder. In practice, we often need to find a solution that satisfies a given set of constraints or at least check that such a solution is possible. Once we have a candidate for the solution, we can feasibly check whether this candidate indeed satisfies all the constraints. In theoretical computer science, "feasibly" is usually interpreted as computable in polynomial time, i.e., in time bounded by a polynomial of the length of the input.

A problem of checking whether a given set of constraints has solution is called a *problem of the class NP* if we can check, in polynomial time, whether a given candidate is a solution; see, e.g., [10]. Examples of such problem includes checking whether a given graph can be colored in 3 colors, checking whether a given propositional formula – i.e., formula of the type

$$(v_1 \vee \neg v_2 \vee v_3) \& (v_4 \vee \neg v_2 \vee \neg v_5) \& \ldots$$

is satisfiable, i.e., whether this formula is true by some combination of the propositional variables v_i.

Each problem from the class NP can be algorithmically solved by trying all possible candidates. For example, we can check whether a graph can be colored by trying all possible assignments of colors to different vertices of a graph, and we can check whether a given propositional formula is satisfiable by trying all 2^n possible combinations of true-or-false values $v_1, \ldots, v_n$. Such exhaustive search algorithms require computation time like 2^n, time that grows exponentially with n. For medium-size inputs, e.g., for $n \approx 300$, the resulting time is larger than the lifetime of the Universe. So, these exhaustive search algorithms are not practically feasible.

It is not known whether problems from the class NP can be solved feasibly (i.e., in polynomial time): this is a famous open problem P$\stackrel{?}{=}$NP. It is known, however, there are problems in the class NP which are *NP-complete* in the sense that every problem from the class NP can be reduced to this problem. Reduction means, in particular, that if we can find a way to efficiently solve one NP-complete problem, then, by reducing other problems from the class NP to this problem, we can thus efficiently solve all the problems from the class NP.

So, it is very important to be able to efficiently solve even one NP-complete problem. (By the way, both above example of NP problems – checking whether a graph can be colored in 3 colors and whether coloring a propositional formula is satisfiable – are NP-complete.)

How to Represent Instances of an NP-Complete Problem. For each NP-complete problem $\mathcal{P}$, its instances are sequences of symbols. In the computer, each such sequence is represented as a sequence of 0s and 1s. Thus, we can append 1 in front of this sequence and interpret the resulting sequence as a binary code of a natural number i (we need to add 1 in front, so that different sequences transform into different numbers, otherwise 0 and 00 will lead to the same number).

In principle, not all natural numbers i correspond to instances of a problem $\mathcal{P}$; we will denote the set of all natural numbers which correspond to such instances by $S_{\mathcal{P}}$. For each $i \in S_{\mathcal{P}}$, the correct answer (true or false) to the i-th instance of the problem $\mathcal{P}$ will be denoted by $s_{\mathcal{P},i}$.

Easier-to-Solve and Harder-to-Solve NP-Complete Problems. For some easier-to-solve problems, there are feasible algorithms which solve "almost all" instances, in the sense that for each n, the proportion of instance $i \leq n$ for which the problem is solved by this algorithm tends to 1. In this case, while the worst-case complexity is still exponential, in practice, almost all problems can be feasibly solved.

A more challenging case are harder-to-solve NP-complete problems, for which no feasible algorithm is known that would solve almost all instances. In this section, we show that our method works on all NP-complete problems, both easier-to-solve and harder-to-solve ones.

What We Mean by Using Physical Observations in Computations. We assume that the sequence ω_i comes from observations. In addition to performing computations, our computational device can, given a natural number i, use the result ω_i of the corresponding i-th observation in its computations. In other words, given an integer i, we can produce ω_i.

In precise theory-of-computation terms, this means computations that use the sequence ω as an oracle; see, e.g., [10].

Comment. Since we are interested in feasible (= polynomial time) computations, the code should be set up in such a way that the overall time of an experiment does not exceed a polynomial of the length of the number i. This can be done, e.g., if we explicitly add maximum waiting time into the description of the experiment, by adding as many 0 s as the time that we plan to wait.

Definition 4. *By a $[\underline{p}, \overline{p}]$-algorithm $\mathcal{A}$, we mean an algorithm which uses, as an oracle, a sequence ω_i which is random with respect to a probability measure determined by a sequence p_i for which $p_i \in [\underline{p}, \overline{p}]$ for all i and which has no other properties.*

Notation. The result of applying an algorithm $\mathcal{A}$ using ω_i to an input i will be denoted by $\mathcal{A}(\omega, i)$.

Definition 5. *Let $\mathcal{P}$ be an NP-complete problem. We say that a feasible $[\underline{p}, \overline{p}]$-algorithm $\mathcal{A}$ solves almost all instances of $\mathcal{P}$ if for every $\varepsilon > 0$ and $\delta > 0$ and for every integer n, there exists an integer $N \geq n$ for which, with probability $\geq 1-\delta$, the proportion of the instances $i \leq N$ of the problem $\mathcal{P}$ which are correctly solved by $\mathcal{A}$ is greater than $1 - \varepsilon$:*

$$\text{Prob}\left(\frac{\#i\{i \leq N : i \in S_\mathcal{P} \ \& \ \mathcal{A}(\omega, i) = s_{\mathcal{P}, i}\}}{\#i\{i \leq N : i \in S_\mathcal{P}\}} > 1 - \varepsilon \right) \geq 1 - \delta.$$

Comment. The restriction to sufficiently long inputs $N \geq n$ makes perfect sense: for short inputs, NP-completeness is not an issue: we can perform exhaustive search of all possible bit sequences of length 10, 20, and even 30. The challenge starts when the length of the input is high.

Proposition 3. *For every NP-complete problem $\mathcal{P}$, there exists a feasible $[\underline{p}, \overline{p}]$-algorithm $\mathcal{A}$ that solves almost all instances of $\mathcal{P}$.*

Comment. In other words, we show that if there is objective interval uncertainty, then, *theoretically*, the use of the corresponding physical observations makes all NP-complete problems easier-to-solve (in the above-described sense).

Of course, as we have mentioned earlier, this does not mean that we already have an efficient algorithm for solving NP-complete problems – but this theoretical possibility is encouraging, and we hope that it will eventually lead to efficient algorithms.

Proof. We know that for every i, the probability p_i that $\omega_i = 1$ is in between $\underline{p}$ and $\overline{p}$. Thus, for every two numbers $N \gg N'$, the proportion of values ω_i ($i = N, N + 1, \ldots, N' - 1$) which are equal to 1, should be either within the interval $[\underline{p}, \overline{p}]$ or at least close to this interval. Let us use this property to design the desired algorithm $\mathcal{A}$.

A value p from the interval $[\underline{p}, \overline{p}]$ is:

– closer to $\overline{p}$ if it is larger than the midpoint $\widetilde{p} \stackrel{\text{def}}{=} \dfrac{\underline{p} + \overline{p}}{2}$ and
– closer to $\underline{p}$ if p is smaller than the midpoint.

The midpoint itself is equidistant from both endpoints $\underline{p}$ and $\overline{p}$.

Let us therefore select an increasing sequence $N_1 < N_2 < \ldots$, and take:

- $\mathcal{A}(\omega, i) = 1$ if the proportion of values $\omega_i = 1$ between N_i and N_{i+1} is greater than or equal to the midpoint $\widetilde{p}$, and
- $\mathcal{A}(\omega, i) = 0$ if this proportion is smaller than $\widetilde{p}$.

Let us prove that, for an appropriate sequence N_i, this algorithm indeed solves almost all instances of the given problem $\mathcal{P}$.

The proposition states that for very $\varepsilon > 0$, $\delta > 0$, and n, there exists an integer $N \geq n$ for which the above inequality holds. To prove the existence of such an N, let us consider the set T of all sequences p for which, for all $N \geq n$, this inequality *does not* hold. We will show that this set T is definable, closed, and nowhere dense. By definition of a sequence that has no other properties (Definition 3), this would imply that the actual sequence p *does not* belong to this set T – and thus, there exists the desired value N, which is exactly what the proposition claims.

Definability is easy: we just had defined this set. Closeness is also rather easy to prove; it can be proven similarly to a similar closeness proof in [5].

The non-trivial part is nowhere density. To prove that the set T is nowhere dense, it is sufficient, for each finite starting sequence $p_1, \ldots, p_n$, to produce an infinite extension p for which the desired integer $N \geq n$ exists (and which, thus, does not belong to the set T).

We will take a sequence p_i all whose elements are either equal to the lower endpoint $\underline{p}$ or to the upper endpoint $\overline{p}$. Specifically, for all the values between N_i and N_{i+1}, we will take:

- $p_i = \overline{p}$ if $s_{\mathcal{P},i} = 1$,
- $p_i = \underline{p}$ if $s_{\mathcal{P},i} = 0$, and
- any of these two values if $i \notin S_{\mathcal{P}}$.

Let us show that for an appropriate choice of the sequence N_i, with probability $\geq 1 - \delta$, for all the values i from n to $N = \dfrac{n}{\varepsilon}$, we will have $\mathcal{A}(\omega, i) = s_{\mathcal{P}, i}$. This will imply that the proportion of such i is indeed greater than $1 - \varepsilon$ with probability $\geq 1 - \delta$.

For each i, we consider the arithmetic average of $k_i \overset{\text{def}}{=} (N_{i+1} - 1) - N_i$ independent 0-1 random values each of which is equal to 1 with some probability p (namely, either with probability p or with probability $\overline{p}$). It is known that this arithmetic average is, in the limit $\overline{k}_i \to \infty$, normally distributed – this fact is a particular case of the Central Limit Theorem. The mean value of this average is equal to the corresponding probability p, and the standard deviation decreases, with k_i, as $\dfrac{1}{\sqrt{k_i}}$. Let us use these facts to estimate the probability that with $p = 0$ we will have $\mathcal{A}(\omega, i) = 1$ or vice versa. In other words, we are interested in the probability that the average differs from its expected values by at least the half-width $w \overset{\text{def}}{=} \dfrac{\overline{p} - p}{2}$. For a normal distribution with mean μ and standard deviation σ, asymptotically, this probability is proportional to $\exp\left(-\dfrac{w^2}{2\sigma^2}\right)$, i.e., to $\exp(-\text{const} \cdot k_i)$. If we select k_i in such a way that $\exp(-\text{const} \cdot k_i) \leq \dfrac{1}{i^2}$, i.e., $k_i = \text{const} \cdot \ln(i)$, then the probability that this happens for one of the values i cannot exceed the sum of the probabilities corresponding to different i, and is, thus, smaller than the sum $\sum\limits_{i=n}^{\infty} \dfrac{1}{i^2}$. Thus, the sum tends to 0 and is, therefore, smaller than δ for all sufficiently large n.

So, we get the desired property if we find N_i for which $k_i \approx N_{i+1} - N_i \sim \text{const} \cdot \ln(i)$. This approximate equality is true if we take $N_i = i \cdot \ln(i)$.

For this choice of N_i, computing $\mathcal{A}(\omega, i)$ requires $N_{i+1} - N_i \sim \ln(i)$ calls to the oracle – a number which is a linear function of the bit length of an integer i. Thus, this algorithm is indeed feasible. The proposition is proven.

Acknowledgments. This work was supported in part by the National Science Foundation grants HRD-0734825, HRD-1242122, and DUE-0926721. The authors are thankful to all the participants of the 16th International Symposium on Scientific Computing, Computer Arithmetic, and Validated Numerics SCAN'2014 (Würzburg, German, September 21–26, 2014) for valuable discussions, and to the anonymous referees for valuable suggestions.

References

1. Cheu, D., Longpré, L.: Towards the possibility of objective interval uncertainty in physics. Reliable Comput. **15**(1), 43–49 (2011)
2. Feynman, R., Leighton, R., Sands, M.: The Feynman Lectures on Physics. Addison Wesley, Boston (2005)
3. Gorban, I.I.: Theory of Hyper-Random Phenomena. Ukrainian National Academy of Sciences Publ., Kyiv (2007). in Russian

4. Gorban, I.I.: Hyper-random phenomena: definition and description. Inf. Theor. Appl. **15**(3), 203–211 (2008)
5. Kosheleva, O., Zakharevich, M., Kreinovich, V.: If many physicists are right and no physical theory is perfect, then by using physical observations, we can feasibly solve almost all instances of each NP-complete problem. Math. Struct. Model. **31**, 4–17 (2014)
6. Kreinovich, V., Longpré, L.: Pure quantum states are fundamental, mixtures (composite states) are mathematical constructions: an argument using algorithmic information theory. Int. J. Theoret. Phys. **36**(1), 167–176 (1997)
7. Li, M., Vitányi, P.: An Introduction to Kolmogorov Complexity and Its Applications. Springer, Heidelberg (2008)
8. Longpré, L., Kreinovich, V.: When are two wave functions distinguishable: a new answer to Pauli's question, with potential application to quantum cosmology. Int. J. Theoret. Phys. **47**(3), 814–831 (2008)
9. Oxtoby, J.C.: Measure and Category: A Survey of the Analogies Between Topological and Measure Spaces. Springer, Heidelberg (1980)
10. Papadimitriou, C.: Computational Complexity. Addison Welsey, Reading (1994)
11. Wester, M.E., et al.: Atomic weights of the elements 2011 (IUPAC Technical Report). Pure Appl. Chem. **85**(5), 1047–1078 (2013)

How Much for an Interval? a Set? a Twin Set? a p-Box? A Kaucher Interval? Towards an Economics-Motivated Approach to Decision Making Under Uncertainty

Joe Lorkowski and Vladik Kreinovich[✉]

University of Texas at El Paso, El Paso, TX 79968, USA
lorkowski@computer.org, vladik@utep.edu

Abstract. A natural idea of decision making under uncertainty is to assign a fair price to different alternatives, and then to use these fair prices to select the best alternative. In this paper, we show how to assign a fair price under different types of uncertainty.

Keywords: Decision making · Interval uncertainty · Set uncertainty · p-box

1 Decision Making Under Uncertainty: Formulation of the Problem

In many practical situations, we have several alternatives, and we need to select one of these alternatives. For example:

- a person saving for retirement needs to find the best way to invest money;
- a company needs to select a location for its new plant;
- a designer must select one of several possible designs for a new airplane;
- a medical doctor needs to select a treatment for a patient, etc.

Decision making is the easiest if we know the exact consequences of selecting each alternative. Often, however, we only have an incomplete information about consequences of different alternative, and we need to select an alternative under this uncertainty.

Traditional decision theory (see, e.g., [8,12]) assumes that for each alternative a, we know the probability $p_i(a)$ of different outcomes i. It can be proven that preferences of a rational decision maker can be described by *utilities* u_i so that an alternative a is better if its expected utility $u(a) \stackrel{\text{def}}{=} \sum_i p_i(a) \cdot u_i$ is larger.

Often, we do not know the probabilities $p_i(a)$. As a result, we do not know the exact value of the gain u corresponding to each alternative. How can we then make a decision?

M. Nehmeier et al. (Eds.): SCAN 2014, LNCS 9553, pp. 66–76, 2016.
DOI: 10.1007/978-3-319-31769-4_6

For the case when we only know the interval $[\underline{u}, \overline{u}]$ containing the actual (unknown) value of the gain u, a possible solution was proposed in the 1950s by a future Nobelist L. Hurwicz [5, 8]: we should select an alternative that maximizes the value $\alpha_H \cdot \overline{u}(a) + (1 - \alpha_H) \cdot \underline{u}(a)$. Here, the parameter $\alpha_H \in [0, 1]$ described the optimism level of a decision maker:

- $\alpha_H = 1$ means optimism;
- $\alpha_H = 0$ means pessimism;
- $0 < \alpha_H < 1$ combines optimism and pessimism.

Hurwicz's approach is widely used in decision making, but it is largely a heuristic, and it is not clear how to extend it other types of uncertainty. It is therefore desirable to develop more theoretically justified recommendations for decision making under uncertainty, recommendations that would be applicable to different types of uncertainty.

In this paper, we propose such recommendations by explaining how to assign a fair price to each alternative, so that we can select between several alternatives by comparing their fair prices.

The structure of this paper is as follows: in Sect. 2, we recall how to describe different types of uncertainty; in Sect. 3, we describe the fair price approach; in the following sections, we show how the fair price approach can be applied to different types of uncertainty.

Comment. Our result for the case of interval uncertainty has been previously described in [9]; other results are new.

2   How to Describe Uncertainty

When we have a full information about a situation, then we can express our desirability of each possible alternative by declaring a price that we are willing to pay for this alternative. Once these prices are determined, we simply select the alternative for which the corresponding price is the highest. In this full information case, we know the exact gain u of selecting each alternative.

In practice, we usually only have partial information about the gain u: based on the available information, there are several possible values of the gain u. In other words, instead of the exact gain u, we only know a *set S* of possible values of the gain.

We usually know lower and bounds for this set, so this set is *bounded*. It is also reasonable to assume that the set S is *closed*: indeed, if we have a sequence of possible values $u_n \in S$ that converges to a number u_0, then, no matter how accurately we measure the gain, we can never distinguish between the limit value u_0 and a sufficiently close value u_n. Thus, we will never be able to conclude that the limit value u_0 is not possible – and thus, it is reasonable to consider it possible, i.e., to include the limit point u_0 into the set S of possible values.

In many practical situations, if two gain values $u < u'$ are possible, then all intermediate values $u'' \in (u, u')$ are possible as well. In this case, the bounded closed set S is simply an *interval* $[\underline{u}, \overline{u}]$.

However, sometimes, some intermediate numbers u'' cannot be possible values of the gain. For example, if we buy an obscure lottery ticket for a simple prize-or-no-prize lottery from a remote country, we either get the prize or lose the money. In this case, the set of possible values of the gain consists of two values. To account for such situations, we need to consider general bounded closed sets.

In addition to knowing which gain values are possible, we may also have an information about which of these values are more probable and which values are less probable. Sometimes, this information has a *qualitative* nature, in the sense that, in addition to the set S of possible gain values, we also know a (closed) subset $s \subseteq S$ of values which are more probable (so that all the values from the difference $S - s$ are less probable). In many cases, the set s also contains all its intermediate values, so it is an interval; an important particular case is when this interval s consists of a single point. In other cases, the set s may be different from an interval.

Often, we have a *quantitative* information about the probability (frequency) of different values $u \in S$. A universal way to describe a probability distribution on the real line is to describe its cumulative distribution function (cdf) $F(u) \overset{\text{def}}{=}$ Prob$(U \leq u)$. In the ideal case, we know the exact cdf $F(u)$. In practice, we usually only know the values of the cdf with uncertainty. Typically, for every u, we may only know the bounds $\underline{F}(u)$ and $\overline{F}(u)$ on the actual (unknown) values $F(u)$. The corresponding interval-valued function $[\underline{F}(u), \overline{F}(u)]$ is known as a *p-box* [2,3].

All this classification relates to the usual *passive* uncertainty, uncertainty over which we have no control. Sometimes, however, we have *active* uncertainty. As an example, let us consider two situations in which we need to minimize the amount of energy E used to heat the building. For simplicity, let us assume that cooling by 1 degree requires 1 unit of energy.

In the first situation, we simply know the interval $[\underline{E}, \overline{E}]$ that contains the actual (unknown) value of the energy E: for example, we know that $E \in [20, 25]$ (and we do not control this energy). In the second situation, we know that the outside temperature is between 50 F and 55 F, and we want to maintain the temperature 75 F. In this case, we also conclude that $E \in [20, 25]$, but this time, we ourselves (or, alternatively, the heating system programmed by us) set up the appropriate amount of energy.

The distinction between the usual (passive) uncertainty and a different (active) type of uncertainty can be captured by considering *improper intervals* first introduced by Kaucher, i.e., intervals $[\underline{u}, \overline{u}]$ in which we may have $\underline{u} > \overline{u}$ see, e.g., [7,13]. For example, in terms of these Kaucher intervals, our first (passive) situation is described by the interval $[15, 20]$, while the second (active) situation is described by an improper interval $[20, 15]$.

In line with this classification of different types of uncertainty, in the following text, we will first consider the simplest (interval) uncertainty, then the general set-valued uncertainty, then uncertainty described by a pair of embedded sets (in particular, by a pair of embedded intervals). After that, we consider situations with known probability distribution, situations with a known p-box, and finally, situations described by Kaucher intervals.

3   Fair Price Approach: Main Idea

When we have full information, we can express our desirability of each possible situation by declaring a price that we are willing to pay to get involved in this situation. To make decisions under uncertainty, it is therefore desirable to assign a fair price to each uncertain situation: e.g., to assign a fair price to each interval and/or to each set.

There are reasonable restrictions on the function that assigns the fair price to each type of uncertainty. First, the fair price should be *conservative*: if we know that the gain is always larger than or equal to $\underline{u}$, then the fair price corresponding to this situation should also be greater than or equal to $\underline{u}$. Similarly, if we know that the gain is always smaller than or equal to $\overline{u}$, then the fair price corresponding to this situation should also be smaller than or equal to $\overline{u}$.

Another natural property is *monotonicity*: if one alternative is clearly better than the other, then its fair price should be higher (or at least not lower).

Finally, the fair price should be *additive* in the following sense. Let us consider the situation when we have two consequent independent decisions. In this case, we can either consider two decision processes separately, or we can consider a single decision process in which we select a pair of alternatives:

- the 1st alternative corresponding to the 1st decision, and
- the 2nd alternative corresponding to the 2nd decision.

If we are willing to pay the amount u to participate in the first process, and we are willing to pay the amount v to participate in the second decision process, then it is reasonable to require that we should be willing to pay $u + v$ to participate in both decision processes.

On the examples of the above-mentioned types of uncertainty, let us describe the formulas for the fair price that can be derived from these requirements.

4   Case of Interval Uncertainty

We want to assign, to each interval $[\underline{u}, \overline{u}]$, a number $P([\underline{u}, \overline{u}])$ describing the fair price of this interval. Conservativeness means that the fair price $P([\underline{u}, \overline{u}])$ should be larger than or equal to $\underline{u}$ and smaller than or equal to $\overline{u}$, i.e., that the fair price of an interval should be located in this interval:

$$P([\underline{u}, \overline{u}]) \in [\underline{u}, \overline{u}].$$

Let us now apply monotonicity. Suppose that we keep the lower endpoint $\underline{u}$ intact but increase the upper bound. This means that we keep all the previous possibilities, but we also add new possibilities, with a higher gain. In other words, we are improving the situation. In this case, it is reasonable to require that after this addition, the fair price should either increase or remain the same, but it should definitely not decrease:

$$\text{if } \underline{u} = \underline{v} \text{ and } \overline{u} < \overline{v} \text{ then } P([\underline{u}, \overline{u}]) \leq P([\underline{v}, \overline{v}]).$$

Similarly, if we dismiss some low-gain alternatives, this should increase (or at least not decrease) the fair price:

$$\text{if } \underline{u} < \underline{v} \text{ and } \overline{u} = \overline{v} \text{ then } P([\underline{u}, \overline{u}]) \leq P([\underline{v}, \overline{v}]).$$

Finally, let us apply additivity. In the case of interval uncertainty, about the gain u from the first alternative, we only know that this (unknown) gain is in $[\underline{u}, \overline{u}]$. Similarly, about the gain v from the second alternative, we only know that this gain belongs to the interval $[\underline{v}, \overline{v}]$.

The overall gain $u + v$ can thus take any value from the interval

$$[\underline{u}, \overline{u}] + [\underline{v}, \overline{v}] \stackrel{\text{def}}{=} \{u + v : u \in [\underline{u}, \overline{u}], v \in [\underline{v}, \overline{v}]\}.$$

It is easy to check that (see, e.g., [6,10]):

$$[\underline{u}, \overline{u}] + [\underline{v}, \overline{v}] = [\underline{u} + \underline{v}, \overline{u} + \overline{v}].$$

Thus, for the case of interval uncertainty, the additivity requirement about the fair prices takes the form

$$P([\underline{u} + \underline{v}, \overline{u} + \overline{v}]) = P([\underline{u}, \overline{u}]) + P([\underline{v}, \overline{v}]).$$

So, we arrive at the following definition:

Definition 1. *By a fair price under interval uncertainty, we mean a function $P([\underline{u}, \overline{u}])$ for which:*

- $\underline{u} \leq P([\underline{u}, \overline{u}]) \leq \overline{u}$ *for all $\underline{u}$ and $\overline{u}$ (conservativeness);*
- *if $\underline{u} = \underline{v}$ and $\overline{u} < \overline{v}$, then $P([\underline{u}, \overline{u}]) \leq P([\underline{v}, \overline{v}])$ (monotonicity);*
- *(additivity) for all $\underline{u}$, $\overline{u}$, $\underline{v}$, and $\overline{v}$, we have*

$$P([\underline{u} + \underline{v}, \overline{u} + \overline{v}]) = P([\underline{u}, \overline{u}]) + P([\underline{v}, \overline{v}]).$$

Proposition 1. [9] *Each fair price under interval uncertainty has the form*

$$P([\underline{u}, \overline{u}]) = \alpha_H \cdot \overline{u} + (1 - \alpha_H) \cdot \underline{u} \text{ for some } \alpha_H \in [0, 1].$$

Comment. We thus get a new justification of Hurwicz optimism-pessimism criterion.

Proof.

1°. Due to monotonicity, $P([u, u]) = u$.

2°. Also, due to monotonicity, $\alpha_H \stackrel{\text{def}}{=} P([0, 1]) \in [0, 1]$.

3°. For $[0, 1] = [0, 1/n] + \ldots + [0, 1/n]$ (n times), additivity implies $\alpha_H = n \cdot P([0, 1/n])$, so $P([0, 1/n]) = \alpha_H \cdot (1/n)$.

4°. For $[0, m/n] = [0, 1/n] + \ldots + [0, 1/n]$ (m times), additivity implies

$$P([0, m/n]) = \alpha_H \cdot (m/n).$$

$5°$. For each real number r, for each n, there is an m such that $m/n \leq r \leq (m+1)/n$. Monotonicity implies

$$\alpha_H \cdot (m/n) = P([0, m/n]) \leq P([0, r]) \leq P([0, (m+1)/n]) = \alpha_H \cdot ((m+1)/n).$$

When $n \to \infty$, $\alpha_H \cdot (m/n) \to \alpha_H \cdot r$ and $\alpha_H \cdot ((m+1)/n) \to \alpha_H \cdot r$, hence $P([0, r]) = \alpha_H \cdot r$.

$6°$. For $[\underline{u}, \overline{u}] = [\underline{u}, \underline{u}] + [0, \overline{u} - \underline{u}]$, additivity implies $P([\underline{u}, \overline{u}]) = \underline{u} + \alpha_H \cdot (\overline{u} - \underline{u})$. The proposition is proven.

5   Case of Set-Valued Uncertainty

Intervals are a specific case of bounded closed sets. We already know how to assign fair price to intervals. So, we arrive at the following definition.

Definition 2. *By a fair price under set-valued uncertainty, we mean a function P that assigns, to every bounded closed set S, a real number $P(S)$, for which:*

- $P([\underline{u}, \overline{u}]) = \alpha_H \cdot \overline{u} + (1 - \alpha_H) \cdot \underline{u}$ *(conservativeness)*;
- $P(S + S') = P(S) + P(S')$, *where* $S + S' \stackrel{\text{def}}{=} \{s + s' : s \in S, s' \in S'\}$ *(additivity)*.

Proposition 2. *Each fair price under set uncertainty has the form* $P(S) = \alpha_H \cdot \sup S + (1 - \alpha_H) \cdot \inf S$.

Proof. It is easy to check that each bounded closed set S contains its infimum $\underline{S} \stackrel{\text{def}}{=} \inf S$ and supremum $\underline{S} \stackrel{\text{def}}{=} \sup S$: $\{\underline{S}, \overline{S}\} \subseteq S \subseteq [\underline{S}, \overline{S}]$. Thus,

$$[2\underline{S}, 2\overline{S}] = \{\underline{S}, \overline{S}\} + [\underline{S}, \overline{S}] \subseteq S + [\underline{S}, \overline{S}] \subseteq [\underline{S}, \overline{S}] + [\underline{S}, \overline{S}] = [2\underline{S}, 2\overline{S}].$$

So, $S + [\underline{S}, \overline{S}] = [2\underline{S}, 2\overline{S}]$. By additivity, we conclude that $P(S) + P([\underline{S}, \overline{S}]) = P([2\underline{S}, 2\overline{S}])$. Due to conservativeness, we know the fair prices $P([\underline{S}, \overline{S}])$ and $P([2\underline{S}, 2\overline{S}])$. Thus, we can conclude that

$$P(S) = P([2\underline{S}, 2\overline{S}]) - P([\underline{S}, \overline{S}]) = (\alpha_H \cdot (2\overline{S}) + (1 - \alpha_H) \cdot (2\underline{S})) - (\alpha_H \cdot \overline{S} + (1 - \alpha_H) \cdot \underline{S}),$$

hence indeed $P(S) = \alpha_H \cdot \overline{S} + (1 - \alpha_H) \cdot \underline{S}$. The proposition is proven.

6   Case of Embedded Sets

In addition to a set S of possible values of the gain u, we may also know a subset $s \subseteq S$ of more probable values u. To describe a fair price assigned to such a pair (S, s), let us start with the simplest case when the original set S is an interval $S = [\underline{u}, \overline{u}]$, and the subset s is a single "most probable" value u_0 within this interval. Such pairs are known as *triples*; see, e.g., [1] and references therein. For triples, addition is defined component-wise:

$$([\underline{u}, \overline{u}], u_0) + ([\underline{v}, \overline{v}], v_0) = ([\underline{u} + \underline{v}, \overline{u} + \overline{v}], u_0 + v_0).$$

Thus, the additivity requirement about the fair prices takes the form

$$P([\underline{u} + \underline{v}, \overline{u} + \overline{v}], u_0 + v_0) = P([\underline{u}, \overline{u}], u_0) + P([\underline{v}, \overline{v}], v_0).$$

Definition 3. *By a fair price under triple uncertainty, we mean a function* $P([\underline{u}, \overline{u}], u_0)$ *for which:*

- $\underline{u} \leq P([\underline{u}, \overline{u}], u_0) \leq \overline{u}$ *for all* $\underline{u} \leq u \leq \overline{u}$ *(conservativeness);*
- *if* $\underline{u} \leq \underline{v}$, $u_0 \leq v_0$, *and* $\overline{u} \leq \overline{v}$, *then* $P([\underline{u}, \overline{u}], u_0) \leq P([\underline{v}, \overline{v}], v_0)$ *(monotonicity);*
- *(additivity) for all* $\underline{u}$, $\overline{u}$, u_0 $\underline{v}$, $\overline{v}$, *and* v_0, *we have*

$$P([\underline{u} + \underline{v}, \overline{u} + \overline{v}], u_0 + v_0) = P([\underline{u}, \overline{u}], u_0) + P([\underline{v}, \overline{v}], v_0).$$

Proposition 3. *Each fair price under triple uncertainty has the form*

$$P([\underline{u}, \overline{u}], u_0) = \alpha_L \cdot \underline{u} + (1 - \alpha_L - \alpha_U) \cdot u_0 + \alpha_U \cdot \overline{u}, \text{ where } \alpha_L, \alpha_U \in [0, 1].$$

Proof. In general, we have

$$([\underline{u}, \overline{u}], u_0) = ([u_0, u_0], u_0) + ([0, \overline{u} - u], 0) + ([\underline{u} - u, 0], 0).$$

So, due to additivity:

$$P([\underline{u}, \overline{u}], u_0) = P([u_0, u_0], u_0) + P([0, \overline{u} - u_0], 0) + P([\underline{u} - u_0, 0], 0).$$

Due to conservativeness, $P([u_0, u_0], u_0) = u_0$.

Similarly to the interval case, we can prove that $P([0, r], 0) = \alpha_U \cdot r$ for some $\alpha_U \in [0, 1]$, and that $P([r, 0], 0) = \alpha_L \cdot r$ for some $\alpha_L \in [0, 1]$. Thus,

$$P([\underline{u}, \overline{u}], u_0) = \alpha_L \cdot \underline{u} + (1 - \alpha_L - \alpha_U) \cdot u_0 + \alpha_U \cdot \overline{u}.$$

The proposition is proven.

The next simplest case is when both sets S and $s \subseteq S$ are intervals, i.e., when, inside the interval $S = [\underline{u}, \overline{u}]$, instead of a "most probable" value u_0, we have a "most probable" subinterval $[\underline{m}, \overline{m}] \subseteq [\underline{u}, \overline{u}]$. The resulting pair of intervals is known as a "twin interval" (see, e.g., [4,11]).

For such twin intervals, addition is defined component-wise:

$$([\underline{u}, \overline{u}], [\underline{m}, \overline{m}]) + ([\underline{v}, \overline{v}], [\underline{n}, \overline{n}]) = ([\underline{u} + \underline{v}, \overline{u} + \overline{v}], [\underline{m} + \underline{n}, \overline{m} + \overline{n}]).$$

Thus, the additivity requirement about the fair prices takes the form

$$P([\underline{u} + \underline{v}, \overline{u} + \overline{v}], [\underline{m} + \underline{n}, \overline{m} + \overline{n}]) = P([\underline{u}, \overline{u}], [\underline{m}, \overline{m}]) + P([\underline{v}, \overline{v}], [\underline{n}, \overline{n}]).$$

Definition 4. *By a fair price under twin uncertainty, we mean a function* $P([\underline{u}, \overline{u}], [\underline{m}, \overline{m}])$ *for which:*

- $\underline{u} \leq P([\underline{u}, \overline{u}], [\underline{m}, \overline{m}]) \leq \overline{u}$ *for all* $\underline{u} \leq \underline{m} \leq \overline{m} \leq \overline{u}$ *(conservativeness);*
- *if* $\underline{u} \leq \underline{v}$, $\underline{m} \leq \underline{n}$, $\overline{m} \leq \overline{n}$, *and* $\overline{u} \leq \overline{v}$, *then* $P([\underline{u}, \overline{u}], [\underline{m}, \overline{m}]) \leq P([\underline{v}, \overline{v}], [\underline{n}, \overline{n}])$ *(monotonicity);*
- *for all* $\underline{u} \leq \underline{m} \leq \overline{m} \leq \overline{u}$ *and* $\underline{v} \leq \underline{n} \leq \overline{n} \leq \overline{v}$, *we have additivity:*

$$P([\underline{u} + \underline{v}, \overline{u} + \overline{v}], [\underline{m} + \underline{n}, \overline{m} + \overline{m}]) = P([\underline{u}, \overline{u}], [\underline{m}, \overline{m}]) + P([\underline{v}, \overline{v}], [\underline{n}, \overline{n}]).$$

Proposition 4. *Each fair price under twin uncertainty has the following form, for some* $\alpha_L, \alpha_u, \alpha_U \in [0, 1]$:

$$P([\underline{u}, \overline{u}], [\underline{m}, \overline{m}]) = \underline{m} + \alpha_u \cdot (\overline{m} - \underline{m}) + \alpha_U \cdot (\overline{u} - \overline{m}) + \alpha_L \cdot (\underline{u} - \underline{m}).$$

Proof. In general, we have

$$([\underline{u}, \overline{u}], [\underline{m}, \overline{m}]) = ([\underline{m}, \underline{m}], [\underline{m}, \underline{m}]) + ([0, \overline{m} - \underline{m}], [0, \overline{m} - \underline{m}]) +$$

$$([0, \overline{u} - \overline{m}], [0, 0]) + ([\underline{u} - \underline{m}, 0], [0, 0]).$$

So, due to additivity:

$$P([\underline{u}, \overline{u}], [\underline{m}, \overline{m}]) = P([\underline{m}, \underline{m}], [\underline{m}, \underline{m}]) + P([0, \overline{m} - \underline{m}], [0, \overline{m} - \underline{m}]) +$$

$$P([0, \overline{u} - \overline{m}], [0, 0]) + P([\underline{u} - \underline{m}, 0], [0, 0]).$$

Due to conservativeness, $P([\underline{m}, \underline{m}], [\underline{m}, \underline{m}]) = \underline{m}$. Similarly to the interval case, we can prove that:

- $P([0, r], [0, r]) = \alpha_u \cdot r$ for some $\alpha_u \in [0, 1]$,
- $P([0, r], [0, 0]) = \alpha_U \cdot r$ for some $\alpha_U \in [0, 1]$;
- $P([r, 0], [0, 0]) = \alpha_L \cdot r$ for some $\alpha_L \in [0, 1]$.

Thus,

$$P([\underline{u}, \overline{u}], [\underline{m}, \overline{m}]) = \underline{m} + \alpha_u \cdot (\overline{m} - \underline{m}) + \alpha_U \cdot (\overline{u} - \overline{m}) + \alpha_L \cdot (\underline{u} - \underline{m}).$$

The proposition is proven.

Finally, let us consider the general case.

Definition 5. *By a fair price under embedded-set uncertainty, we mean a function P that assigns, to every pair of bounded closed sets (S, s) with $s \subseteq S$, a real number $P(S, s)$, for which:*

- $P([\underline{u}, \overline{u}], [\underline{m}, \overline{m}]) = \underline{m} + \alpha_u \cdot (\overline{m} - \underline{m}) + \alpha_U \cdot (\overline{U} - \overline{m}) + \alpha_L \cdot (\underline{u} - \underline{m})$
  (conservativeness);
- $P(S + S', s + s') = P(S, s) + P(S', s')$ *(additivity).*

Proposition 5. *Each fair price under embedded-set uncertainty has the form*

$$P(S, s) = \inf s + \alpha_u \cdot (\sup s - \inf s) + \alpha_U \cdot (\sup S - \sup s) + \alpha_L \cdot (\inf S - \inf s).$$

Proof. Similarly to the proof of Proposition 2, we can conclude that

$$(S, s) + ([\inf S, \sup S], [\inf s, \sup s]) = ([2 \cdot \inf S, 2 \cdot \sup S], [2 \cdot \inf s, 2 \cdot \sup s]).$$

By additivity, we conclude that

$$P(S, s) + P([\inf S, \sup S], [\inf s, \sup s]) =$$

$$P([2 \cdot \inf S, 2 \cdot \sup S], [2 \cdot \inf s, 2 \cdot \sup s]),$$

hence

$$P(S, s) = P([2 \cdot \inf S, \cdot \sup S], [2 \cdot \inf s, 2 \cdot \sup s]) - P([\inf S, \sup S], [\inf s, \sup s]).$$

Due to conservativeness, we know the fair prices

$$P([2 \cdot \inf S, 2 \cdot \sup S], [2 \cdot \inf s, 2 \cdot \sup s]) \text{ and } P([\inf S, \sup S], [\inf s, \sup s]).$$

Subtracting these expressions, we get the desired formula for $P(S, s)$. The proposition is proven.

7 Cases of Probabilistic and p-Box Uncertainty

Suppose that for some financial instrument, we know the corresponding probability distribution $F(u)$ on the set of possible gains u. What is the fair price P for this instrument?

Due to additivity, the fair price for n copies of this instrument is $n \cdot P$. According to the Large Numbers Theorem, for large n, the average gain tends to the mean value $\mu = \int u \, dF(u)$.

Thus, the fair price for n copies of the instrument is close to $n \cdot \mu$: $n \cdot P \approx n \cdot \mu$. The larger n, the closer the averages. So, in the limit, we get $P = \mu$.

So, the fair price under probabilistic uncertainty is equal to the average gain $\mu = \int u \, dF(u)$.

Let us now consider the case of a p-box $[\underline{F}(u), \overline{F}(u)]$. For different functions $F(u) \in [\underline{F}(u), \overline{F}(u)]$, values of the mean μ form an interval $[\underline{\mu}, \overline{\mu}]$, where $\underline{\mu} = \int u \, d\overline{F}(u)$ and $\overline{\mu} = \int u \, d\underline{F}(u)$. Thus, the price of a p-box is equal to the price of an interval $[\underline{\mu}, \overline{\mu}]$.

We already know that the fair price of this interval is equal to

$$\alpha_H \cdot \overline{\mu} + (1 - \alpha_H) \cdot \underline{\mu}.$$

Thus, we conclude that the fair price of a p-box $[\underline{F}(u), \overline{F}(u)]$ is $\alpha_H \cdot \overline{\mu} + (1 - \alpha_H) \cdot \underline{\mu}$, where $\underline{\mu} = \int u \, d\overline{F}(u)$ and $\overline{\mu} = \int u \, d\underline{F}(u)$.

8 Case of Kaucher (Improper) Intervals

For Kaucher intervals, addition is also defined component-wise; in particular, for all $\underline{u} < \overline{u}$, we have

$$[\underline{u}, \overline{u}] + [\overline{u}, \underline{u}] = [\underline{u} + \overline{u}, \underline{u} + \overline{u}].$$

Thus, additivity implies that

$$P([\underline{u}, \overline{u}]) + P([\overline{u}, \underline{u}]) = P([\underline{u} + \overline{u}, \underline{u} + \overline{u}]).$$

We know that $P([\overline{u}, \underline{u}]) = \alpha_H \cdot \underline{u} + (1 - \alpha_H) \cdot \overline{u}$ and $P([\underline{u} + \overline{u}, \underline{u} + \overline{u}]) = \underline{u} + \overline{u}$. Hence:

$$P([\underline{u}, \overline{u}]) = (\underline{u} + \overline{u}) - (\alpha_H \cdot \underline{u} + (1 - \alpha_H) \cdot \overline{u}).$$

Thus, the fair price $P([\underline{u}, \overline{u}])$ of an improper interval $[\underline{u}, \overline{u}]$, with $\underline{u} > \overline{u}$, is equal to $P([\underline{u}, \overline{u}]) = \alpha_H \cdot \overline{u} + (1 - \alpha_H) \cdot \underline{u}$.

9   Summary and Conclusions

In this paper, for different types of uncertainty, we derive the formulas for the fair prices under reasonable conditions of conservativeness, monotonicity, and additivity.

In the simplest case of interval uncertainty, when we only know the interval $[\underline{u}, \overline{u}]$ of possible values of the gain u, the fair price is equal to

$$P([\underline{u}, \overline{u}]) = \alpha_H \cdot \overline{u} + (1 - \alpha_H) \cdot \underline{u},$$

for some parameter $\alpha_H \in [0, 1]$. Thus, the fair price approach provides a justification for the formula originally proposed by a Nobelist L. Hurwicz, in which α_H describes the decision maker's optimism degree: $\alpha_H = 1$ corresponds to pure optimism, $\alpha_H = 0$ to pure pessimism, and intermediate values of α_H correspond to a realistic approach that takes into account both best-case (optimistic) and worst-case (pessimistic) scenarios.

In a more general situation, when the set S of possible values of the gain u is not necessarily an interval, the fair price is equal to

$$P(S) = \alpha_H \cdot \sup S + (1 - \alpha_H) \cdot \inf(S).$$

If, in addition to the set S of possible values of the gain u, we also know a subset $s \subseteq S$ of "most probable" gain values, then the fair price takes the form

$$P(S, s) = \inf s + \alpha_u \cdot (\sup s - \inf s) + \alpha_U \cdot (\sup S - \sup s) + \alpha_L \cdot (\inf S - \inf s),$$

for some values α_u, α_L, and α_U from the interval $[0, 1]$. In particular, when both sets S and s are intervals, i.e., when $S = [\underline{u}, \overline{u}]$ and $s = [\underline{m}, \overline{m}]$, the fair price takes the form

$$P([\underline{u}, \overline{u}], [\underline{m}, \overline{m}]) = \underline{m} + \alpha_u \cdot (\overline{m} - \underline{m}) + \alpha_U \cdot (\overline{u} - \overline{m}) + \alpha_L \cdot (\underline{u} - \underline{m}).$$

When the interval s consists of a single value u_0, this formula turns into

$$P([\underline{u}, \overline{u}], u_0) = \alpha_L \cdot \underline{u} + (1 - \alpha_L - \alpha_U) \cdot u_0 + \alpha_U \cdot \overline{u}.$$

When, in addition to the set S, we also know the cumulative distributive function (cdf) $F(u)$ that describes the probability distribution of different possible values u, then the fair price is equal to the expected value of the gain

$$P(F) = \int u \, dF(u).$$

In situations when for each u, we only know the interval $[\underline{F}(u), \overline{F}(u)]$ of possible values of the cdf $F(u)$, then the fair price is equal to

$$P([\underline{F}, \overline{F}]) = \alpha_H \cdot \int u \, d\overline{F}(u) + (1 - \alpha_H) \cdot \int u \, d\underline{F}(u).$$

Finally, when uncertainty is described by an improper interval $[\underline{u}, \overline{u}]$ with $\underline{u} > \overline{u}$, the fair price is equal to

$$P([\underline{u}, \overline{u}]) = \alpha_H \cdot \overline{u} + (1 - \alpha_H) \cdot \underline{u}.$$

Acknowledgments. This work was supported in part by the National Science Foundation grants HRD-0734825 and HRD-1242122 (Cyber-ShARE Center of Excellence) and DUE-0926721, by Grant 1 T36 GM078000-01 and 1R43TR000173-01 from the National Institutes of Health, and by grant N62909-12-1-7039 from the Office of Naval Research.

The authors are thankful to all the participants of the 16th International Symposium on Scientific Computing, Computer Arithmetic, and Validated Numerics SCAN'2014 (Würzburg, German, September 21–26, 2014) for their interest, and to the anonymous referees for valuable suggestions.

References

1. Cole, A.J., Morrison, R.: Triplex: a system for interval arithmetic. Softw. Pract. Experience **12**(4), 341–350 (1982)
2. Ferson, S.: Risk Assessment with Uncertainty Numbers: RiskCalc. CRC Press, Boca Raton, Florida (2002)
3. Ferson, S., Kreinovich, V., Oberkampf, W., Ginzburg, L.: Experimental Uncertainty Estimation and Statistics for Data Having Interval Uncertainty, Sandia National Laboratories, Report SAND2007-0939 (2007)
4. Gardefies, E., Trepat, A., Janer, J.M.: SIGLA-PL/1: development and applications. In: Nickel, K.L.E. (ed.) Interval Mathematics 1980, pp. 301–315. Academic Press, New York (1980)
5. Hurwicz, L.: Optimality criteria for decision making under ignorance, Cowles Commission Discussion Paper, Statistics, No. 370 (1951)
6. Jaulin, L., Kieffer, M., Didrit, O., Walter, E.: Applied Interval Analysis, with Examples in Parameter and State Estimation. Robust Control and Robotics. Springer, London (2001)
7. Kaucher, E.: Über Eigenschaften und Anwendungsmöglichkeiten der erweiterten Intervallrechnung und des hyperbolische Fastköpers über R. Comput. Suppl. **1**, 81–94 (1977)
8. Luce, R.D., Raiffa, R.: Games and Decisions: Introduction and Critical Survey. Dover, New York (1989)
9. McKee, J., Lorkowski, J., Ngamsantivong, T.: Note on fair price under interval uncertainty. J. Uncertain Syst. **8**(3), 186–189 (2014)
10. Moore, R.E., Kearfott, R.B., Cloud, M.J.: Introduction to Interval Analysis. SIAM Press, Philadelphia, Pennsylviania (2009)
11. Nesterov, V.M.: Interval and twin arithmetics. Reliable Comput. **3**(4), 369–380 (1997)
12. Raiffa, H.: Decision Analysis: Introductory Lectures on Choices Under Uncertainty. Mcgraw-Hill, New York (1997)
13. Sainz, M.A., Armengol, J., Calm, R., Herrero, P., Jorba, L., Vehi, J.: Modal Interval Analysis. Springer, Berlin, Heidelberg, New York (2014)

Sliding Mode Approaches Considering Uncertainty for Reliable Control and Computation of Confidence Regions in State and Parameter Estimation

Luise Senkel[✉], Andreas Rauh, and Harald Aschemann

Chair of Mechatronics, University of Rostock, 18059 Rostock, Germany
{Luise.Senkel,Andreas.Rauh,Harald.Aschemann}@uni-rostock.de

Abstract. Robust control procedures are essential for a reliable functionality of technical applications. Therefore, firstly, the mathematical description of the system and, secondly, bounded as well as stochastic disturbances play a major role in control engineering. Bounded uncertainty occurs due to lack of knowledge about system parameters, manufacturing tolerances and measurement inaccuracy. Stochastic disturbances, namely process and measurement noise, play further a very important role in system dynamics. Both classes of uncertainty are considered in the presented control and estimation purposes by using interval arithmetics, where the estimator is necessary to reconstruct non-measurable system states. Interval representations of uncertain variables provide the possibility to stabilize dynamic (nonlinear) systems in a robust way. This is necessary because parameters and measured data are typically only known within given tolerance bounds. Therefore, this paper combines interval arithmetics with the advantages of sliding mode approaches for control and estimation of states and parameters taking into account also stochastic disturbances. The efficiency of these approaches is shown in terms of an application describing the longitudinal dynamics of a vehicle.

Keywords: Sliding mode techniques · Uncertainty · Interval arithmetics

1 Introduction

A challenging task in control theory is to find a unique concept that can be applied to a large number of systems, that provides sufficiently accurate results, stabilizes the system dynamics in a robust way, and can cope with uncertainty (unknown parameters) as well as random effects (e.g. non-modeled friction, measurement inaccuracies). Unfortunately, the applicability of common numerical calculations is limited in terms of rounding errors, truncation errors, and input errors [4]. These problems can be overcome by calculations with intervals describing the range of a variable instead of its scalar value. Then, statements about the dynamics of a system described by parameter ranges can be proposed without

© Springer International Publishing Switzerland 2016
M. Nehmeier et al. (Eds.): SCAN 2014, LNCS 9553, pp. 77–96, 2016.
DOI: 10.1007/978-3-319-31769-4_7

the need of statistical methods (e.g. Monte-Carlo methods) [3]. Nevertheless, interval arithmetic has to cope with the problem of overestimation due to the dependency problem and the wrapping effect [5]. In combination with robust control and estimation purposes, a reliable possibility to deal with uncertainty and to quantify their worst-case influence in the system dynamics can be provided while robustness and stability of the error dynamics are guaranteed.

Sliding mode techniques, so-called variable-structure approaches, are well known for their robust performance and the possibility to handle bounded uncertainties and disturbances in a more efficient way than other approaches as for example state-of-the-art back-stepping procedures [12]. Commonly, the user defines a manifold (called sliding surface) that is assumed to be reached by the system in a stable manner. Once this surface is reached, the system will not diverge anymore and remains in the near surrounding area of this stable mode. However, a large problem in sliding mode approaches is still the so-called chattering that occurs inevitably due to noise that affects the switching function due to discretization in computer implementations [12]. In recent years, second-order and other higher-order sliding modes have been developed to reduce chattering. Nevertheless, common sliding modes are limited in their applicability due to quite restrictive matching conditions [7]. For that reason, this paper presents sliding mode techniques that make use of intervals to reduce chattering by taking into account bounded and stochastic uncertainty of parameters and states. In general, the applicability of these approaches is not limited to a special class of systems [7]. Therefore, the Itô differential operator is applied for an online calculation of the variable-structure gain (called switching amplitude) in each time step despite stochastic disturbances instead of the usual offline computation.

This paper is structured as follows: Sect. 2 describes a scenario for which the sliding mode approaches, firstly a control procedure (Sect. 3) and secondly an observer for estimation of parameters and states (Sect. 4) has been applied. The results are shown in Sect. 5 before this paper is concluded and finalized with an outlook on further work in Sect. 6.

2   Application Scenario

The control and estimation procedures presented in the following are implemented on a laboratory test rig available at the Chair of Mechatronics at the University of Rostock which can be interpreted as the longitudinal dynamics of a vehicle, cf. Fig. 1. The following assumptions have been made: all mass moments of inertia of the motor, the brake as well as of the drive and load side shafts are summarized into one mass moment of inertia J. Moreover, a velocity-proportional friction coefficient d (friction occurs inevitably in all bearings, between the deflector rolls and the toothed belt) and the requirement that static friction is compensated by an underlying motor control are considered. Then, the system model is described by $\mathbf{f}(\mathbf{x}(t), \mathbf{p}, \mathbf{u}(t)) = \dot{\mathbf{x}}(t) = \left[\dot{x}_1(t), \dot{x}_2(t)\right]^T$ with two system states $x_1(t) = \varphi_M(t)$ (angle on the drive side) and $x_2(t) = \dot{x}_1(t) = \dot{\varphi}_M(t) = \omega_M(t)$ (corresp. angular velocity) according to

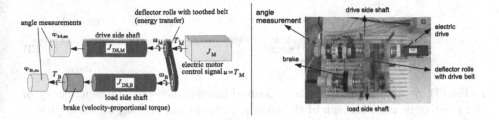

Fig. 1. Schematic visualization and photo of the available test rig.

$$\dot{\mathbf{x}}(t) = \mathbf{A} \cdot \mathbf{x}(t) + \mathbf{b} \cdot u(t) = \begin{bmatrix} 0 & 1 \\ 0 & \alpha \end{bmatrix} \begin{bmatrix} x_1(t) \\ x_2(t) \end{bmatrix} + \begin{bmatrix} 0 \\ \beta \end{bmatrix} u(t) \;,\; y(t) = x_1(t) \;. \quad (1)$$

In Eq. (1), two parameters $\alpha = -\frac{d}{J} \in [\alpha]$ and $\beta = \frac{1}{J} \in [\beta]$ are included. Both parameters influence the system dynamics in a significant way due to the multiplicative coupling with the time-varying system states and the input (motor control signal). Because of this and due to the fact that both parameters are a-priori unknown, it is necessary to determine these and the system states simultaneously. Note that only the first state $y = x_1$ is measurable. However, for the control strategy in the next section, knowledge about both states is necessary.

3 Sliding Mode Techniques for Control Purposes

In recent years, sliding mode approaches for control purposes have reached wide attention due to their inherent robustness despite uncertainties and disturbances. Both can be compensated in terms of the variable structure part in order to stabilize the system's tracking error efficiently. In the following, the classical approach is described which is then extended to an interval-based formulation and stochastic processes.

3.1 Classical Control Approach

The classical sliding mode controller for single-input single-output systems is based on a Lyapunov function [9] with $V(t) = \frac{1}{2}s^2(t)$ and its time derivative $\dot{V}(t) = s(t) \cdot \dot{s}(t)$ with the sliding variable chosen as a Hurwitz polynomial of order $n - 1$ (in the following, all time arguments are omitted)

$$s = \kappa_0 \cdot (x^{(0)} - x_d^{(0)}) + \kappa_1 \cdot (x^{(1)} - x_d^{(1)}) + \dots + (x^{(n-1)} - x_d^{(n-1)}) \;. \quad (2)$$

Here, $x^{(0)} = x_1$ denotes the first state of a system which is given in nonlinear controller canonical form, and $x^{(i)}$ with $i = \{1, ..., n-1\}$ the derivatives up to the order $n - 1$. The terms $x_d, .., x_d^{(n-1)}$ denote the desired trajectory and its time derivatives, resp. The condition for asymptotic stability $\dot{V} < 0$ for all $s \neq 0$ can be fulfilled by using the switching amplitude η and the definition of the absolute value of the sliding variable $|s| = s \cdot \text{sign}(s)$ according to

$$\dot{V} \overset{!}{\leq} -\eta \cdot |s| \Rightarrow s\dot{s} \overset{!}{\leq} -\eta \cdot s \cdot \text{sign}(s) \Rightarrow \dot{s} + \eta \cdot \text{sign}(s) \overset{!}{\leq} -\epsilon \cdot \text{sign}(s)$$

$$\Rightarrow \dot{s} + \underbrace{(\eta + \epsilon)}_{:=\tilde{\eta} > 0} \cdot \text{sign}(s) \overset{!}{\leq} 0 . \tag{3}$$

In Eq. (3), $\tilde{\eta}$ can be interpreted as a desired convergence rate. In dependency of the positive or negative sign of the sliding variable, the signum function yields

$$\text{sign}(s) = \begin{cases} 1, & \text{if } s > 0 \\ -1, & \text{if } s < 0 \\ 0, & \text{else} . \end{cases} \tag{4}$$

For the described application scenario, the sliding surface has the form $s = \kappa_0 \cdot (x_1 - x_{1,d}) + (x_2 - x_{2,d})$. The control law results after inserting the 2nd state equation of (1) into the time derivative $\dot{s} = \kappa_0(\dot{x}_1 - \dot{x}_{1,d}) + \kappa_1(\dot{x}_2 - \dot{x}_{2,d})$ in

$$u = \frac{-\kappa_0(x_2 - x_{2,d}) - (\alpha x_2 - \dot{x}_{2,d}) - \tilde{\eta} \cdot \text{sign}(s)}{\beta} . \tag{5}$$

In Eq. (5), $\kappa_0 > 0$ and $\tilde{\eta} > 0$ have to be chosen such that the condition for asymptotic stability $\dot{V} < 0$ holds. The disadvantage of this approach is that the results are often too conservative, especially in case of uncertain parameters so that unnecessary large switching amplitudes defined by the user are necessary. Moreover, noise and actuator wear may occur [9]. Therefore, this approach is extended in the following such that uncertainty as well as stochastic processes are taken into consideration. Moreover, the scalar switching amplitude $\tilde{\eta}$ is generalized to a time-varying vector of the same dimension as the state vector $\mathbf{x}$.

3.2   Extension to Uncertain and Stochastic Processes

In the following, an interval-based sliding mode controller (ISMC) is described [10]. Consider an uncertain system of order n described by

$$d\mathbf{x} = \mathbf{f}(\mathbf{x}, \mathbf{p}, u)dt + \mathbf{G}_p \, d\mathbf{w}_p \quad \text{and} \quad \mathbf{y} = \mathbf{C}_C(\mathbf{p}) \cdot \mathbf{x} + \mathbf{G}_m \, d\mathbf{w}_m \tag{6}$$

for all $\mathbf{p} \in [\mathbf{p}]$ and $\mathbf{x} \in [\mathbf{x}]$ with the standard Brownian motion of the process $d\mathbf{w}_p \in \mathbb{R}^{r_p}$ and of the measurement $d\mathbf{w}_m \in \mathbb{R}^{r_m}$ as well as the corresponding matrices of standard deviations $\mathbf{G}_p \in \mathbb{R}^{n \times r_p}$ and $\mathbf{G}_m \in \mathbb{R}^{n_y \times r_m}$, and the output matrix $\mathbf{C}_C$. In the linear case[1], it is assumed, that the function $\mathbf{f}(\mathbf{x}, \mathbf{p}, u)$ is given in state-space notation with the system and input matrices $\mathbf{A}_C$ and $\mathbf{B}_C$ according to $\mathbf{f}(\mathbf{x}, \mathbf{p}, u) = \mathbf{A}_C \mathbf{x} + \mathbf{B}_C u$ for $\mathbf{p} \in [\mathbf{p}]$. Requiring the existence of a desired trajectory and its derivatives for all states, the vector of desired trajectories is denoted by $\mathbf{x}_d$. Including intervals for system parameters, the following control law[2] can be defined according to

---

[1] Linearity in the state vector $\mathbf{x}$ which is assumed in this application scenario.

[2] Here, only a system with a single input $\dim(u) = 1$ is described. The approach is also applicable for a system with more than one input. In this case, the vector of switching amplitudes becomes a matrix. This also holds for the linear controller gain $\mathbf{K}$.

$$u = u_{FF} - \mathbf{k}^T \cdot \mathbf{x} + \boldsymbol{\eta}^T \cdot \text{sign}(\mathbf{x} - \mathbf{x}_d) \ . \tag{7}$$

In Eq. (7), a (static or a dynamic) feedforward control term u_{FF} is included. An underlying state feedback term is denoted by $\mathbf{k}^T \cdot \mathbf{x}$, where the controller gain $\mathbf{k}$ is calculated by e.g. pole assignment for one parameter vector $\mathbf{p} \in [\mathbf{p}]$. As already mentioned, the scalar switching amplitude from the classical approach becomes a vector $\boldsymbol{\eta}$ with $\dim(\boldsymbol{\eta}) = \dim(\mathbf{x}) = n \times 1$ (system order n). To calculate the switching amplitude vector, the Itô differential operator [6] is applied while further taking into consideration the uncertain system

$$[\mathbf{f}] = \mathbb{A}_C \mathbf{x} + \mathbb{B}_C u \ . \tag{8}$$

Note, that the terms $\mathbb{A}_C$, $\mathbb{B}_C$ are interval evaluations according to $\mathbb{A}_C \in \mathbb{A}_C := \mathbb{A}_C(\mathbf{x}, [\mathbf{p}])$ and $\mathbb{B}_C \in \mathbb{B}_C := \mathbb{B}_C(\mathbf{x}, [\mathbf{p}])$. Consequently, the Itô differential operator according to the definition

$$L(V_C) = \frac{\partial V_C}{\partial t} + \left(\frac{\partial V_C}{\partial \tilde{\mathbf{x}}}\right)^T \cdot ([\mathbf{f}] - \dot{\mathbf{x}}_d) + \frac{1}{2}\text{trace}\left\{\mathbf{G}_p^T \frac{\partial^2 V_C}{\partial \tilde{\mathbf{x}}^2}\mathbf{G}_p\right\} \tag{9}$$

is used. It is evaluated taking into account a control error interval $[\varDelta \mathbf{x}_c]$ in $\mathbf{x} \in [\mathbf{x}] = \mathbf{x} + [\varDelta \mathbf{x}_c]$ to consider also control errors as well as parameter uncertainty specified as range bounds in terms of intervals $\mathbf{p} \in [\mathbf{p}]$. A suitable candidate for a Lyapunov function is given by $V_C = \frac{1}{2}\tilde{\mathbf{x}}^T \mathbf{P}_C \tilde{\mathbf{x}}$, with the definition of the vector-valued sliding variable $\tilde{\mathbf{x}} := \mathbf{x} - \mathbf{x}_d = [x_1 - x_{1,d}, \ x_2 - x_{2,d}, \ ... , x_n - x_{n,d}]^T$. As it can be seen, there is no explicit time dependency in V_C so that $\frac{\partial V_C}{\partial t} = 0$. By employing the condition $L(V_C) \overset{!}{<} -\mathbf{q}_C^T \text{abs}(\tilde{\mathbf{x}})$ with a user-defined element-wise non-negative convergence rate vector $\mathbf{q}_C$, the components $i \in \{1, ..., n\}$ of the switching amplitude vector are calculated by

$$\eta_i = \begin{cases} \sup\left([\mathbf{M}]_i^+ \cdot \left(-[\dot{V}_{a,C}] - \mathbf{q}_C^T \text{abs}([\tilde{\mathbf{x}}]) - T\right)\right) + \mu, & \text{if } [\mathbf{M}]_i^+ < 0 \\ \inf\left([\mathbf{M}]_i^+ \cdot \left(-[\dot{V}_{a,C}] - \mathbf{q}_C^T \text{abs}([\tilde{\mathbf{x}}]) - T\right)\right) - \mu, & \text{if } [\mathbf{M}]_i^+ > 0 \\ 0, & \text{else} \ . \end{cases} \tag{10}$$

In Eq. (10), the abbreviation $[\mathbf{M}] := [\mathbb{B}_C]^T \mathbf{P}_C \|[\tilde{\mathbf{x}}]\|$ is used and the absolute values are given by $\text{abs}([\tilde{\mathbf{x}}]) = [|[x_1] - x_{1,d}| \ ... \ |[x_n] - x_{n,d}|]$. Moreover, the matrix of the absolute values $\|[\tilde{\mathbf{x}}]\| \in \mathbb{R}^{n \times n}$ with $[\tilde{x}_i] = ([x_i] - x_{i,d})$ (for all $i \in \{1, ..., n\}$) and the sign function are defined as

$$\|[\tilde{\mathbf{x}}]\| = \begin{bmatrix} [\tilde{x}_1] \cdot \text{sign}([\tilde{x}_1]) & [\tilde{x}_1] \cdot \text{sign}([\tilde{x}_2]) & ... & [\tilde{x}_1] \cdot \text{sign}([\tilde{x}_n]) \\ [\tilde{x}_2] \cdot \text{sign}([\tilde{x}_1]) & [\tilde{x}_2] \cdot \text{sign}([\tilde{x}_2]) & ... & [\tilde{x}_2] \cdot \text{sign}([\tilde{x}_n]) \\ \vdots & \vdots & \ddots & \vdots \\ [\tilde{x}_n] \cdot \text{sign}([\tilde{x}_1]) & [\tilde{x}_n] \cdot \text{sign}([\tilde{x}_2]) & ... & [\tilde{x}_n] \cdot \text{sign}([\tilde{x}_n]) \end{bmatrix} \quad \text{and} \tag{11}$$

$$\text{sign}([\tilde{x}_i]) = \begin{cases} 1, & \text{if } \inf([\tilde{x}_i]) > 0 \\ -1, & \text{if } \sup([\tilde{x}_i]) < 0 \\ 0, & \text{else} \ . \end{cases} \tag{12}$$

Furthermore, $[\dot{V}_{a,C}] = [\tilde{\mathbf{x}}]^T \mathbf{P}_C([\mathbb{A}_C] - [\mathbb{B}_C]\mathbf{K})[\mathbf{x}] + [\tilde{\mathbf{x}}]^T \mathbf{P}_C[\mathbb{B}_C]u_{FF} - [\tilde{\mathbf{x}}]^T \mathbf{P}_C\dot{\mathbf{x}}_d$
and the trace of the stochastic processes $T = \frac{1}{2}\text{trace}\left\{\mathbf{G}_p^T \frac{\partial^2 V_C}{\partial \tilde{\mathbf{x}}^2}\mathbf{G}_p\right\}$ hold. The left
pseudo inverse[3] of the matrix $[\mathbf{M}]$ is calculated by $[\mathbf{M}]^+ = \left([\mathbf{M}]^T \cdot [\mathbf{M}]\right)^{-1}\cdot[\mathbf{M}]^T$.
In case that not all system states are measurable, it is necessary to reconstruct
them by an observer (see next section). Then, $\mathbf{x}$ in the control law is replaced
by the estimated states $\hat{\mathbf{x}}$.

For this stability proof, the sign of $L(V_C)$ is relevant; if $L(V_C) < 0$ the
system is stable. Note that the boundary $L(V_C) = 0$ of the provable stability
domain is commonly shaped like an ellipsoid whose volume should be as small as
possible to reduce the non-stabilizable area [10]. Currently, the simulation of this
approach works fine also in combination with the observer in Subsects. 4.2 and
4.3 (feedback of the estimated states). In future work, it will be implemented on
the described test rig in experiment and connected with the approach described
in Subsect. 4.4.

4   Sliding Mode Techniques for Estimation Purposes

Due to the fact that measurements in real applications are often limited, knowledge of the non-measurable states is necessary to implement an efficient control strategy. In the presented case, in addition to state estimation, uncertain
parameters will be identified simultaneously. Therefore, a classical sliding mode
observer approach is extended such that the real-time capability is ensured.
Often, parameter identification procedures are evaluated offline by using measured data which results in a set of parameters that are assumed to be constant
for the complete identification horizon. In the following, it is assumed that parameters can change in each time step (index k) within a defined range. Especially
for parameters containing uncertain or random effects as static or sliding friction, this assumption is reasonable. Therefore, intervals defining suitable range
bounds are considered in Subsects. 4.2, 4.3 and 4.4. Moreover, the presented
extension of a classical observer (Subsect. 4.1) is able to deal with process as
well as measurement noise but still guaranteeing the system's stability in each
time step.

4.1   A Classical Observer Approach

As it has been shown in [9], assume a dynamic system $\dot{\mathbf{x}} = \mathbf{f}(\mathbf{x}, \mathbf{u})$ whose state
equations can be written in the form

$$\dot{\mathbf{x}} = \mathbf{f} = \mathbf{A} \cdot \mathbf{x} + \mathbf{B} \cdot \mathbf{u} + \mathbf{S} \cdot \mathbf{w}(\mathbf{x}, \mathbf{u}) \ , \quad \mathbf{y} = \mathbf{C} \cdot \mathbf{x} \ . \tag{13}$$

In (13), $\mathbf{x}$ denotes the state vector, $\mathbf{A}$ as well as $\mathbf{B}$ denote the constant system
and input matrices, and $\mathbf{u}$ is a vector-valued control signal. The matrix $\mathbf{S} \in \mathbb{R}^{n\times q}$
includes influences of a-priori unknown terms on the system dynamics which have

---

[3] For single-input systems, $[\mathbf{M}]$ turns out to be a column vector.

to fulfill the condition $\|\mathbf{w}(\mathbf{x}, \mathbf{u})\| \leq \rho(\mathbf{u})$ with a fixed upper bound for the vector norm $\rho(\mathbf{u})$ [1]. Consequently, the product $\mathbf{S} \cdot \mathbf{w}(\mathbf{x}, \mathbf{u})$ contains all nonlinearities included in the system model that do not fit in the linear part of the set of state equations. Moreover, the constant output matrix $\mathbf{C}$ links the system states in a linear way to the output $\mathbf{y}$. Now, a variable-structure observer described by a set of ordinary differential equations (ODEs) [1]

$$\dot{\hat{\mathbf{x}}} = \hat{\mathbf{A}} \cdot \hat{\mathbf{x}} + \hat{\mathbf{B}} \cdot \mathbf{u} + \rho(\mathbf{u}) \cdot \tilde{\mathbf{S}} \cdot \tilde{\mathbf{e}} + \mathbf{H}_p \cdot (\mathbf{y}_m - \hat{\mathbf{y}}) \tag{14}$$

with the output equation $\hat{\mathbf{y}} = \hat{\mathbf{C}}\hat{\mathbf{x}}$ can be defined. In Eq. (14), $\rho(\mathbf{u})$ is a factor for the variable-structure part which has to be chosen such that the approximation $\mathbf{S} \cdot \mathbf{w} \approx \rho(\mathbf{u}) \cdot \tilde{\mathbf{S}} \cdot \tilde{\mathbf{e}}$ holds. The observer gain matrix $\mathbf{H}_p$ is used for the stabilization of the error dynamics of the linear part and is usually determined by pole assignment. The vector $\mathbf{y}_m$ includes all measured system outputs. All remaining terms in (14) denoted by the symbol $\hat{}$ characterize system, input and output matrices as well as the estimated state vector of the observer parallel model that replicates the original one. Moreover, the error vector is defined as $\tilde{\mathbf{e}} = \frac{\mathbf{S}^T \mathbf{P}(\mathbf{x}-\hat{\mathbf{x}})}{\|\mathbf{S}^T \mathbf{P}(\mathbf{x}-\hat{\mathbf{x}})\|}$ and accounts for deviations between the true (original) and estimated system states. There, the positive definite matrix $\mathbf{P} = \mathbf{P}^T$ results from solving the Lyapunov equation $(\hat{\mathbf{A}} - \mathbf{H}_p \cdot \hat{\mathbf{C}}) \cdot \mathbf{P} + \mathbf{P} \cdot (\hat{\mathbf{A}} - \mathbf{H}_p \cdot \hat{\mathbf{C}})^T + \mathbf{Q} = \mathbf{0}$ with a weighting matrix $\mathbf{Q} > 0$. Note that the sliding mode observer (as well as any other observer) is only applicable if the pair $\left(\hat{\mathbf{A}}, \hat{\mathbf{C}}\right)$ is observable. As soon as a change of sign occurs in the term $\mathbf{e}_m = \mathbf{y}_m - \hat{\mathbf{y}}$, the term $\mathbf{w}$ in (13) is reproduced approximately by $\tilde{\mathbf{e}}$ in the switching part of the observer. According to the (matching) conditions

$$\mathbf{S} \cdot \mathbf{w} \approx \rho(\mathbf{u}) \cdot \tilde{\mathbf{S}} \cdot \tilde{\mathbf{e}} \approx \rho(\mathbf{u}) \cdot \tilde{\mathbf{S}} \cdot \operatorname{sign}(\mathbf{e}_m) \tag{15}$$

the matrices $\mathbf{S}$ and $\tilde{\mathbf{S}}$ have an identical structure. Considering this condition, the observer differential equation is adapted to

$$\dot{\hat{\mathbf{x}}} = \hat{\mathbf{A}} \cdot \hat{\mathbf{x}} + \hat{\mathbf{B}} \cdot \mathbf{u} + \mathbf{H}_p \cdot \mathbf{e}_m + h_s \cdot \tilde{\mathbf{S}} \cdot \operatorname{sign}(\mathbf{e}_m) \tag{16}$$

with the sign function depending component-wise on the difference $\mathbf{e}_m$. By this reformulation of Eq. (16), it is now possible to handle uncertainty by means of the variable structure term because the model does not only include a locally valid model anymore. The stabilization of the error dynamics becomes now possible in spite of nonlinearities in the system representation [9]. In the following, this extended sliding mode observer will be modified further such that it is possible to estimate multiplicatively coupled states and parameters simultaneously by means of the modification

$$\dot{\hat{\mathbf{x}}} = \underbrace{\hat{\mathbf{A}} \cdot \hat{\mathbf{x}} + \hat{\mathbf{B}} \cdot \mathbf{u}}_{\hat{\mathbf{f}}(\hat{\mathbf{x}}, \mathbf{u})} + \mathbf{H}_p \cdot \mathbf{e}_m + \mathbf{P}^+ \cdot \hat{\mathbf{C}}^T \cdot \mathbf{H}_s \cdot \operatorname{sign}(\mathbf{e}_m) \tag{17}$$

with $\hat{\mathbf{A}} := \hat{\mathbf{A}}(\hat{\mathbf{x}})$, $\hat{\mathbf{B}} := \hat{\mathbf{B}}(\hat{\mathbf{x}})$. As it can be seen, the quite restrictive matching condition (15) is removed. The variable structure gain, called switching amplitude vector, is a diagonal matrix according to $\mathbf{H}_s = \text{diag}(\mathbf{h}_s)$ with $\dim(\mathbf{h}_s) = n_y \times 1$ and has to be defined such that the Lyapunov function candidate $V = \frac{1}{2}(\mathbf{x} - \hat{\mathbf{x}})^T \mathbf{P}(\mathbf{x} - \hat{\mathbf{x}}) > 0$ is greater than zero and its time derivative less than zero for asymptotic stability according to $\dot{V} = (\mathbf{x} - \hat{\mathbf{x}})^T \mathbf{P}(\dot{\mathbf{x}} - \dot{\hat{\mathbf{x}}}) < 0$.

In classical sliding mode approaches, the variable-structure gain is defined as a constant value. In Subsect. 4.2, a possibility is shown that calculates this gain in each time step to reduce chattering and to avoid actuator saturations. Moreover, uncertainty of parameters and states will be included as well as stochastic processes. The aim is to estimate states and to identify parameters simultaneously by a cascaded observer structure which will be shown in Subsect. 4.3.

4.2 Extended Approach Considering Uncertainty and Stochastic Processes - Estimation of Point Values

In this section, the dual problem to control tasks — namely state estimation — is solved by using sliding mode techniques taking into account parameter uncertainty, sensor inaccuracies and noise processes as already described in [7] in detail. Here, the calculated number of the switching amplitudes is equal to the number of measurable states in order to guarantee stability and to reach the sliding surface as fast as possible. Moreover, by a direct calculation of the switching amplitude vector, its values can be adapted automatically in each time step instead of predefining a suitable constant gain by intelligent guessing.

As for control design, a suitable Lyapunov function candidate is necessary, which is chosen as $V_O = \frac{1}{2}\mathbf{e}^T \cdot \mathbf{P}_O \cdot \mathbf{e}$ with the deviation $\mathbf{e} = \mathbf{x} - \hat{\mathbf{x}}$ between the true and estimated system states. The matrix $\mathbf{P}_O$ results from solving the Lyapunov equation $\tilde{\mathbf{A}}_O \mathbf{P}_O^T + \tilde{\mathbf{A}}_O^T \mathbf{P}_O + \mathbf{Q} = \mathbf{0}$ with $\tilde{\mathbf{A}}_O = \hat{\mathbf{A}} - \mathbf{H}_p \hat{\mathbf{C}}$ that holds for one working point ($\mathbf{P}_O = \mathbf{P}_O^T > 0$). Moreover, stochastic disturbances that affect the system dynamics as process or measurement noise can again be considered by applying the Itô differential operator

$$L(V_O) = \frac{\partial V_O}{\partial t} + \left(\frac{\partial V_O}{\partial \mathbf{e}}\right)^T \cdot [\bar{\mathbf{f}}] + \frac{1}{2}\text{trace}\left\{\mathbf{G}^T \frac{\partial^2 V_O}{\partial \mathbf{e}^2}\mathbf{G}\right\} \tag{18}$$

with $[\bar{\mathbf{f}}] = [\mathbf{f}] - [\hat{\bar{\mathbf{f}}}([\hat{\mathbf{x}}], [\mathbf{p}], \mathbf{u})]$ and the interval extension of the system ODEs (8). The term $\hat{\bar{\mathbf{f}}}(\hat{\mathbf{x}}, [\mathbf{p}], \mathbf{u})$ describes an observer parallel model by

$$[\hat{\bar{\mathbf{f}}}(\hat{\mathbf{x}}, [\mathbf{p}], \mathbf{u})] := \underbrace{\hat{\mathbb{A}}(\hat{\mathbf{x}}, [\mathbf{p}]) \cdot [\hat{\mathbf{x}}] + \hat{\mathbb{B}}(\hat{\mathbf{x}}, [\mathbf{p}]) \cdot \mathbf{u} + \mathbf{H}_p \cdot [\mathbf{e}_m]}_{\hat{\mathbf{f}}(\mathbf{x}, [\mathbf{p}], \mathbf{u})}$$

$$+ \mathbf{P}_O^+ \hat{\mathbb{C}}^T \cdot \mathbf{H}_s \cdot \text{sign}(\mathbf{e}_m + [\Delta \mathbf{y}_m]), \quad [\hat{\mathbf{y}}] := \hat{\mathbb{C}} \cdot [\hat{\mathbf{x}}] . \tag{19}$$

In Eq. (19)[4], $\mathbf{e}_m \in [\mathbf{e}_m] = \mathbf{y}_m - \hat{\mathbf{y}} + [\Delta \mathbf{y}_m]$ denotes the component-wise defined interval measurement error vector and accounts for bounded uncertainty

---

[4] Note that time discretization errors are neglected and the procedure is not limited to a special class of systems. Here, a SISO system is considered.

that becomes noticeable in deviations between the measured $\mathbf{y}_m$ and estimated system outputs $\hat{\mathbf{y}} \in [\hat{\mathbf{y}}]$ with the measurement error interval $[\Delta\mathbf{y}_m]$. Moreover, $\hat{\mathbb{A}}$, $\hat{\mathbb{B}}$ as well as $\hat{\mathbb{C}}$ are the interval evaluations of the system, input and output matrices $\hat{\mathbf{A}}(\hat{\mathbf{x}}, [\mathbf{p}])$, $\hat{\mathbf{B}}(\hat{\mathbf{x}}, [\mathbf{p}])$ and $\hat{\mathbf{C}}(\hat{\mathbf{x}}, [\mathbf{p}])$. The matrix $\mathbf{G}$ contains the standard deviation of both process $\mathbf{G}_p$ as well as measurement noise $\mathbf{G}_m$ according to $\mathbf{G} = [\mathbf{G}_p \quad -\mathbf{H}_p\mathbf{G}_m]$; the matrix $\mathbf{H}_p$ stabilizes the linear part of the observer denoted by $\hat{\mathbf{f}}(\hat{\mathbf{x}}, \mathbf{p}, \mathbf{u})$ for one working point $\hat{\mathbf{x}} \in [\hat{\mathbf{x}}]$ and $\mathbf{p} \in [\mathbf{p}]$. This observer gain $\mathbf{H}_p$ can be determined by pole assignment, minimizing a quadratic cost function (both for one special operating point) or by solving linear matrix inequalities (valid for the whole defined operating range) [10].

Reaching the sliding surface of all estimated states means that the difference between the system states themselves and the estimated ones becomes small so that switching around the sliding surface occurs. To prevent the observer from unnecessary switchings in regions where the sign of $\mathbf{e}_m$ cannot be evaluated, the interval error vector $[\mathbf{e}] = [\mathbf{x}] - [\hat{\mathbf{x}}]$ is introduced with $[\hat{\mathbf{x}}] = \hat{\mathbf{x}} + [\Delta\mathbf{x}_e]$ (estimation error interval $[\Delta\mathbf{x}_e]$). For measurable states, the estimation interval is replaced by the measurement interval according to the sensor model $[\Delta\mathbf{y}_m] = \hat{\mathbf{C}} \cdot [\Delta\mathbf{x}_e]$. Due to the interval extension of $\mathbf{e}$ to $[\mathbf{e}]$, the sliding surface becomes an area leading to smaller switching amplitudes $\mathbf{H}_s$.

Applying the condition $L(V_O) \overset{!}{<} -\mathbf{q}_O^T \|[\mathbf{e}_m]\|$ with a user-defined convergence rate $\mathbf{q}_O$, the expression $[\mathbf{e}]^T\mathbf{P}_O \cdot [\bar{\mathbf{f}}] + \frac{1}{2}\text{trace}\left\{\mathbf{G}^T\frac{\partial^2 V_O}{\partial \mathbf{e}^2}\mathbf{G}\right\} < -\mathbf{q}_O^T\|[\mathbf{e}_m]\|$ leads to

$$[\mathbf{e}]^T\mathbf{P}_O \cdot \left(\mathbb{A} \cdot [\hat{\mathbf{x}}] + \mathbb{B} \cdot \mathbf{u} - \hat{\mathbb{A}} \cdot [\hat{\mathbf{x}}] - \hat{\mathbb{B}} \cdot \mathbf{u} - \mathbf{H}_p \cdot [\mathbf{e}_m]\right) -$$

$$-[\mathbf{e}]^T\mathbf{P}_O \cdot \left(\mathbf{P}_O^+\hat{\mathbf{C}}^T\mathbf{H}_s \cdot \text{sign}(\mathbf{e}_m)\right) + \frac{1}{2}\text{trace}\left\{\mathbf{G}^T\frac{\partial^2 V_O}{\partial \mathbf{e}^2}\mathbf{G}\right\} < -\mathbf{q}_O^T\|[\mathbf{e}_m]\| . \tag{20}$$

Note this inequality needs to be valid for all possible configurations inside the specified intervals for states and parameters. Because of the positive definiteness of the matrix $\mathbf{P}_O$, the term $\mathbf{P}_O\mathbf{P}_O^+ = \mathbf{I}^{n\times n}$ is equal to the identity matrix $\mathbf{I}$ with $\mathbf{P}_O = \mathbf{P}_O^T$. With this simplification, the components of the diagonal matrix $\mathbf{H}_s = \text{diag}(\mathbf{h}_s)$ with $\dim(\mathbf{H}_s) = n_y \times n_y$ can be calculated component-wise from Eq. (20). Therefore, the term $[\mathbf{e}]^T\mathbf{P}_O \cdot \left(\mathbf{P}_O^+\hat{\mathbf{C}}^T\mathbf{H}_s \cdot \text{sign}([\mathbf{e}_m])\right) = [\mathbf{e}]^T \cdot \left(\hat{\mathbf{C}}^T\mathbf{H}_s \cdot \text{sign}([\mathbf{e}_m])\right) = h_{s,1} \cdot [e_{m,1}] \cdot \text{sign}([e_{m,1}]) + ... + h_{s,n_y} \cdot [e_{m,n_y}] \cdot \text{sign}([e_{m,n_y}])$ can be reformulated using $[e_{m,i}] \cdot \text{sign}([e_{m,i}]) = \|[e_m]\|$ $(i \in 1,...,n_y)$ according to

$$[\mathbf{e}]^T\mathbf{P}_O \cdot \left(\mathbf{P}_O^+\hat{\mathbf{C}}^T\mathbf{H}_s \cdot \text{sign}([\mathbf{e}_m])\right) = \mathbf{h}_s^T\|[\mathbf{e}_m]\| . \tag{21}$$

Note that the output matrix $\hat{\mathbf{C}} := \hat{\mathbb{C}}$ is independent of states or parameters in this case. The absolute value of the interval measurement error $\|[\mathbf{e}_m]\|$ yields

$$|[e_{m,i}]| = |[\underline{e}_{m,i}; \overline{e}_{m,i}]| = \begin{cases} [-\overline{e}_{m,i} ; -\underline{e}_{m,i}] , & \text{for } \overline{e}_{m,i} \leq 0 , \\ [\underline{e}_{m,i} ; \overline{e}_{m,i}] , & \text{for } \underline{e}_{m,i} \geq 0 , \\ [0 ; \max\{|\underline{e}_{m,i}|, |\overline{e}_{m,i}|\}] & \text{else .} \end{cases} \quad (22)$$

From Eq. (18) with (21), the switching amplitude vector can be calculated by

$$\mathbf{h}_s = \begin{cases} \mathbf{0}, & \text{if } [\delta_O] \subseteq [\mathbf{e}_m]^T [\mathbf{e}_m] \\ \sup\left(|[\mathbf{e}_m]|^+ \cdot \left([\dot{V}_{a,O}] + \frac{1}{2} \cdot \text{trace} \left\{ \mathbf{G}^T \frac{\partial^2 V_O}{\partial \mathbf{e}^2} \mathbf{G} \right\} \right) + \mathbf{q}_O^T \right), & \text{else .} \end{cases} \quad (23)$$

In Eq. (23), $[\dot{V}_{a,O}] = [\mathbf{e}]^T \mathbf{P}_O \cdot (\mathbb{A}_C \cdot [\mathbf{x}] + \mathbb{B}_C \cdot \mathbf{u} - \hat{\mathbf{A}} \cdot [\hat{\mathbf{x}}] - \hat{\mathbb{B}} \cdot \mathbf{u} - \mathbf{H}_p \cdot [\mathbf{e}_m])$ holds with the interval error vector $[\mathbf{e}]$. To prevent a division by zero, a small interval $[\delta_O]$ around zero is included in the calculation of the switching amplitude to reduce on the one hand unnecessary chattering and to reduce the value of the switching amplitude on the other hand [10]. The interval pseudo inverse in Eq. (23) is defined as $|[\mathbf{e}_m]|^+ = \left(|[\mathbf{e}_m]|^T |[\mathbf{e}_m]| \right)^{-1} \cdot |[\mathbf{e}_m]|^T$. A stability proof can be done by evaluating Eq. (18) in each time step.

4.3   Cascaded Structure of the Observer

In the following, the approach described in the previous section is applied to the simplified model for the longitudinal dynamics of a vehicle (see Sect. 2) in terms of a cascaded structure. To estimate all system states and to identify the system parameters, two subsystems are necessary due to multiplicative couplings of states and parameters in the system model. The states are estimated by the first subsystem and serve as virtual measurements for the second subsystem that identifies the parameters in each time step by using integrator disturbance models. The physical background of those is, that the parameters are assumed to be located within specified intervals in which they are allowed to vary between two time steps. Figure 2 shows the definitions of matrices, the models of both subsystems as well as the corresponding observer parallel models. From the latter ones, the linear structure of the interval-based sliding mode observer (ISMO) can be derived by factorization of $\hat{\mathbf{f}}^{(i)}$ for both subsystems $i \in \{S_1, S_2\}$. Note the representation by the system matrix $\hat{\mathbf{A}}^{(S_2)}$ and the input vector $\hat{\mathbf{b}}^{(S_2)}$ of the second subsystem is evaluated for one operating point and consequently this also holds for $\mathbf{H}_p^{(S_2)}$ and $\mathbf{P}_O^{(S_2)}$. However, the stability proof is still valid because the variable-structure part stabilizes the error dynamics of the complete system.

4.4   Extended Approach Considering Uncertainty and Stochastic Processes - Estimation of Confidence Intervals

In the previous subsection, point values for system states and parameters were estimated such that the system's stability can be guaranteed. Due to calculation with intervals, these point values are just one solution of others. Therefore, the following strategies provide the possibility to estimate all possible solutions

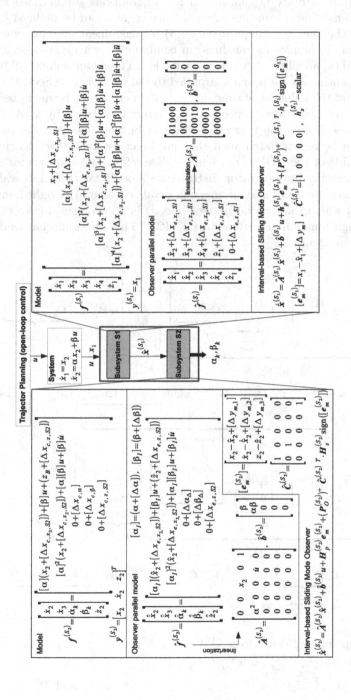

Fig. 2. Structure of the cascaded observer, definitions of the nominal system models $\mathbf{f}^{(i)}$, observer parallel models $\hat{\mathbf{f}}^{(i)}$, (both necessary for the calculation of the switching amplitude) and the interval-based sliding mode observers $\dot{\hat{\mathbf{x}}}^{(i)}$ for the estimation of states and parameters that both are guaranteed to be located in specified range bounds by adapting the switching amplitude in each time step, $i \in \{S_1, S_2\}$, cf. [7].

which results in the calculation of confidence regions (enclosures) for parameters and states. Note that stability is guaranteed for all solutions summarized by the confidence regions. The calculation of confidence regions can be realized by using (1) Müller's theorem, (2) cooperativity, (3) a quasi-linear system representation (with real or complex eigenvalues) in combination with a state-space transformation, and (4) an affine system representation. These four points will be described in the following and, if possible, applied to the scenario (see Sect. 2) in simulation. Note that the influence of time discretization errors can be neglected, because the considered uncertainty represented by intervals is larger by multiple orders of magnitudes. Before the four strategies can be implemented, the two subsystems need to be reformulated such that no direct interval dependencies in the right-hand side of the ODEs are included. Therefore, the state vector is extended in the following by integrator disturbance models for these intervals (in the presented scenario, only measurement intervals occur on the right-hand sides of the ODEs, where the interval in $\text{sign}([\Delta \mathbf{e}_m])$ is no problem due to the definition of the sign function according to Eq. (12)). This reformulation leads to the extended subsystems

$$
\underbrace{\begin{bmatrix} \dot{\hat{x}}_1 \\ \dot{\hat{x}}_2 \\ \dot{\hat{x}}_3 \\ \dot{\hat{x}}_4 \\ \dot{\hat{z}}_1 \\ [\Delta \dot{y}_m] \end{bmatrix}}_{\dot{\mathbf{x}}_{ext}^{(S_1)}} = \underbrace{\left(\begin{bmatrix} \hat{\mathbf{A}}^{(S_1)}, \ \mathbf{H}_p^{(S_1)} \\ \mathbf{0}^{1\times 6} \end{bmatrix} - \underbrace{\begin{bmatrix} \mathbf{H}_p^{(S_1)} \\ 0 \end{bmatrix}}_{\mathbf{H}_{p,ext}^{(S_1)}} \cdot \underbrace{[\hat{\mathbf{C}}^{(S_1)} \ 0]}_{\hat{\mathbf{C}}_{ext}^{(S_1)}} \right)}_{\mathbf{A}_{O,ext}^{(S_1)}} \underbrace{\begin{bmatrix} \hat{x}_1 \\ \hat{x}_2 \\ \hat{x}_3 \\ \hat{x}_4 \\ \hat{z}_1 \\ [\Delta y_m] \end{bmatrix}}_{\hat{\mathbf{x}}_{ext}^{(S_1)}} + \mathbf{H}_{p,ext}^{(S_1)} \cdot \underbrace{\hat{\mathbf{C}}_{ext}^{(S_1)} \cdot \mathbf{x}_{ext}^{(S_1)}}_{y_{m,ext}^{(S_1)}}
$$

$$
+ \mathbf{b}_{ext}^{(S_1)} \cdot u + (\mathbf{P}_{O,ext}^{(S_1)})^+ \cdot \hat{\mathbf{C}}_{ext}^{(S_1)} \cdot \mathbf{H}_s^{(S_1)} \cdot \underbrace{\text{sign}(y_m - \hat{y} + [\Delta y_m])}_{[e_m^{(S_1)}]} \quad \text{and} \qquad (24)
$$

$$
\underbrace{\hspace{4cm}}_{\text{scalar point-value}}
$$

$$
\underbrace{\begin{bmatrix} \dot{\hat{\mathbf{x}}}^{(S_2)} \\ [\Delta \dot{y}_{m,1}] \\ [\Delta \dot{y}_{m,2}] \\ [\Delta \dot{y}_{m,3}] \end{bmatrix}}_{\dot{\mathbf{x}}_{ext}^{(S_2)}} = \underbrace{\left(\begin{bmatrix} \hat{\mathbf{A}}^{(S_2)}, \ \mathbf{H}_p^{(S_2)} \\ \mathbf{0}^{3\times 8} \end{bmatrix} - \underbrace{\begin{bmatrix} \mathbf{H}_p^{(S_2)} \\ \mathbf{0}^{3\times 3} \end{bmatrix}}_{\mathbf{H}_{p,ext}^{(S_2)}} \cdot \underbrace{[\hat{\mathbf{C}}^{(S_2)} \ \mathbf{0}^{3\times 3}]}_{\hat{\mathbf{C}}_{ext}^{(S_2)}} \right)}_{\mathbf{A}_{O,ext}^{(S_2)}} \underbrace{\begin{bmatrix} \hat{\mathbf{x}}^{(S_2)} \\ [\Delta y_{m,1}] \\ [\Delta y_{m,2}] \\ [\Delta y_{m,3}] \end{bmatrix}}_{\hat{\mathbf{x}}_{ext}^{(S_2)}} + \underbrace{\begin{bmatrix} \mathbf{b}^{(S_2)} \\ 0 \\ 0 \\ 0 \end{bmatrix}}_{\mathbf{b}_{ext}^{(S_2)}} \cdot u
$$

$$
+ \mathbf{H}_{p,ext}^{(S_2)} \cdot \underbrace{\hat{\mathbf{C}}_{ext}^{(S_2)T} \cdot \mathbf{x}_{ext}^{(S_2)}}_{y_{m,ext}^{(S_2)}} + (\mathbf{P}_{O,ext}^{(S_2)})^+ \cdot \hat{\mathbf{C}}_{ext}^{(S_2)} \cdot \mathbf{H}_s^{(S_2)} \cdot \underbrace{\text{sign}(y_m - \hat{y} + [\Delta y_m])}_{[e_m^{(S_2)}]} \ .
$$

$$
(25)
$$

with the vector and matrix definitions $\mathbf{x}_{ext}^{(S_1)} = [x_1 \ x_2 \ x_3 \ x_4 \ z_1 \ [\Delta y_m]]^T$, $\mathbf{b}_{ext}^{(S_1)} = [0 \ 0 \ 0 \ 0 \ 0 \ 0]^T$, $\mathbf{C}_{ext}^{(S_1)} = [\mathbf{C}_{ext}^{(S_1)} \ 0]^T$, $\mathbf{P}_{O,ext}^{(S_1)} = [\mathbf{P}_O^{(S_1)} \ \mathbf{0}^{5\times 1} \ ; \ \mathbf{0}^{1\times 6}]$, $\hat{\mathbf{x}}^{(S_2)} = [\hat{x}_2 \ \hat{x}_3 \ \hat{\alpha} \ \hat{\beta} \ \hat{z}_2]^T$, $\dot{\hat{\mathbf{x}}}^{(S_2)} = [\dot{\hat{x}}_2 \ \dot{\hat{x}}_3 \ \dot{\hat{\alpha}} \ \dot{\hat{\beta}} \ \dot{\hat{z}}_2]^T$, and $\mathbf{P}_{O,ext}^{(S_2)} = [\mathbf{P}_O^{(S_2)} \ \mathbf{0}^{5\times 3} \ ; \ \mathbf{0}^{3\times 6}]$ (see also Fig. 2).

Müller's Theorem. Müller's Theorem can always be taken into account, even if the proof of cooperativity of a system fails. In this case, interval enclosures of a system described by ODEs can be calculated as follows. According to Müller [2], the enclosures of all time-varying system states can be computed, if lower and upper functions can be found that bound the right-hand side of the system ODEs $\dot{\mathbf{x}} = \mathbf{f}(\mathbf{z}, \mathbf{p}, u)$ for all uncertain states $\mathbf{z} \in [\mathbf{z}]$ and parameters $\mathbf{p} \in [\mathbf{p}]$. These functions are denoted by $\mathbf{f}_\chi$ (lower function) and $\mathbf{f}_\psi$ (upper function). Then, the bounding system can be defined by the worst-case enclosures

$$\chi_i \leq x_i \leq \psi_i \tag{26}$$

for all $i = \{1, ..., n\}$ component-wise (system order n). Analogously, the system equations can be determined according to the differential inequalities

$$f_{\chi,i} \leq \dot{x}_i \leq f_{\psi,i} \tag{27}$$

for all possible states, parameters and inputs. Then, a system of order $2n$ has to be solved to find the lower and upper enclosures for all states in a numerical way by verified solution algorithms. Therefore, the lower and upper bounds result from minimization and maximization

$$f_{\chi,i} \leq \min\{f_i(\mathbf{z}, \mathbf{p}, u) | \chi_j \leq z_j \leq \psi_j \text{ with } z_i \equiv \chi_i \text{ and } \underline{p}_k \leq p_k \leq \overline{p}_k\} , \tag{28}$$

$$f_{\psi,i} \geq \max\{f_i(\mathbf{z}, \mathbf{p}, u) | \chi_j \leq z_j \leq \psi_j \text{ with } z_i \equiv \psi_i \text{ and } \underline{p}_k \leq p_k \leq \overline{p}_k\} \tag{29}$$

for $j \in \{1, ..., i-1, i+1, ..., n\}$ and $k \in \{1, ..., n_p\}$. For the application example in Sect. 2, this procedure leads do unstable solutions that are too conservative and include unphysical combinations due to overestimation.

Enclosures for Cooperative Systems. A given system $\mathbf{f}(\mathbf{x}, \mathbf{p}, u)$ is called *cooperative system* if it is monotone concerning its initial values. The domain for all possible states can be calculated by evaluating the system for all corner points of intervals containing the initial values. In general, the calculation of confidence regions by taking into account the cooperativity of a system is a special case of Müller's theorem. Sufficient conditions for a cooperative system are that

- all states are non-negative for all times according to $\dot{x}_i = f_i(x_1, ..., x_{i-1}, 0, x_{i+1,...,x_{n_x}}) \geq 0$ (parameters and inputs are not denoted explicitly),
- all entries of the Jacobian fulfill the condition $\frac{\partial f_i}{\partial x_j} \geq 0$ for all $i, j \in \{1, ..., n\}$ $(i \neq j)$.

If these conditions are fulfilled, guaranteed lower and upper bounds can be calculated by evaluating the system ODEs at all corner points instead of using the whole range of x_i as it is done with Müller's theorem. This decreases the computational effort significantly. All corner points are used to find the infimum (lower bound) and supremum (upper bound) of the enclosures in each time step. For this initial value problem, verified numerical solution algorithms are used to avoid errors due to computational accuracy, rounding, and time discretization [8,13]. For the given application scenario, enclosures for the system states and the two system parameters are of interest. Therefore, this algorithm needs to be implemented for both subsystems. This leads for the first subsystem to $2^6 \cdot 6$ and for the second subsystem to $2^8 \cdot 8$ state equations containing unphysical parameter combinations. Unfortunately, overestimation due to this large number of configurations leads to unstable solutions for both subsystems.

Quasi-Linear System Representation with State-Space Transformation.
Another possibility to calculate enclosures for states and parameters can be found by reformulating the system into a quasi-linear representation. In order to reduce overestimation caused by the dependency problem, the system states need to be decoupled. This generates a state-space transformation into new coordinates (by using Jordan or Metzler matrices) and allows a recursive evaluation of the new system equations. Afterwards, a backward transformation into original coordinates is necessary. Here, the Jordan matrix is taken into account, in which the (real or complex) eigenvalues of the system are located on the diagonal of the matrix and zeros on all other entries. The algorithm presented in the following needs to be implemented for both subsystems in order to get the enclosures of parameters and states. The state-space transformation follows with $\hat{\mathbf{x}}_{ext} = \mathcal{V} \cdot \mathbf{z}$ and $\dot{\hat{\mathbf{x}}}_{ext} = \mathcal{V} \cdot \dot{\mathbf{z}}$ for the single-input single-output system (S_1) and the single-input multi-output system (S_2) to

$$
\begin{aligned}
[\dot{\mathbf{z}}] &= \mathrm{inv}(\mathcal{V}) \cdot (\mathbf{A}_{O,ext}[\hat{\mathbf{x}}_{ext}] + \mathbf{b}_{ext}u + \mathbf{H}_{p,ext}\mathbf{y}_m + (\mathbf{P}_{O,ext})^+ \mathbf{C}_{ext}^T \cdot \mathbf{H}_s \mathrm{sign}([\mathbf{e}_m])) \\
&= \mathrm{inv}(\mathcal{V}) \cdot (\mathbf{A}_{O,ext}\mathcal{V}[\mathbf{z}] + \mathbf{b}_{ext}u + \mathbf{H}_{p,ext}\mathbf{y}_m + (\mathbf{P}_{O,ext})^+ \mathbf{C}_{ext}^T \cdot \mathbf{H}_s \mathrm{sign}([\mathbf{e}_m])) \\
&= \mathbf{J}[\mathbf{z}] + \mathrm{inv}(\mathcal{V}) \cdot (\mathbf{b}_{ext}u + \mathbf{H}_{p,ext}\mathbf{y}_m + (\mathbf{P}_{O,ext})^+ \mathbf{C}_{ext}^T \cdot \mathbf{H}_s \mathrm{sign}([\mathbf{e}_m])) \ .
\end{aligned}
\tag{30}
$$

The relation between the Jordan canonical form and the matrix of eigenvectors $\mathcal{V}$ is given by $\mathbf{J} = \mathrm{diag}(\lambda_i) = \mathrm{inv}(\mathcal{V}) \cdot \mathbf{A}_{O,ext} \cdot \mathcal{V}$ where λ_i are real or complex eigenvalues according to $\lambda_i = \sigma_i \pm j \cdot \omega_i$ with multiplicity one. Both matrices result directly from the Matlab command $[\mathcal{V}, \mathbf{J}] = \mathrm{eig}(\mathbf{A}_{O,ext})$ for both subsystems separately. The state-space transformation is the prerequisite for the algorithm in Fig. 3 that is based on the Euler method. Due to the fact that the system matrix of subsystem S_2 significantly depends on the positive or negative sign of the time derivative of the input signal $\dot{u}$ (see Fig. 2), the algorithm needs to be divided into 4 cases where for S_1 only a single case is necessary. This is justified by the fact that this system matrix represents an integrator chain state estimator which provides virtual measurements for subsystem S_2. The following algorithm is executed in each time step, where two subsequent time steps t_k (index k) and t_{k+1} (index $k+1$) are taken into account.

Additionally, in Fig. 3 the terms

$$
\gamma_+ = \mathbf{b} \cdot u + \mathbf{H}_{p,ext,+} \cdot \mathbf{y}_m + (\mathbf{P}_{O,ext,+})^+ \cdot \hat{\mathbf{C}}_{ext}^T \cdot \mathrm{sign}(\mathbf{y}_m - \hat{\mathbf{y}} + [\Delta\mathbf{y}_m]) \tag{31}
$$

$$
\gamma_- = \mathbf{b} \cdot u + \mathbf{H}_{p,ext,-} \cdot \mathbf{y}_m + (\mathbf{P}_{O,ext,-})^+ \cdot \hat{\mathbf{C}}_{ext}^T \cdot \mathrm{sign}(\mathbf{y}_m - \hat{\mathbf{y}} + [\Delta\mathbf{y}_m]) \tag{32}
$$

Explicit Euler Method for S_2 (only one case for S_1)			
		Positive or negative sign of $\dot{u}_k$ and $\dot{u}_{k+1}$	
Case 1	Case 2	Case 3	Case 4
$(\dot{u}_k < 0)$ && $(\dot{u}_{k+1} > 0)$	$(\dot{u}_k > 0)$ && $(\dot{u}_{k+1} < 0)$	$(\dot{u}_k < 0)$ && $(\dot{u}_{k+1} < 0)$	$(\dot{u}_k > 0)$ && $(\dot{u}_{k+1} > 0)$
$[\mathbf{z}] = \mathbf{V}_+^{-1}\mathbf{V}_-[\mathbf{z}_k]$ $\underline{z} = inf([\mathbf{z}])$ $\overline{z} = sup([\mathbf{z}])$	$[\mathbf{z}] = \mathbf{V}_-^{-1}\mathbf{V}_+[\mathbf{z}_k]$ $\underline{z} = inf([\mathbf{z}])$ $\overline{z} = sup([\mathbf{z}])$	$\underline{z} = inf([\mathbf{z}_k])$ $\overline{z} = sup([\mathbf{z}_k])$	$\underline{z} = inf([\mathbf{z}_k])$ $\overline{z} = sup([\mathbf{z}_k])$
$\underline{z}_{k+1} =$ $\underline{z} + T \cdot (\mathbf{J}_+\underline{z} + \gamma_+)$ $\overline{z}_{k+1} =$ $\overline{z} + T \cdot (\mathbf{J}_+\overline{z} + \gamma_+)$	$\underline{z}_{k+1} =$ $\underline{z} + T \cdot (\mathbf{J}_-\underline{z} + \gamma_-)$ $\overline{z}_{k+1} =$ $\overline{z} + T \cdot (\mathbf{J}_-\overline{z} + \gamma_-)$	$\underline{z}_{k+1} =$ $\underline{z} + T \cdot (\mathbf{J}_-\underline{z} + \gamma_-)$ $\overline{z}_{k+1} =$ $\overline{z} + T \cdot (\mathbf{J}_-\overline{z} + \gamma_-)$	$\underline{z}_{k+1} =$ $\underline{z} + T \cdot (\mathbf{J}_+\underline{z} + \gamma_+)$ $\overline{z}_{k+1} =$ $\overline{z} + T \cdot (\mathbf{J}_+\overline{z} + \gamma_+)$

Fig. 3. Structure diagram: quasi-linear system representation, $T = t_{k+1} - t_k$.

with the pseudo inverses $(.)^+$ are used. In dependency of the possible change of sign from $\dot{u}_k$ to $\dot{u}_{k+1}$, diagonal matrices of eigenvalues $\mathbf{J}_+$ and $\mathbf{J}_-$, matrices of eigenvectors $\mathbf{V}_+$ and $\mathbf{V}_-$, linear observer gains $\mathbf{H}_{p,ext,+}$ or $\mathbf{H}_{p,ext,-}$, and matrices $\mathbf{P}_{O,ext,+}$ or $\mathbf{P}_{O,ext,-}$ are used. These are evaluated offline for two fixed operating points of $\dot{u}$ in the system matrix $\hat{\mathbf{A}}_{ext}^{(S_2)}$ (index $+$ if $\dot{u}_{k+1} > 0$ and index $-$ if $\dot{u}_{k+1} < 0$). Then, a backward transformation from $[\mathbf{z}_{k+1}] = [\underline{z}_{k+1}, \overline{z}_{k+1}]$ follows with $[\mathbf{x}_{ext,k+1}] = \mathbf{V}_+ \cdot [\mathbf{z}_{k+1}]$ or $[\mathbf{x}_{ext,k+1}] = \mathbf{V}_- \cdot [\mathbf{z}_{k+1}]$ in dependency of the sign of $\dot{u}$. The goal of this procedure is to get intervals for parameters and states that comprise all possible configurations where stability can be guaranteed. Figure 4 shows this in a graphical way: Starting with a large initial interval $[\mathbf{x}_0]$, at the end of the simulation a smaller interval $[\mathbf{x}_f]$ results. Unfortunately, this algorithm provides only good results for subsystem S_1. The results of subsystem S_2 are not feasible due to overestimation and rotation of interval boxes (leading to non-negligible wrapping effect) caused by the importance of the positive or negative sign of $\dot{u}$.

Affine System Representation. In the previous subsection, only overestimation due to the dependency problem could be reduced but not overestimation due to the wrapping effect so that the algorithm failed for subsystem S_2. Therefore, a new system representation is shown that makes it possible to directly map states and parameters to their initial intervals in each time step. This reduces computational effort, neither state-space nor backward transformation are necessary (because there are no dependencies of the states due to constant initial intervals) and no interval box rotations occur. The algorithm is depicted in Fig. 5 and based on the Euler forward discretization (sample time $T = t_{k+1} - t_k$,

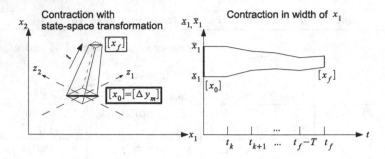

Fig. 4. Contracting interval box over time for a system with two states in two points of view, where the considered interval is only defined in x_1-direction (comparable to the measurement interval of subsystem S_1).

identity matrix $\mathbf{I}$). As it can be seen from the previous four cases in Fig. 3, now only two cases for S_2 are necessary (for S_1 again only one case is used). A new matrix $\mathbf{M}_{k+1}$ and a vector ρ_{k+1} using Eqs. (31) and (32) are calculated in each time step which then represent the new enclosure vector $[\mathbf{x}_{k+1}]$ still referring to the constant initial interval vector $[\mathbf{x}_0]$. This procedure is successful for both subsystems, where the results for S_1 from the affine system representation are equal to the results from the state-space transformation procedure. Figure 6 shows the connection between the sliding mode observer estimating point values and the observer estimating confidence regions. Note that currently both observers are implemented with an open-loop control. In future work, a closed-loop control will be performed by using the states and parameters estimated from the observer calculating confidence regions with the affine system representation in simulation (dashed line in Fig. 6).

Affine system representation for S_2, dim(S_2) = 8 (only one case for S_1, dim(S_1) = 6)	
Initialization: $\mathbf{x}_0 \in \mathbb{R}^{n \times 1}$, $\mathbf{x}_0 \in [\mathbf{x}_0]$, $\mathbf{M} = \mathbf{M}_0 := \mathbf{I} \in \mathbb{R}^{n \times n}$, $\rho = \rho_0 := \mathbb{0} \in \mathbb{R}^{n \times 1}$	
	Positive or negative sign of $\dot{u}_k$
Case 1: $\dot{u}_k < 0$	Case 2: $\dot{u}_k \geq 0$
$\mathbf{M}_- := \mathbf{I} + T \cdot \mathbf{A}_{O,ext,-}$	$\mathbf{M}_+ := \mathbf{I} + T \cdot \mathbf{A}_{O,ext,+}$
$\mathbf{M}_{k+1} := \mathbf{M}_- \mathbf{M}_k$	$\mathbf{M}_{k+1} := \mathbf{M}_+ \mathbf{M}_k$
$\rho_{k+1} := \mathbf{M}_- \rho_k + T \cdot \gamma_{-,k}$	$\rho_{k+1} := \mathbf{M}_+ \rho_k + T \cdot \gamma_{+,k}$
Update step $[\mathbf{x}_{k+1}] = \mathbf{M}_{k+1} [\mathbf{x}_0] + \rho_{k+1}$	

Fig. 5. Structure diagram: Affine system representation.

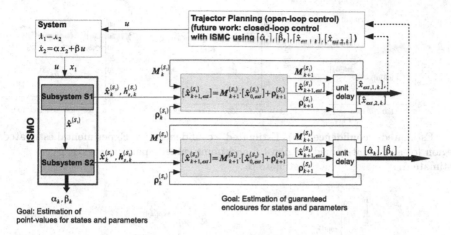

Fig. 6. Overview of point-value observer and confidence region estimator.

5   Results

In this section, the results of the point valued sliding mode observer and the extension to the estimation of confidence regions are shown. Note that the following visualizations result from open-loop control. The connection to closed-loop control in experiments is subject of future work. In Fig. 7, the confidence regions for the parameters Fig. 7(a) and (b) and the states Fig. 7(d) and (e) are visualized. Obviously, the intervals become smaller over time which was proposed in Fig. 4. Additionally, the results of the point valued observer for the parameter estimates can be seen in Fig. 7(a) and (b) labeled by $\hat{\alpha}_k$ and $\hat{\beta}_k$. In this context, Fig. 7(c) presents the estimated point valued parameters from experimental data. These estimates are located inside the confidence regions depicted in Fig. 7(a) and (b). Moreover in Fig. 7(f) and (g), the errors between the estimated states and the desired state trajectories are shown which provide good results taking into consideration that the states are affected by noise processes. In order to validate the results of the parameter identification in each time step from the second subsystem, a comparable procedure is necessary [11]. Therefore, a least-squares parameter identification has been performed. There, the parameter estimates are assumed to be constant for time intervals of 8 s taking into account the second system equation $\dot{x}_{2,m} = \hat{\alpha}_k x_{2,m} + \hat{\beta}_k u_m$ which has to be valid also for measurements (index m). From this, the parameters can be calculated for several measurements over 8 s according to

$$
\begin{bmatrix} \alpha \\ \beta \end{bmatrix} = \left(\begin{bmatrix} x_{2,m}(t_1) & u_m(t_1) \\ \vdots & \vdots \\ x_{2,m}(t_f) & u_m(t_f) \end{bmatrix} \right)^{+} \begin{bmatrix} \dot{x}_{2,m}(t_1) \\ \vdots \\ \dot{x}_{2,m}(t_f) \end{bmatrix}
\tag{33}
$$

where $t_f - t_1 = 8$ s. The time derivatives $x_{2,m}$ and $\dot{x}_{2,m}$ of the measurable state x_1 have been calculated by first-order low pass filtered derivative approximations.

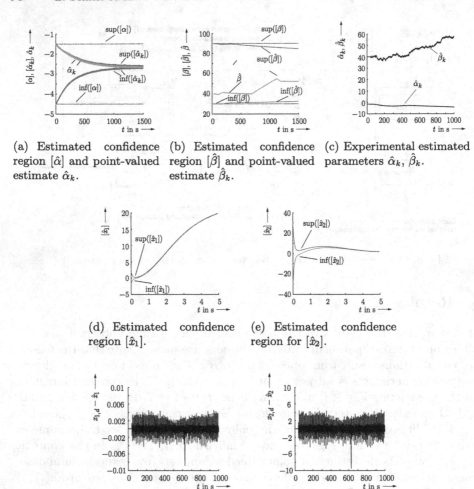

(a) Estimated confidence region $[\hat{\alpha}]$ and point-valued estimate $\hat{\alpha}_k$.

(b) Estimated confidence region $[\hat{\beta}]$ and point-valued estimate $\hat{\beta}_k$.

(c) Experimental estimated parameters $\hat{\alpha}_k$, $\hat{\beta}_k$.

(d) Estimated confidence region $[\hat{x}_1]$.

(e) Estimated confidence region for $[\hat{x}_2]$.

(f) Estimation error for x_1.

(g) Estimation error for x_2.

Fig. 7. Confidence regions for the system parameters (a), (b) with ($[\Delta y_{m,1}] = [-0.1; 0.1]$, $[\Delta y_{m,2}] = [-0.1; 0.1]$, $[\Delta y_{m,3}] = [-0.01; 0.01]$, $\mathbf{h}_{s,max} = 400$) including simulative point-valued estimates $\hat{\alpha}_k$ and $\hat{\beta}_k$ from ISMO; (c) experimental estimated parameters of the ISMO; (d), (e) confidence regions for the two system states; (f) and (g) estimation errors of the reconstructed point-valued states using ISMO.

The measurements for the motor torque u_m have also been low-pass filtered. With the estimated parameter values of both, the sliding mode observer as well as the least-squares method, the system (1) described in Sect. 2 has been simulated again separately including the parameter estimates of each method. The results can be seen in Fig. 8, which shows the root mean square errors for $\Delta_{x_1} = x_1 - x_{1,m}$ and $\Delta_{x_2} = x_2 - x_{2,m}$ ($x_{2,m}$ results from differentiation of $x_{1,m}$) of both methods. It becomes obvious, that the ISMO provides better estimates due to smaller root

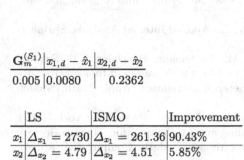

$G_m^{(S_1)}$	$x_{1,d} - \hat{x}_1$	$x_{2,d} - \hat{x}_2$
0.005	0.0080	0.2362

	LS	ISMO	Improvement
x_1	$\Delta_{x_1} = 2730$	$\Delta_{x_1} = 261.36$	90.43%
x_2	$\Delta_{x_2} = 4.79$	$\Delta_{x_2} = 4.51$	5.85%

Fig. 8. Results: Standard deviations of the estimation errors (subsystem S_1) compared to the standard deviation of the simulated measurement noise (top); validation of the parameter identification by least-squares (LS) and interval-based sliding mode observer (ISMO) (bottom).

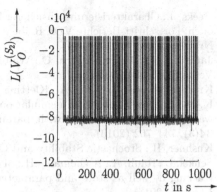

Fig. 9. Stability proof for S_2.

mean square errors than the least-squares estimation. Moreover, Fig. 9 shows the successful stability proof for the second subsystem, because $L(V_O^{(S_2)})$ is negative for the complete time horizon.

6 Conclusions and Outlook

In this paper, sliding mode approaches for control and estimation tasks were shown under consideration of uncertainty as well as stochastic processes. Non-measurable states and unknown parameters are estimated by observer strategies, firstly with point valued results and secondly by estimating confidence intervals using the interval toolbox INTLAB, the C++ library C-XSC and s-functions in MATLAB/SIMULINK. In future work, these strategies will be coupled as closed-loop control and evaluated online on the described test rig. The real-time capability is already secured. Moreover, the influence of time discretization errors will be investigated. The control procedure will especially be applied to other mechanical and thermodynamic systems.

References

1. Engell, S.: Entwurf nichtlinearer Regelungen. R. Oldenbourg Verlag, München (1995)
2. Gennat, M., Tibken, B.: Guaranteed bounds for uncertain systems: methods using linear lyapunov-like functions, differential inequalities and a midpoint method. In: 12th GAMM - IMACS International Symposium on Scientific Computing, Computer Arithmetic and Validated Numerics (2006)

3. Heeks, J.: Charakterisierung unsicherer Systeme mit intervallarithmetischen Methoden. Fortschritt-Berichte VDI, Reihe 8 Mess-, Steuerungs- und Regelungstechnik, Nr. 919 (2001). (in German)
4. Jaulin, L., Kieffer, M., Didrit, O., Walter, É.: Applied Interval Analysis. Springer-Verlag, London (2001)
5. Krasnochtanova, I., Rauh, A., Kletting, M., Aschemann, H., Hofer, E.P., Schoop, K.-M.: Interval methods as a simulation tool for the dynamics of biological wastewater treatment processes with parameter uncertainties. Appl. Math. Model. **34**(3), 744–762 (2010)
6. Kushner, H.: Stochastic Stability and Control. Academic Press, New York (1967)
7. Senkel, L., Rauh, A., Aschemann, H.: Sliding mode techniques for robust trajectory tracking as well as state and parameter estimation. Math. Comput. Sci. **8**(3–4), 543–561 (2014)
8. Rauh, A.: Theorie und Anwendung von Intervallmethoden für Analyse und Entwurf robuster und optimaler Regelungen dynamischer Systeme. Fortschritt-Berichte VDI, Reihe 8 Mess-, Steuerungs- und Regelungstechnik, Nr. (1148) (2008). (in German)
9. Rauh, A., Aschemann, H.: Interval-based sliding mode control and state estimation of uncertain systems. In: Proceedings of IEEE International Conference on Methods and Models in Automation and Robotics MMAR, Miedzyzdroje, Poland (2012)
10. Senkel, L., Rauh, A., Aschemann, H.: Robust sliding mode techniques for control and state estimation of dynamic systems with bounded and stochastic uncertainty. In: Second International Conference on Vulnerability and Risk Analysis and Managemet, ICVRAM 2014 (2014)
11. Senkel, L., Rauh, A., Aschemann, H.: Robust sliding mode techniques for control and state estimation of dynamic systems with bounded and stochastic uncertainty. In: Proceedings of 18th European Conference on Mathematics for Industry, ECMI (2014)
12. Shtessel, Y., Edwards, C., Fridman, L., Levant, A. (eds.): Sliding Mode Control and Observation. Springer, Birkhäuser, Heidelberg, Boston (2014)
13. Smith, H.L.: Monotone Dynamical Systems: An Introduction to the Theory of Competitive and Cooperative Systems. Mathematical Surveys and Monographs, vol. 41. American Mathematical Society, Providence (1995)

Linear Algebra

Efficiency of Reproducible Level 1 BLAS

Chemseddine Chohra[1,2]($\boxtimes$), Philippe Langlois[1,2], and David Parello[1,2]

[1] Digits, Architectures et Logiciels Informatiques,
Univ. Perpignan Via Domitia, 66860 Perpignan, France
{chemseddine.chohra,philippe.langlois,david.parello}@univ-perp.fr
[2] Laboratoire d'Informatique Robotique et de Microélectronique de Montpellier,
Univ. Montpellier II, UMR 5506, CNRS, 34095 Montpellier, France

Abstract. Numerical reproducibility failures appear in massively parallel floating-point computations. One way to guarantee this reproducibility is to extend the IEEE-754 correct rounding to larger computing sequences, *e.g.* to the BLAS. Is the extra cost for numerical reproducibility acceptable in practice? We present solutions and experiments for the level 1 BLAS and we conclude about their efficiency.

1 Introduction

Numerical reproducibility is an open question for current high performance computing platforms. Dynamic scheduling and non-deterministic reduction on multithreaded systems affect the operation order. This leads to non-reproducible results because the floating-point addition is not associative. Numerical reproducibility is important for debugging and for validating results, particularly if legal agreements require the bitwise reproduction of the execution results. Failures have been reported in numerical simulation for energy science, dynamic weather forecasting, atomic or molecular dynamic, fluid dynamic – entries in [8].

Solutions provided at the middleware level forbid the dynamic behavior and so impact the performances — see [11] for TBB, [15] for OpenMP or Intel MKL. A first algorithmic solution has been recently proposed in [6]. Their summation algorithms, *ReprodSum* and *FastReprodSum*, guarantee the reproducibility independently from the computation order. They return about the same accuracy as the performance optimized algorithm only running a small constant times slower.

Correctly rounded results ensure numerical reproducibility. IEEE754–2008 floating-point arithmetic is correctly rounded in its four rounding modes [1]. We propose to extend this property to the level 1 routines of the BLAS that depend on the summation order: asum, dot and nrm2, respectively the sum of the absolute values, the dot product and the vectorial Euclidean norm. Recent algorithms that compute the correctly rounded sum of n floating-point values allow us to implement such reproducible parallel computation. The main issue is to investigate whether the running-time overhead of these reproducible routines remains reasonable enough in practice. In this paper we present experimental

© Springer International Publishing Switzerland 2016
M. Nehmeier et al. (Eds.): SCAN 2014, LNCS 9553, pp. 99–108, 2016.
DOI: 10.1007/978-3-319-31769-4_8

answers to this question. Our experimental framework is significant of the current computing practice: it consists in a shared memory parallel system with several sockets of multicore x86 processing units. We apply standard optimization techniques to implement efficient sequential and parallel level 1 routines. We show that for large vectors, reproducible and accurate routines introduces almost no overhead compared to their original counterparts in a performance-optimized library, Intel MKL [7]. For shorter ones, reasonable overheads are measured and presented in Table 2. Since Level 1 BLAS performance is mainly dominated by the memory transfers, additional computation does not significantly increase the running time, especially for large vectors.

The paper is organized as follows. In Sect. 2, we briefly present some accurate summation algorithms, their optimizations and an experimental performance analysis to decide how to efficiently implement the level 1 BLAS. The experimental framework used throughout the paper is also described in this part. Section 3 is devoted to the performance analysis of the sequential implementation of the level 1 BLAS routines. Section 4 describes their parallel implementations and the measure of their efficiency; we gather plots for all previously presented sequential and parallel algorithms. We conclude describing the future developments of this ongoing project towards efficient and reproducible BLAS.

2   Choice of Optimized Floating-Point Summation

Level 1 BLAS subroutines mainly rely on floating-point sums. It exists several correctly rounded summation algorithms. Our first step aims to derive optimized implementations of such algorithms and to choose the most efficient ones. In the following, we briefly describe these accurate algorithms and then, how to optimize and to compare them.

All floating-point computations satisfy the IEEE754–2008. Let $fl(\sum p_i)$ be the computed sum of a length n floating-point vector p. The relative error of the classical accumulation is of the order of $u \cdot n \cdot cond(\sum p_i)$, where $cond(\sum p_i) = \sum |p_i|/|\sum p_i|$ is the condition number of the sum. u is the machine precision that equals 2^{-53} for IEEE754 binary64.

2.1   Some Accurate or Reproducible Summation Algorithms

Algorithm SumK [10] reduces the previous relative error bound as if the classical accumulation is performed in K times the working precision:

$$\frac{|SumK(p) - \sum p_i|}{|\sum p_i|} \leq \frac{(n \cdot u)^K}{1 - (n \cdot u)^K} \cdot cond(\sum p_i) + u. \tag{2.1}$$

SumK replaces the floating-point add by Knuth's TwoSum algorithm that computes both the sum and its rounding error [9]. *SumK* iteratively accumulates these rounding errors to enhance the final result accuracy. The correct rounding could be achieved by choosing a large enough K to vanish the effect of the condition number in Eq. (2.1) — but in practice this latter is usually unknown.

Algorithm iFastSum [18] repeats *SumK* to error-free transform the entry vector. As [2], this distillation process terminates returning a correctly rounded result thanks to a dynamic control of the error.

Algorithms AccSum [14] *and FastAccSum* [13] also rely on error-free transformations of the entry vector. They split the summands, relatively to $\max |p_i|$ and n, such that their higher order parts are then exactly accumulated. This split-and-accumulate steps are iterated to enhance the accuracy up to return a faithfully rounded sum. These algorithms return the correctly rounded sum and *FastAccSum* requires 25 % less floating-point operations than *AccSum*.

HybridSum [18] *and OnlineExact sum* [19] exploit the short range of the floating-point number exponents. These algorithms accumulate the summands with a same exponent in a specific way to produce a short vector with no rounding error. The length of the output vector of this error-free transform step is the exponent range. *HybridSum* splits the summands such that floating-point numbers can be used as error-free accumulators. *OnlineExact* uses two floating-point numbers to simulate a double length accumulator. These algorithms then apply *iFastSum* to evaluate the correctly rounded sum of the error-free short vector(s).

ReprodSum and FastReprodSum [6] respectively rely on *AccSum* and *FastAccSum* to compute not fully accurate but reproducible sums independently of the summation order. So numerical reproducibility of parallel sums is ensured for every number of computing units.

2.2   Experimental Framework and Implementation

Table 1 describes our experimental framework. Aggressive compiler options as -ffast-math are disabled to prevent the modification of the sensitive floating-point properties of these algorithms.

Rounding intermediate results to the binary64 format (53 bit mantissa) and value safe optimizations are provided with -fp-model double and -fp-model strict options. For a fair comparison, all algorithms are manually optimized by a best effort process. AVX vectorization, data prefetching and loop unrolling are carefully combined to pull out the best implementation of each algorithm. Optimization details are presented in Appendix A of [3]. Source code is available at [12].

Runtimes are measured in cycles with the hardware counters thanks to the RDTSC assembly instruction. We display the minimum cycle measures over more than fifty runs for each data. Condition dependant data are computed with the dot product generator from [10].

2.3   Test and Results

Figure 1a and b present the runtime measured in cycles divided by the vector size (y-axis). Vector lengths vary between 2^{10} and 2^{25} (x-axis) and two condition numbers are considered : 10^8 and 10^{32}.

Table 1. Experimental framework

Software	
Compiler (language)	ICC 14.0.2 (C99)
Options	`-O3 -axCORE-AVX-I -fp-model double -fp-model strict -funroll-all-loops`
Parallel library	OpenMP 4.0
BLAS library	Intel MKL 11
Hardware	
Processor	Xeon E5 2660 (Sandy Bridge) at 2.2 GHz
Cache	L1: 32 KB, L2: 256 KB, shared L3 for each socket: 20 MB
Bandwidth	51.2 GB/s
#cores	2 × 8 cores (hyper-threading disabled)

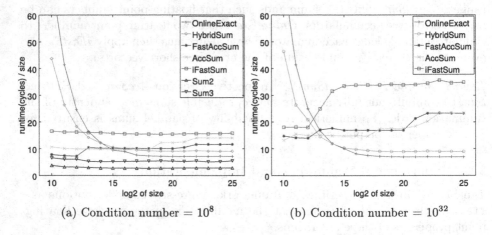

(a) Condition number = 10^8 (b) Condition number = 10^{32}

Fig. 1. Runtime/size ratio for optimized summation algorithms

It is not a surprise that *HybridSum* and *OnlineExact* are interesting for larger size vectors. These algorithms produce one or two short vectors (length = 2048 in binary64) whose distillation is of constant time compared to the linear times of the data preprocessing step (exponent extraction) or also, of the successive error free transformations in the other algorithms. Moreover they are very less sensitive to the conditioning of the entry vector. Shorter size vectors benefit from the other algorithms, especially from *FastAccSum* while their conditioning remains small.

In the following we take advantage of these different behaviors according to the size of the entry vector. We call it a "mixed solution". In practice for the level 1 BLAS routines, *FastAccSum* or *iFastSum* are useful for short vectors while larger ones benefit from *HybridSum* or *OnlineExact* as we will explain it.

3   Sequential Level 1 BLAS

Now we focus on the sum of the absolute value vector (asum), the dot product (dot) and the 2-norm (nrm2). Note that other level 1 BLAS subroutines do not suffer neither of accuracy nor of reproducibility failures. In this section, we start with sequential algorithms detailing our implementations and their efficiency.

3.1   Sum of Absolute Values

The condition number of asum equals 1. So $SumK$ is enough to efficiently get a correctly rounded result. According to Eq. (2.1), K is chosen such that $n \leq u^{1/K-1}$.

Figure 2a exhibits that the correctly rounded asum costs less than $2 \times$ the optimized MKL dasum. Indeed $K = 2$ applies for the considered sizes. Note that $K = 3$ is enough until $n \leq 2^{35}$, $i.e.$ until 256 Terabyte of data.

3.2   Dot Product

The dot product of two n-vectors is transformed into a sum of a $2n$-vector with Dekker's TwoProd [5]. This sum is correctly rounded using a "mixed solution". Short vectors are correctly rounded with $FastAccSum$. For large n, we avoid to build and read this intermediate $2n$-vector: the two TwoProd results are directly exponent-driven accumulated into the short vectors of $OnlineExact$. This explains why this latter is interesting for shorter dot products than what we can expect from Sect. 2.3.

Figure 2c shows this runtime divided by the input vector size — the condition number is 10^{32}. Despite the previous optimizations, the extra cost ratio compared to MKL dot is between 3 and 6. This is essentially justified by the additional computations (memory transfers are unchanged). If a fused-multiply-and-add unit (FMA) is available, the 2MultFMA algorithm [9] that only costs 2 FMA (compared to the TwoProd's 17 flop) certainly improves these values.

3.3   Euclidean Norm

It is not difficult to implement an efficient and reproducible Euclidean norm. Reproducibility is ensured by the correct rounding of the sum of the squares and then by the correct rounding of the IEEE-754 square root. Of course this reproducible 2-norm is only faithfully rounded. Hence a "mixed solution" is similar to the dot one.

Here the MKL nrm2 is not used as the comparison reference since we measure very disappointing runtime for it. We implement a non-reproducible simple and efficient 2-norm with the optimized MKL dot (cblas_ddot). We named it nOrm2.

The memory transfer cost dominates the computing one for dot and nOrm2: compared to dot, nOrm2 halves the memory transfer volume, performs the same number of floating-point operations and runs twice faster, see Fig. 2c and e.

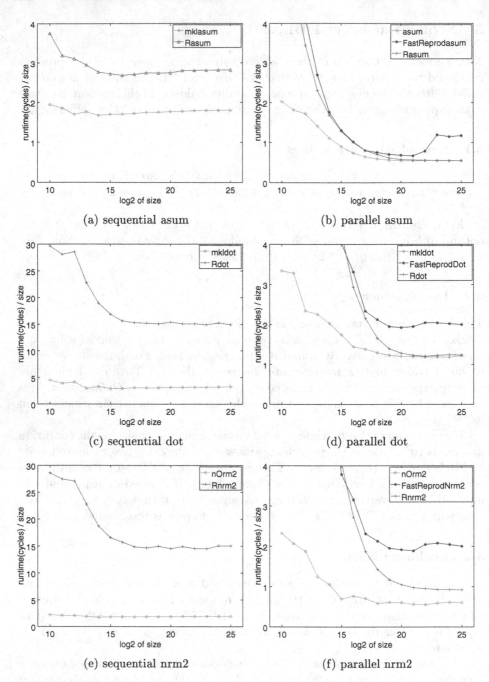

(a) sequential asum

(b) parallel asum

(c) sequential dot

(d) parallel dot

(e) sequential nrm2

(f) parallel nrm2

Fig. 2. Runtime/size ratio of sequential and parallel level 1 BLAS (cond $= 10^{32}$)

As previously mentioned, the "mixed solution" dot product is still computation-dominated. This justifies that the previous dot ratios prohibitively double for our sequential nrm2.

4   Reproducible Parallel Level 1 BLAS

Now we consider the parallel implementations. As in the previous section, parallel asum relies on parallel $SumK$ while parallel dot and nrm2 derive from a parallel version of a "mixed solution" for the dot product. We start introducing these two parallel algorithms. Then we derive the parallel reproducible level 1 BLAS and perform its performance analysis.

4.1   From Parallel Sums to Reproducible Level 1 BLAS

Parallel SumK. It derives from the sequential version and has already been introduced in [17]. It consists in 2 steps. Step 1 applies the $SumK$ algorithm on the local data without the final error compensation for every K iterations. Hence it returns a K-length vector S such that $(S_j)_{j=1,K}$ is the sum of the jth layer in $SumK$ applied to the local subvector. Step 2 gathers these K-length vectors to the master unit and applies the sequential $SumK$.

Parallel dot "mixed solution". Every n-length entry vector is split within P threads (or computing units) and N denotes the length of these local subvectors. The key point is to perform efficient error-free transformations of these N-vectors until the last reduction step. This consists in a 4 step process presented with Fig. 3 for $P = 2$. Steps 1 and 2 are processed by the P threads with local private vectors. Step 1 is similar to the sequential case and produces one vector of $size = 2N$ or 2048 or 4096: TwoProd transforms short N-vectors into a $2N$-one while this latter is not built for larger entries but directly exponent-driven accumulated into the $size$-length vector as for *HybridSum* or *OnlineExact*. Step 2: the $size$-length vector is distilled (as for iFastSum) into a smaller vector of non overlapping floating-point numbers. Step 3: every thread fuses this small vector into a global shared one. Step 4 is performed by the master thread that computes the correctly rounded result of the global vector with *FastAccSum*.

Let us remark that the small vector issued from Step 2 is at most of length 40 in binary64. Hence the distillation certainly benefits from cache effect. The next fusing step moves across the computing units these vectors of length 40 in the worst case. This induces a communication over-cost especially for distributed memory environments. Nevertheless it introduces no more reduction step than a classic parallel summation.

The Reproducible Parallel Level 1 BLAS. The reproducible parallel Rasum derives from parallel $SumK$ as in Sect. 3.1. The parallel dot "mixed solution" gives reproducible parallel Rdot and Rnrm2. In practice, the parallel implementation of the Step 1 differs from the sequential one as follows.

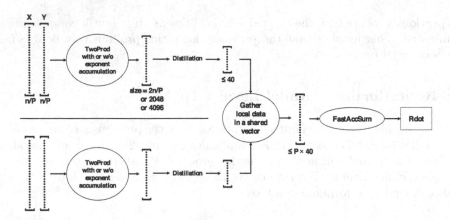

Fig. 3. Parallel dot "mixed solution"

For shorter vectors, *iFastSum* is preferred to *FastAccSum* to minimize the Step 3 communications. For medium sized vectors, *HybridSum* is preferred to *OnlineExact* for Rdot to minimize the Step 2 distillation cost. Otherwise *OnlineExact* is chosen to minimize the exponent extraction cost.

4.2   Test and Results

The experimental framework is unchanged. Each physical core runs at most one thread thanks to the KMP_AFFINITY variable. For every routine, we run from 1 to 16 threads on 16 cores to select the most efficient configurations with respect to the vector size. This optimal number of threads is given in parentheses in Table 2 except when it corresponds to the maximum possible resources (16). Intel MKL's (hidden) choice is denoted with a $\star$.

For the next performance comparisons, optimized parallel routines are necessary as references. We use the MKL parallel dot and we implement asum and nrm2 parallel versions. Our parallel asum runs up to 16 MKL dasum and performs a final reduction. Our parallel nOrm2 derives similarly from the sequential nOrm2 introduced in Sect. 3.3. These implementations exhibit the best performances in Fig. 2. As in Sect. 2.3, our implementations of *ReprodSum* and

Table 2. Runtime extra cost for the reproducibility of parallel level 1 BLAS

Vector size	10^3		10^4		10^5	10^6	10^7
Rasum/asum	2.0	(1/1)	1.5	(4/2)	1.3	1.1	1
Rdot/mkldot	6.4	(8/$\star$)	3.8	(8/$\star$)	1.6	1.1	1
Rnrm2/nOrm2	9.1	(8/$\star$)	7.1	(8/$\star$)	3.4	1.6	1.5
Rasum/FastReprodasum	0.9	(1/1)	0.9	(4/4)	1.0	0.8	0.5
Rdot/FastReprodDot	1.5	(8/1)	1.5	(8/8)	0.9	0.7	0.6
Rnrm2/FastReprodNrm2	1.7	(8/1)	1.5	(8/8)	0.9	0.5	0.4

FastReprodSum are optimized in a fair way using again AVX vectorization, data prefetching and loop unrolling. The latter one is selected for the sequel.

We compare our reproducible Rasum, Rdot and Rnrm2, to the optimized but non-reproducible reference implementations, and to the one derived from *FastReprodSum*. Results are presented in Fig. 2 and Table 2.

Our reproducible Rasum compares very well to the optimized asum: the initial $2 \times$ extra cost tends to 1 for n about 10^6, see Fig. 2. Compared to the sequential cases and since it operates now on $16 \times$ smaller local vectors, our reproducible Rdot and Rnrm2 reach their optimal linear performance for larger entry sizes. Nevertheless the reproducible Rdot runs less than $2 \times$ slower than the MKL reference for vector size up to 10^5, see Fig. 2d. For the same reasons as in the sequential case (Sect. 3.3), our reproducible Rnrm2 is not enough efficient to exhibit the same optimal tendency. Nevertheless the Rnmr2 overhead now reduces to the more convincing ratios compared to nOrm2, see Fig. 2f.

Finally our fully accurate reproducible level 1 routines compare quite favorably to those derived from the reproducible *FastReprodSum*, especially for large vectors: see Fig. 2. Those latest algorithms read twice the entry vector and thus suffer from cache effects for large vectors. It is not the case for our algorithms. On the other hand, the additional computation required by *OnlineExact* or *Hybrid-Sum* benefit from the floating-point unit availability.

5   Conclusion and Future Developments

This experimental work illustrates that reproducible level 1 BLAS can be implemented with a reasonable extra cost compare to the performance-optimized non-reproducible routines. Moreover our implementations offer full accuracy almost for free compared to the existing reproducible solutions.

Indeed the floating-point peak performance of current machines is far to be exploited by level 1 BLAS. So the additional floating-point operations required by our accuracy enhancement do not significantly increase their execution time.

Of course these results are quantitatively linked to the experimental framework. Nevertheless the same tendencies should be observed in other current computing contexts. Work is ongoing to benefit from FMA within dot and nrm2, to validate an hybrid OpenMP + MPI implementation on larger HPC cluster, to port and optimize this approach to accelerators (as Intel Xeon Phi) and to compare it to the expansions and software long accumulator of [4].

Finally there is alas no reason to be optimistic for the BLAS level 3 where the floating-point units have no space left for extra computation. Reproducible solutions need to be implemented from scratch, for example following [16].

References

1. IEEE 754–2008, Standard for Floating-Point Arithmetic. Institute of Electrical and Electronics Engineers, New York (2008)
2. Bohlender, G.: Floating-point computation of functions with maximum accuracy. IEEE Trans. Comput. **C-26**(7), 621–632 (1977)
3. Chohra, C., Langlois, P., Parello, D.: Implementation and Efficiency of Reproducible Level 1 BLAS (2015). http://hal-lirmm.ccsd.cnrs.fr/lirmm-01179986
4. Collange, C., Defour, D., Graillat, S., Iakimchuk, R.: Reproducible and accurate matrix multiplication in ExBLAS for high-performance computing. In: SCAN 2014, Würzburg, Germany (2014)
5. Dekker, T.J.: A floating-point technique for extending the available precision. Numer. Math. **18**, 224–242 (1971)
6. Demmel, J.W., Nguyen, H.D.: Fast reproducible floating-point summation. In: Proceedings of 21th IEEE Symposium on Computer Arithmetic. Austin, Texas, USA (2013)
7. Intel Math Kernel Library. http://www.intel.com/software/products/mkl/
8. Jézéquel, F., Langlois, P., Revol, N.: First steps towards more numerical reproducibility. ESAIM: Proc. **45**, 229–238 (2013)
9. Muller, J.M., Brisebarre, N., de Dinechin, F., Jeannerod, C.P., Lefèvre, V., Melquiond, G., Revol, N., Stehlé, D., Torres, S.: Handbook of Floating-Point Arithmetic. Birkhäuser, Boston (2010)
10. Ogita, T., Rump, S.M., Oishi, S.: Accurate sum and dot product. SIAM J. Sci. Comput. **26**(6), 1955–1988 (2005)
11. Reinders, J.: Intel Threading Building Blocks, 1st edn. O'Reilly & Associates Inc., Sebastopol (2007)
12. http://webdali.univ-perp.fr/ReproducibleSoftware
13. Rump, S.M.: Ultimately fast accurate summation. SIAM J. Sci. Comput. **31**(5), 3466–3502 (2009)
14. Rump, S.M., Ogita, T., Oishi, S.: Accurate floating-point summation - part I: faithful rounding. SIAM J. Sci. Comput. **31**(1), 189–224 (2008)
15. Story, S.: Numerical reproducibility in the Intel Math Kernel Library. Salt Lake City, November 2012
16. Van Zee, F.G., van de Geijn, R.A.: BLIS: a framework for rapidly instantiating BLAS functionality. ACM Trans. Math. Software **41**(3), 14:1–14:33 (2015)
17. Yamanaka, N., Ogita, T., Rump, S., Oishi, S.: A parallel algorithm for accurate dot product. Parallel Comput. **34**(68), 392–410 (2008)
18. Zhu, Y.K., Hayes, W.B.: Correct rounding and hybrid approach to exact floating-point summation. SIAM J. Sci. Comput. **31**(4), 2981–3001 (2009)
19. Zhu, Y.K., Hayes, W.B.: Algorithm 908: online exact summation of floating-point streams. ACM Trans. Math. Softw. **37**(3), 37:1–37:13 (2010)

Tight Bounds on the Radius of Nonsingularity

David Hartman[1,2]($\boxtimes$) and Milan Hladík[1]

[1] Department of Applied Mathematics, Faculty of Mathematics and Physics,
Charles University, Malostranské nám. 25, 11800 Prague, Czech Republic
{hartman,hladik}@kam.mff.cuni.cz
[2] Institute of Computer Science Academy of Sciences,
18207 Prague 8, Czech Republic
hartman@cs.cas.cz

Abstract. Radius of nonsingularity of a square matrix is the minimal distance to a singular matrix in the maximum norm. Computing the radius of nonsingularity is an NP-hard problem. The known estimations are not very tight; one of the best one has the relative error $6n$. We propose a randomized approximation method with a constant relative error 0.7834. It is based on a semidefinite relaxation. Semidefinite relaxation gives the best known approximation algorithm for MAXCUT problem, and we utilize similar principle to derive tight bounds on the radius of nonsingularity. This gives us rigorous upper and lower bounds despite randomized character of the algorithm.

Keywords: Radius of nonsingularity · Bounds · Semidefinite programming

1 Introduction

It is well known that (non)singularity of the matrix is a determinative property for many phenomena in optimization or system theory. Computationally (non)singularity of a real value matrix is not a hard problem. In real-word, however, systems under study are analysed via data that are often subject to uncertainties and measurement errors. For these reasons it is beneficial in many situations to be able to detect existence of a singular matrix for range of possible values of its corresponding elements, which starts to be more complex problem – as we will see later even NP-hard.

Let be more specific for a while to show motivational example. We may assume that there is a matrix A as a result of series of measurements with elements a_{ij}, where the corresponding elements are subject to uncertainty. Let moreover assume that we can ensure that there exists a precise bound for the measurement precision represented by δ. This in fact means that the resulting

D. Hartman—Supported by grants 13-17187S and 13-10660S of the Czech Science Foundation.
M. Hladík—Supported by grant 13-10660S of the Czech Science Foundation.

© Springer International Publishing Switzerland 2016
M. Nehmeier et al. (Eds.): SCAN 2014, LNCS 9553, pp. 109–115, 2016.
DOI: 10.1007/978-3-319-31769-4_9

values are in intervals $[a_{ij} - \delta, a_{ij} + \delta]$ for all i, j. Mentioned results can be therefore represented as an interval matrix [8,9]:

$$A = [\underline{A}, \overline{A}] = \{A \in \mathbb{R}^{n \times n}; \underline{A} \leq A \leq \overline{A}\}$$

Determining a specific property of an interval matrix is usually represented by determining this property for all realizations from the corresponding intervals. In the same way, non-singularity of interval matrix can be seen as a search for singular matrix within specified interval. The determination of radius of non-singularity, on the other hand, goes even further and searches for the maximal interval within which there is no singular matrix. More formally, an interval matrix is called singular if it contains a singular matrix. Then the following decision problem can be considered [2]:

Instance: Let us have a square interval matrix A, where both matrices $\underline{A}$ and $\overline{A}$ are rational

Question: Is A singular?

Except these computational motivations there is at least one scientific field that may use results about radius of non-singularity. This connection is represented in notion of structured singular value [11] which is closely related with radius of nonsingularity [2]. This term has relatively long history in automation and control and its main use is in analysis of feedback systems with uncertainties, especially concerning robustness and stability [10].

The determination of radius of non-singularity or, more precisely, bounds of this characteristic seems to be however easier using tools from ordinary matrix theory. Let us therefore concentrate on ordinary matrices. Given a matrix $A \in \mathbb{R}^{n \times n}$, the radius of nonsingularity [1–6] is defined by

$$d(A) := \inf\{\varepsilon > 0; \ (\exists \text{ singular } B)(\forall i, j) : |a_{ij} - b_{ij}| \leq \varepsilon\}.$$

In other words, it is the minimum distance of A to a singular matrix in the maximum norm. There was also studied generalization [1–4] in the form of

$$d(A, \Delta) := \inf\{\varepsilon > 0; \ (\exists \text{ singular } B)(\forall i, j) : |a_{ij} - b_{ij}| \leq \varepsilon \Delta_{ij}\},$$

where $\Delta \in \mathbb{R}^{n \times n}$ is a given non-negative matrix. Thus, $d(A)$ is a special case of $d(A, \Delta)$ when putting $\Delta := E$, and the matrix E consists of all ones, i.e. all elements $e_{ij} = 1$.

In above form the determination of this characteristic for any matrix A seems to be a hard task. In [2–4] it was shown that

$$d(A) = \frac{1}{\|A^{-1}\|_{\infty,1}}, \tag{1}$$

where $\| \cdot \|_{\infty,1}$ is a matrix norm defined as

$$\|M\|_{\infty,1} := \max\{\|Mx\|_1; \ \|x\|_\infty = 1\} = \max\{\|Mz\|_1; \ z \in \{\pm 1\}^n\}.$$

Since computing $\|\cdot\|_{\infty,1}$ is known to be an NP-hard problem as shown in [2], the same is true for the radius of nonsingularity. Moreover, it has been shown that there is no polynomial time algorithm for approximating $d(A)$ with a relative error at most $\frac{1}{4n^2}$ provided $P \neq NP$; see [3,4]. That is why there were investigated various lower and upper bounds. In [3,4], one has

$$\frac{1}{\rho(|A^{-1}|E)} \leq d(A) \leq \frac{1}{\max\limits_{i=1,\ldots,n} (E|A^{-1}|)_{ii}},$$

where $\rho(A)$ is the spectral radius of A and E, similarly to the above case, is a matrix consisting of all ones.

Rump [5,6] developed the other bounds as

$$\frac{1}{\rho(|A^{-1}|E)} \leq d(A) \leq \frac{6n}{\rho(|A^{-1}|E)}.$$

In both mentioned bounds, the matrix E can be substituted by matrix Δ and the resulting equations then provide bounds for general version $d(A, \Delta)$ of radius of nonsingularity, see e.g. [3].

Except these basic solutions there is one method introduced recently by Kolev [13]. He is dealing with this problem from interval analysis perspective and transforms its solution into search of real maximum magnitude eigenvalue of an associated generalized eigenvalue problem. By iterative solution of the last mentioned problem the original problem can be solved assuming additional conditions on the whole process.

Roughly speaking, these conditions say that all solutions have to be computable and moreover meet some additional sign conditions. Although the resulting algorithm is polynomial, these conditions can represent an obstacle in solution and it is therefore still reasonable to ask for better bounds or approximate solutions.

2   SDP Relaxation

We develop an approximation for $\|M\|_{\infty,1}$. In view of (1), it will give the same accurate approximation for $d(A)$, too. Our approach is based on semi-definite relaxation, and first we proceed in a similar manner as for a randomized 0.878-approximation algorithm for MAXCUT problem [7].

The problem of computing $\|M\|_{\infty,1}$ can be formulated as

$$\max \sum_{i,j=1}^{n} m_{ij} x_i y_j \text{ subject to } x, y \in \{\pm 1\}^n. \tag{2}$$

The semidefinite programming relaxation consists of replacing discrete variables $x_i, y_j \in \{\pm 1\}$, $i, j = 1, \ldots, n$ by unit vectors $u_i, v_j \in \mathbb{R}^n$, $i, j = 1, \ldots, n$, as follows

$$\max \sum_{i,j=1}^{n} m_{ij} u_i^T v_j \text{ subject to } u_i, v_i \in \mathbb{R}^n,$$

$$\|u_i\|_2 = \|v_i\|_2 = 1, \ i = 1, \ldots, n. \tag{3}$$

For any feasible solution x, y to (2), we find a corresponding solution to (3) by putting $u_i := (0, \ldots, 0, x_i)$ and $v_i := (0, \ldots, 0, y_i)$, $i = 1, \ldots, n$. This shows that (3) is a relaxation to the optimization problem (2). That is, the optimal value to (3) is an upper bound on the optimal value to (2).

This already represents problem with real variables. We have to go further to obtain a system for which there are known approximate solutions. Let therefore define the matrix $Z \in \mathbb{R}^{2n \times 2n}$ as

$$Z := U^T U, \tag{4}$$

where U has the columns $u_1, \ldots, u_n, v_1, \ldots, v_n$. The matrix Z is obviously positive semidefinite, denoted by $Z \succeq 0$, and the diagonal entries are equal to 1. Contrary, any positive semidefinite matrix Z can be factorized as in (4), and ones on the diagonal of Z imply that the columns of U are unit vectors. Hence (3) takes the equivalent form of a semidefinite program

$$\max \sum_{i,j=1}^{n} m_{ij} z_{i,j+n} \text{ subject to } Z \succeq 0,$$

$$z_{ii} = 1, \ i = 1, \ldots, 2n. \tag{5}$$

The semidefinite program (5) can be solved in polynomial time with an arbitrary (a priori given) precision, see [12] or in connection with actual relaxation in [7] (verification of the numerically computed solutions were dealt in [14]). Let $\varepsilon > 0$ be a given precision and denote by γ the computed approximate optimal value to (5). Then

$$\gamma \geq \|M\|_{\infty,1} - \varepsilon,$$

or

$$\gamma + \varepsilon \geq \|M\|_{\infty,1}.$$

This gives us an upper bound on the $(\infty, 1)$-norm.

In order to calculate a lower bound on $\|M\|_{\infty,1}$, we utilize an optimal solution of (5) resp. (3) to find a good feasible solution to (2). Let $u_1^*, \ldots, u_n^*, v_1^*, \ldots, v_n^*$ be an (approximative) optimal solution to (3). Compared with the solution of (2) this is highly dimensional. To obtain original lower dimension, we have to adopt some sort of "rounding". Let $p \in \mathbb{R}^n$ be a unit vector and define the mapping

$$w \mapsto \begin{cases} 1 & \text{if } p^T w \geq 0, \\ -1 & \text{otherwise} \end{cases} \tag{6}$$

We can see that the resulting vector is dependent not only on the original, but also on the value of p. This p is meant to be chosen randomly, which results in so called randomized rounding sampling p uniformly distributed within the corresponding ball [7].

We utilize the following lemma from [7].

Lemma 1. *Let $u, v \in \mathbb{R}^n$ be unit vectors. The probability that (6) maps u and v to different values is $\frac{1}{\pi} \arccos u^T v$.*

Consider a feasible solution $x_1^*, \ldots, x_n^*, y_1^*, \ldots, y_n^*$ to (2) determined by images of $u_1^*, \ldots, u_n^*, v_1^*, \ldots, v_n^*$ with respect to the mapping (6). By Lemma 1, the expected objective value of this solution is

$$\sum_{i,j} m_{ij} \left(1 - \frac{1}{\pi} \arccos u_i^{*T} v_j^*\right) - m_{ij} \frac{1}{\pi} \arccos u_i^{*T} v_j^*$$

$$= \sum_{i,j} m_{ij} \left(1 - \frac{2}{\pi} \arccos u_i^{*T} v_j^*\right). \tag{7}$$

Let $\alpha \approx 0.87856723$ be the Goemans-Williamson value characterizing the approximation ratio of their approximation algorithm for MAXCUT [15]. It represents an optimal value of the problem $\min_{z \in [-1,1]} \frac{2 \arccos z}{\pi(1-z)}$. The following lemma gives an auxiliary result for bounding (7) by means of a linear function.

Lemma 2. *For each $z \in [-1, 1]$ we have*

$$\alpha z + \alpha - 1 \leq 1 - \frac{2}{\pi} \arccos z \leq \alpha z + 1 - \alpha.$$

Proof. We prove the first inequality $\alpha z + \alpha - 1 \leq 1 - \frac{2}{\pi} \arccos z$. The second one is a straightforward corollary since $1 - \frac{2}{\pi} \arccos z$ is an odd function.

We want to find values a and b such that $az + b \leq 1 - \frac{2}{\pi} \arccos z$ for every $z \in [-1, 1]$. Since the value of the function $1 - \frac{2}{\pi} \arccos z$ at $z = -1$ is -1, we focus on the line $ax + b$ coming through the point $(-1, -1)$ and supporting the function from below. Thus the line reads $ax + a - 1$ and a is the maximal value such that $az + a - 1 \leq 1 - \frac{2}{\pi} \arccos z$ for every $z \in [-1, 1]$. In other words,

$$a \leq \frac{2 - \frac{2}{\pi} \arccos z}{z + 1}, \quad \forall z \in [-1, 1].$$

Hence

$$a = \min_{z \in [-1,1]} \frac{2 - \frac{2}{\pi} \arccos z}{z + 1} = \min_{z \in [-1,1]} \frac{2\pi - 2 \arccos z}{\pi(z + 1)}.$$

Substituting $w := -z$, we obtain

$$a = \min_{w \in [-1,1]} \frac{2\pi - 2 \arccos(-w)}{\pi(1 - w)} = \min_{w \in [-1,1]} \frac{2\pi - 2(\pi - \arccos w)}{\pi(1 - w)}$$

$$= \min_{w \in [-1,1]} \frac{2 \arccos w}{\pi(1 - w)} = \alpha.$$

$\square$

Now, we use Lemma 2 to find a lower bound to expected value of the solution represented by (7). Let $i, j \in \{1, \ldots, n\}$, then

$$
m_{ij}\left(1 - \frac{2}{\pi} \arccos z\right) \geq \begin{cases} m_{ij}(\alpha z + \alpha - 1) & \text{if } m_{ij} \geq 0, \\ m_{ij}(\alpha z - \alpha + 1) & \text{otherwise,} \end{cases}
$$

or,

$$
m_{ij}\left(1 - \frac{2}{\pi} \arccos z\right) \geq m_{ij}\alpha z + |m_{ij}|(\alpha - 1).
$$

Thus the lower bound to (7) can be established as follows

$$
\sum_{i,j} m_{ij}\left(1 - \frac{2}{\pi} \arccos u_i^{*T} v_j^*\right) \geq \sum_{i,j} m_{ij}\alpha u_i^{*T} v_j^* + |m_{ij}|(\alpha - 1)
$$
$$
= \alpha\gamma + (\alpha - 1)e^T|M|e,
$$

where e is a vector that consists of ones.

Hence we have an expected lower bound on $\|M\|_{\infty,1}$

$$
\|M\|_{\infty,1} \geq \alpha\gamma + (\alpha - 1)e^T|M|e.
$$

The right-hand side depends on the entries of M, but we can employ the estimate $\|M\|_{\infty,1} \leq e^T|M|e$ to obtain

$$
\|M\|_{\infty,1} \geq \alpha\gamma + (\alpha - 1)e^T|M|e \geq \alpha\gamma + (\alpha - 1)\|M\|_{\infty,1},
$$

whence

$$
\|M\|_{\infty,1} \geq \frac{\alpha}{2 - \alpha}\gamma.
$$

This gives us a randomized algorithm with the approximation ratio $\frac{\alpha}{2-\alpha} \approx 0.78343281$.

References

1. Poljak, S., Rohn, J.: Radius of Nonsingularity. Technical report KAM Series (88–117), Department of Applied Mathematics, Charles University, Prague (1988)
2. Poljak, S., Rohn, J.: Checking robust nonsingularity is NP-hard. Math. Control Signals Syst. **6**(1), 1–9 (1993)
3. Rohn, J.: Checking properties of interval matrices. Technical report, Institute of Computer Science, Academy of Sciences of the Czech Republic, Prague, 686 (1996)
4. Kreinovich, V., Lakeyev, A., Rohn, J., Kahl, P.: Computational Complexity and Feasibility of Data Processing and Interval Computations. Kluwer Academic Publishers, Dordrecht (1998)
5. Rump, S.M.: Almost sharp bounds for the componentwise distance to the nearest singular matrix. Linear Multilinear Algebra **42**(2), 93–107 (1997)

6. Rump, S.M.: Bounds for the componentwise distance to the nearest singular matrix. SIAM J. Matrix Anal. Appl. **18**(1), 83–103 (1997)
7. Gärtner, B., Matoušek, J.: Approximation Algorithms and Semidefinite Programming. Springer, Heidelberg (2012)
8. Moore, R.E., Kearfott, R.B., Cloud, M.J.: Introduction to Interval Analysis. SIAM, Philadelphia (2009)
9. Neumaier, A.: Interval Methods for Systems of Equations. Cambridge University Press, Cambridge (1990)
10. Packard, A., Doyle, J.C.: The complex structured singular value. Automatica **29**, 71–109 (1993)
11. Stein, G., Doyle, J.C.: Beyond singular values and loop shapes. J. Guidance Control Dyn. **14**(1), 5–16 (1991)
12. Grötschel, M., Lovász, L., Schrijver, A.: The ellipsoid method and its consequences in combinatorial optimization. Combinatorica **1**, 169–197 (1981)
13. Kolev, L.V.: A method for determining the regularity radius of interval matrices. Reliable Comput. **16**(1), 1–26 (2011)
14. Jansson, C., Chaykin, D., Keil, C.: Rigorous error bounds for the optimal value in semidefinite programming. SIAM J. Numer. Anal. **46**(1), 180–200 (2007)
15. Goemans, M.X., Williamson, D.P.: Improved approximation algorithms for maximum cut and satisfiability problems using semidefinite programming. J. ACM **42**(6), 1115–1145 (1995)

Optimal Preconditioning for the Interval Parametric Gauss–Seidel Method

Milan Hladík[(✉)]

Department of Applied Mathematics, Faculty of Mathematics and Physics,
Charles University, Malostranské nám. 25, 118 00 Prague, Czech Republic
hladik@kam.mff.cuni.cz

Abstract. We deal with an interval parametric system of linear equations, and focus on the problem how to find an optimal preconditioning matrix for the interval parametric Gauss–Seidel method. The optimality criteria considered are to minimize the width of the resulting enclosure, to minimize its upper end-point or to maximize its lower end-point. We show that such optimal preconditioners can be computed by solving suitable linear programming problems. We also show by examples that, in some cases, such optimal preconditioners are able to significantly decrease an overestimation of the results of common methods.

Keywords: Interval computation · Interval parametric system · Preconditioner · Linear programming

1 Introduction

Consider an interval linear system of equations

$$Ax = b, \quad A \in \boldsymbol{A}, \ b \in \boldsymbol{b}, \tag{1}$$

where

$$\boldsymbol{A} := [\underline{A}, \overline{A}] = \{A \in \mathbb{R}^{n \times n}; \ \underline{A} \leq A \leq \overline{A}\},$$
$$\boldsymbol{b} := [\underline{b}, \overline{b}] = \{b \in \mathbb{R}^n; \ \underline{b} \leq b \leq \overline{b}\}$$

are an interval matrix and an interval vector, respectively. The solution set is defined as

$$\Sigma := \{x \in \mathbb{R}^n; \ \exists A \in \boldsymbol{A}, \ \exists b \in \boldsymbol{b} : Ax = b\}.$$

Since it is nonconvex in general, the problem is usually to compute an interval vector enclosing the solution set. Computing the smallest enclosure is an NP-hard problem [1], so the known polynomial-time methods overestimate more-or-less the optimal enclosure. There are, however, plenty of methods varying in time complexity and tightness of the resulting enclosures [1,4,11].

© Springer International Publishing Switzerland 2016
M. Nehmeier et al. (Eds.): SCAN 2014, LNCS 9553, pp. 116–125, 2016.
DOI: 10.1007/978-3-319-31769-4_10

Notation. The midpoint and the radius matrices corresponding to an interval matrix A are defined respectively as

$$A^c := \frac{1}{2}(\underline{A} + \overline{A}), \quad A^\Delta := \frac{1}{2}(\overline{A} - \underline{A}).$$

Similarly we define interval vectors and intervals. The ith column of a matrix $C \in \mathbb{R}^{n \times n}$ is denoted by C_{*i}.

Preconditioning. Many methods for enclosing the solution set use preconditioning. Let $C \in \mathbb{R}^{n \times n}$. Then the interval system (1) preconditioned by C reads

$$A'x = b', \quad A' \in (CA), \; b' \in (Cb),$$

where CA and Cb are calculated by interval arithmetic [11]. The solution set corresponding to the preconditioned system contains the original one as a subset, so by preconditioning we do not miss any solution. Even though the solution set inflates by preconditioning, most of the methods used perform better when the system is preconditioned by a suitable matrix.

It is commonly recommended to use the preconditioner $C = (A^c)^{-1}$ or its numerical approximation. Some theoretical properties justifying this choice were stated by Neumaier [10,11]. This does not mean, however, that the midpoint inverse preconditioner yields the best results for each method and for each input data.

Kearfott [5] initiated a research in constructing an optimal preconditioning matrix [6–8]. The authors investigated the interval Gauss–Seidel method with an application in nonlinear equation solving by the interval Newton method. They showed that the optimal precodnitioner for the interval Gauss–Seidel method can be formulated in terms of a linear programming, so it is polynomially computable. A hybrid preconditioning strategy combining the midpoint inverse and a certain kind of optimal preconditioners was proposed by Gau and Stadtherr [2], and some numerical tests and an application in global optimization were presented by Lin and Stadtherr [9].

The Interval Gauss–Seidel Method. Let us recall the interval Gauss–Seidel method briefly. Let $x \supseteq \Sigma$ be an initial enclosure of the solution set. One interval Gauss–Seidel iteration for the preconditioned system is based on the operations

$$z_i := \frac{1}{(CA)_{ii}} \left((Cb)_i - \sum_{j \neq i} (CA)_{ij} x_j \right),$$

$$x_i := x_i \cap z_i,$$

for $i = 1, \ldots, n$.

Interval Parametric Systems. An interval linear parametric system of equations is a family of systems

$$A(p)x = b(p), \quad p \in \boldsymbol{p},$$

where the constraint matrix and the right-hand side vector linearly depend on parameters $p_1, \ldots, p_K$,

$$A(p) = \sum_{k=1}^{K} A^k p_k, \quad b(p) = \sum_{k=1}^{K} b^k p_k.$$

Herein, $A^1, \ldots, A^K \in \mathbb{R}^{n \times n}$ are given matrices, $b^1, \ldots, b^K \in \mathbb{R}^n$ are given vectors, and $\boldsymbol{p} = (\boldsymbol{p}_1, \ldots, \boldsymbol{p}_K)$ is a given interval vector. The corresponding solution set is defined as

$$\Sigma_p := \{x \in \mathbb{R}^n; \, \exists p \in \boldsymbol{p} : A(p)x = b(p)\}.$$

Methods for computing an enclosure of the solution set were discussed, e.g., in [3, 17]. A parametrized version of interval Gauss–Seidel iteration in particular was addressed in Popova [13]. For parametric systems, preconditioning is applied, too.

In principle, a parametric system can be relaxed and the problem reduced to solving the standard interval system

$$Ax = b, \quad A \in \boldsymbol{A}, \, b \in \boldsymbol{b},$$

where

$$\boldsymbol{A} := \sum_{k=1}^{K} A^k \boldsymbol{p}_k, \quad \boldsymbol{b} := \sum_{k=1}^{K} b^k \boldsymbol{p}_k$$

are evaluated by interval arithmetic. For a preconditioned system by $C \in \mathbb{R}^{n \times n}$, the tightest relaxation is done by evaluating

$$\boldsymbol{A} := \sum_{k=1}^{K} (CA^k) \boldsymbol{p}_k, \quad \boldsymbol{b} := \sum_{k=1}^{K} (Cb^k) \boldsymbol{p}_k.$$

Notice that a relaxation leads to overestimation of the solution set in general since we lose information about dependencies between the interval parameters.

The interval Gauss–Seidel iteration for preconditioned parametric system reads

$$\boldsymbol{z}_i := \frac{1}{\left(\sum_{k=1}^{K} (CA^k)_{ii} \boldsymbol{p}_k\right)} \left(\sum_{k=1}^{K} (Cb^k)_i \boldsymbol{p}_k - \sum_{j \neq i} \left(\sum_{k=1}^{K} (CA^k)_{ij} \boldsymbol{p}_k \right) \boldsymbol{x}_j \right), \quad (2)$$

$$\boldsymbol{x}_i := \boldsymbol{x}_i \cap \boldsymbol{z}_i,$$

for $i = 1, \ldots, n$.

For parametric systems, a residual form of enclosures is often employed. Let $x^0 \in \mathbb{R}^n$, for example the solution of $A(p^c)x = b(p^c)$. Then the residual form enclosure of Σ_p has the form of $x = x^0 + y$, where y encloses the solution set to the parametric system

$$A(p)x = b(p) - A(p)x^0, \quad p \in \boldsymbol{p}.$$

The interval Gauss–Seidel iteration (2) for this system works in the same manner as for the original system, only the vectors b^k are replaced by $b^k - A^k x^0$, $k = 1, \ldots, K$.

Goal. The purpose of this paper is to extend the above mentioned results to interval parametric systems of linear equations by designing an optimal preconditioner for the parametric interval Gauss–Seidel method.

2   Optimal Preconditioners

In this section, we show how to construct an optimal preconditioner for (2). We focus on the direct version only since for the residual form it works analogously.

Since the ith step of (2) depends only on the ith row of C, we will design C row by row. For this purpose, let $i \in \{1, \ldots, n\}$ be fixed, and consider the ith row of C, denoted by c.

Optimality of the preconditioner can be viewed from diverse perspectives; see various criteria surveyed in Kearfott et al. [7,8]. We will be concerned with the following objectives

- minimize the resulting width, that is, the objective is $\min 2z_i^\Delta$,
- minimize the resulting upper bound, that is, the objective is $\min \overline{z}_i$,
- maximize the resulting lower bound, that is, the objective is $\max \underline{z}_i$.

If we apply both the second and the third preconditioners, we obtain the smallest interval as a result after the intersection. This observation relies on standard interval arithmetic. Provided we allow division by zero-containing intervals and utilize generalized arithmetic, then tighter results are possible; see S-preconditioners in [7,8].

In the following, we will discuss the first and the second criteria only since the third criterion is easily reduced to the second one.

2.1   Minimal Width

Now, we deal with the first mentioned criterion – to minimize $2z_i^\Delta$. Suppose that $0 \in \boldsymbol{x}$ and $0 \in \boldsymbol{z}_i$. This is the case, for instance, when we apply the residual form and $x^0 \in \Sigma_p$. However, the resulting preconditioner seems to perform well even if the assumption is not satisfied despite it needn't be optimal.

In order that $\boldsymbol{z}_i$ is bounded, we will assume that the denominator in (2) does not contain the zero. Moreover, we will normalize c such that the denominator has the form of $[1, r]$ for some $r \geq 1$. Then, from our assumptions it follows that the operation in (2) is simplified to

$$\sum_{k=1}^{K} (cb^k) p_k - \sum_{j \neq i} \left(\sum_{k=1}^{K} (cA_{*j}^k) p_k \right) x_j$$

Denote

$$\beta_k := |cb^k|, \quad k = 1, \ldots, K,$$
$$\alpha_{jk} := |cA_{*j}^k|, \quad j = 1, \ldots, n, \ k = 1, \ldots, K,$$
$$\eta_j := \overline{\left(\sum_{k=1}^{K} (cA_{*j}^k) p_k \right) x_j}, \quad j \neq i,$$
$$\psi_j := \underline{\left(\sum_{k=1}^{K} (cA_{*j}^k) p_k \right) x_j}, \quad j \neq i.$$

Then our objective function reads

$$\min \sum_{k=1}^{K} 2p_k^\Delta \beta_k + \sum_{j \neq i} (\eta_j - \psi_j). \tag{3}$$

Now, we set up the constraints. By the definition of β_k, we have

$$\beta_k \geq cb^k, \ \beta_k \geq -cb^k, \quad k = 1, \ldots, K. \tag{4}$$

Since β_k is minimized in the objective function, at least one of the inequalities will hold as equation, whence $\beta_k = |cb^k|$. Similarly for α_{jk} we obtain

$$\alpha_{jk} \geq cA_{*j}^k, \ \alpha_{jk} \geq -cA_{*j}^k, \quad j = 1, \ldots, n, \ k = 1, \ldots, K. \tag{5}$$

The condition that the denominator is has the form of $[1, r]$ is formulated as the equation

$$c \sum_{k=1}^{K} A_{*i}^k p_k^c - \sum_{k=1}^{K} p_k^\Delta \alpha_{ik} = 1. \tag{6}$$

Eventually, we reformulate conditions on η_j and ψ_j. Since $0 \in x_j$, the upper end-point of the interval product in the definition of η_j is attained either by the product of their upper end-points or their lower end-points. Thus, we get

$$\eta_j \geq c \sum_{k=1}^{K} A_{*j}^k p_k^c \underline{x}_j - \sum_{k=1}^{K} p_k^\Delta \underline{x}_j \alpha_{jk}, \quad j \neq i, \tag{7}$$

$$\eta_j \geq c \sum_{k=1}^{K} A_{*j}^k p_k^c \overline{x}_j + \sum_{k=1}^{K} p_k^\Delta \overline{x}_j \alpha_{jk}, \quad j \neq i. \tag{8}$$

Similarly for ψ_j,

$$\psi_j \le c \sum_{k=1}^{K} A_{*j}^{k} p_k^c \underline{x}_j + \sum_{k=1}^{K} p_k^{\Delta} \underline{x}_j \alpha_{jk}, \quad j \ne i, \tag{9}$$

$$\psi_j \le c \sum_{k=1}^{K} A_{*j}^{k} p_k^c \overline{x}_j - \sum_{k=1}^{K} p_k^{\Delta} \overline{x}_j \alpha_{jk}, \quad j \ne i. \tag{10}$$

Since η_j is maximized and ψ_j is minimized in the objective function, at least one of the inequalities is fulfilled as equation. Analogous considerations hold for α_{jk}. Therefore, we gathered all the constraints to formulate the optimization problem.

Optimization Problem. The optimal preconditioner of the first type is found by solving the optimization problem (3) under the constraints (4)–(10). This as a linear programming problem with $Kn + K + 3n - 2$ unknowns c, β_k, α_{jk}, η_j, and ψ_j, and $2Kn + 2K + 4n - 3$ constraints.

Notice that for standard interval linear Eq. (1), our approach would require approximately n^3 variables as there is a quadratic number of parameters. This is more than the linear programming formulation from Kearfott [6,7] using only a linear number of variables. His method, however, cannot be directly extended to parametric systems.

Overall, to determine the optimal preconditioner C, we have to solve n linear programs, which is a polynomial time problem. Moreover, C needn't be calculated in a verified way since any matrix can serve as a preconditioner.

On the other hand, solving n linear programs requires some computational effort, so it would be inefficient to compute an optimal C in each iteration of the interval Gauss–Seidel method. It seems more suitable to call the standard version using midpoint inverse preconditioner (or any other method), and after that to tighten the resulting enclosure by running several iterations with an optimal C.

2.2   Minimal Upper Bound

Herein, the criterion is to minimize $\overline{z}_i$. Suppose first that $\overline{z}_i > 0$. Using definitions of c, β_k, α_{jk}, and ψ_j from the previous section, the objective is formulated as

$$\min c \sum_{k=1}^{K} b^k p_k^c + \sum_{k=1}^{K} p_k^{\Delta} \beta_k - \sum_{j \ne i} \psi_j.$$

The constraints (4)–(6), (9)–(10) are employed in this problem, too. In addition, we have to take into account the remaining two possibilities for which ψ_j can be attain, and hence we involve also the inequalities

$$\psi_j \leq c \sum_{k=1}^{K} A_{*j}^k p_k^c x_j - \sum_{k=1}^{K} p_k^\Delta x_j \alpha_{jk}, \quad j \neq i,$$

$$\psi_j \leq c \sum_{k=1}^{K} A_{*j}^k p_k^c \overline{x}_j + \sum_{k=1}^{K} p_k^\Delta \overline{x}_j \alpha_{jk}, \quad j \neq i.$$

If $\overline{z}_i \leq 0$, then we just replace (6) by the equation

$$c \sum_{k=1}^{K} A_{*i}^k p_k^c + \sum_{k=1}^{K} p_k^\Delta \alpha_{ik} = 1, \tag{11}$$

which normalizes the denominator in (2) to have the form of $[r, 1]$ for some $r \in (0, 1]$. In this case, we have to include the condition $r \geq 0$, which draws

$$c \sum_{k=1}^{K} A_{*i}^k p_k^c - \sum_{k=1}^{K} p_k^\Delta \alpha_{ik} \geq 0.$$

The situation $r = 0$ makes practically no harm (even theoretically by realizing what will be the result if extended arithmetic is used).

The weak point is that we do not know a priori whether $\overline{z}_i > 0$ or not. We recommend to use the condition $\overline{x}_i > 0$ instead. It means, if $\overline{x}_i > 0$, then we use (6), otherwise we use (11). The only possible fail may occur when $\overline{z}_i \leq 0$ and $\overline{x}_i > 0$. In this case, the optimization problem does not find the optimal solution, however, the optimal value would be non-positive. Therefore, the upper bound is reduced substantially (with respect to sign change) from $\overline{x}_i > 0$ to a non-positive value.

The resulting linear program has less variables by $n - 1$ than the previous one from Sect. 2.1, and the number of constraints is the same. That is why the time complexities are almost the same.

3   Examples

The examples below show how optimal preconditoners behave for various initial enclosures, for various optimality criteria and for both versions (direct and residual) of the Gauss–Seidel method. For the residual form of the interval Gauss–Seidel iteration, we employed the minimal width approach (Sect. 2.1), and for the direct version, we used both the minimum upper and maximum lower bounds preconditioners.

The main purpose if the examples is to illustrate that while in some cases an optimal preconditioner makes no improvement, in another cases it may significantly reduce the overestimation. The computations were done in MATLAB with help of the interval toolbox INTLAB v7.1 (see Rump [16]).

Example 1. Consider Example 4 from Popova [12], where

$$A(p) := \begin{pmatrix} 1 & p_1 \\ p_1 & p_2 \end{pmatrix}, \quad b(p) := \begin{pmatrix} p_3 \\ p_3 \end{pmatrix}, \quad p \in \boldsymbol{p} = ([0, 1], -[1, 4], [0, 2])^T.$$

The initial enclosure of Σ_p is obtained by calling the `verifylss` function from Intlab on the relaxed system $A(p)x = b(p)$,

$$x = ([-4.4849, 6.6667], [-5.3334, 4.9697])^T.$$

First, we call the residual form of the interval parametric Gauss–Seidel method. For the center $x^* := x^c$ and the residual interval vector $y := x - x^c$, one iteration yields the same result

$$y^1 = ([-5.3940, 5.3940], [-4.1516, 4.1516])^T$$

for the respectively midpoint inverse and the minimal width preconditioners

$$(A^c)^{-1} = \begin{pmatrix} 0.9091 & 0.1818 \\ 0.1818 & -0.3636 \end{pmatrix}, \quad C = \begin{pmatrix} 1 & 0.2 \\ 0.5 & -1 \end{pmatrix}.$$

The corresponding contracted enclosure is

$$x^1 = ([-4.3031, 6.4850], [-4.3334, 3.9698])^T.$$

In contrast, the direct interval parametric Gauss–Seidel iteration with midpoint inverse preconditioning gives

$$x^2 = ([-4.2668, 6.6668], [-4.3334, 3.3334])^T,$$

which tightened about 13.77% of the interval width on average, whereas the optimal preconditioners yields

$$x^3 = ([-4.0308, 6.6283], [-3.8769, 2.6812])^T,$$

which tightened about 20.38% of the interval width on average. This enclosure was computed by calculating separately the upper and the lower end-points by using respectively the preconditioners

$$C^u = \begin{pmatrix} 1 & 0.2115 \\ 0.5978 & -1 \end{pmatrix}, \quad C^l = \begin{pmatrix} 1 & 0.1889 \\ 0.4022 & -1 \end{pmatrix}.$$

Comparing x^1 and x^3, we see that no one is better than the other one w.r.t. inclusion.

It is interesting to consider the interval hull of the relaxed system, $x^4 = ([0, 4], [-2, 2])^T$, as an initial enclosure, too. For the residual form method, the midpoint inverse preconditioner does not improve this enclosure, but the optimal preconditioner reduces it to

$$x^5 = ([0.0000, 3.8182], [-2.0001, 1.7686])^T.$$

For the direct version, the midpoint inverse preconditioner also fails to tighten x^4, whereas the optimal preconditioner reduces the second component by half to

$$x^6 = ([0, 4], [-2, 0])^T.$$

Example 2. In Example 5.2 from Popova and Krämer [15], a resistive network was considered with uncertain resistances. The output voltage was computed by solving the interval parametric system

$$
A(p) := \begin{pmatrix} 30 & -10 & -10 & -10 & 0 \\ -10 & 10+p_1+p_2 & -p_1 & 0 & 0 \\ -10 & -p_1 & 15+p_1+p_3 & -5 & 0 \\ -10 & 0 & -5 & 15+p_4 & 0 \\ 0 & 0 & -5 & 5 & 1 \end{pmatrix}, \quad b(p) := \begin{pmatrix} 1 \\ 0 \\ 0 \\ 0 \\ 0 \end{pmatrix},
$$

where $p \in \boldsymbol{p} = [8, 12] \times [4, 8] \times [8, 12] \times [8, 12]$.

We will consider the enclosure computed by the residual and the direct interval Gauss–Seidel method with the inverse midpoint preconditioner and initiated by the `verifylss` enclosure for the relaxed system.

The residual form yields the enclosure

$$
\boldsymbol{x}^1 = ([0.0595, 0.0851], [0.0262, 0.0587], [0.0247, 0.0514],
$$
$$
[0.0251, 0.0479], [-0.0352, 0.0499])^T,
$$

which is no further improved by the optimal preconditioner.

The direct form yields in $0.2194\,s$ the enclosure

$$
\boldsymbol{x}^2 = ([0.0575, 0.0871], [0.0268, 0.0660], [0.0247, 0.0557],
$$
$$
[0.0267, 0.0491], [-0.0527, 0.0674])^T.
$$

Using the optimal preconditioner, it takes $1.2535\,s$ to reduce the enclosure radii by 15 % on average, and the resulting enclosure is

$$
\boldsymbol{x}^3 = ([0.0626, 0.0862], [0.0293, 0.0646], [0.0273, 0.0541],
$$
$$
[0.0276, 0.0482], [-0.0359, 0.0573])^T.
$$

For comparison, `verifylss` enclosure for the system preconditioned by the inverse midpoint reads

$$
\boldsymbol{x}^4 = ([0.0576, 0.0871], [0.0187, 0.0662], [0.0202, 0.0558],
$$
$$
[0.0240, 0.0491], [-0.0525, 0.0672])^T.
$$

Hence, our enclosure $\boldsymbol{x}^3$ has by about 22 % (on average) smaller radii than $\boldsymbol{x}^4$.

4   Conclusion

We proposed a linear programming based method to compute an optimal preconditioning matrix for the parametric interval Gauss–Seidel iterations. Even though large numerical studies would be needed, some illustrative examples show that the optimal preconditioner can sometimes reduce the ubiquitous overestimation. Besides that, future research may be addressed to other types of optimality

(S-preconditioners, pivoting preconditioners, and others), or to directly focus on the interval Newton method (as done in Kearfott [5,6,8]). It would be also interesting to investigate optimality of various preconditioners in generalized interval systems, for instance for AE solutions of (non)-parametric interval systems [14].

Acknowledgments. The author was supported by the Czech Science Foundation Grant P402-13-10660S.

References

1. Fiedler, M., Nedoma, J., Ramík, J., Rohn, J., Zimmermann, K.: Linear Optimization Problems with Inexact Data. Springer, New York (2006)
2. Gau, C.-Y., Stadtherr, M.A.: New interval methodologies for reliable chemical process modeling. Comput. Chem. Eng. **26**(6), 827–840 (2002)
3. Hladík, M.: Enclosures for the solution set of parametric interval linear systems. Int. J. Appl. Math. Comput. Sci. **22**(3), 561–574 (2012)
4. Hladík, M.: New operator and method for solving real preconditioned interval linear equations. SIAM J. Numer. Anal. **52**(1), 194–206 (2014)
5. Kearfott, R.B.: Preconditioners for the interval Gauss-Seidel method. SIAM J. Numer. Anal. **27**(3), 804–822 (1990)
6. Kearfott, R.B.: Decomposition of arithmetic expressions to improve the behavior of interval iteration for nonlinear systems. Comput. **47**(2), 169–191 (1991)
7. Kearfott, R.B.: A comparison of some methods for bounding connected and disconnected solution sets of interval linear systems. Comput. **82**(1), 77–102 (2008)
8. Kearfott, R.B., Hu, C., Novoa, M.: A review of preconditioners for the interval Gauss-Seidel method. Interval Comput. **1991**(1), 59–85 (1991)
9. Lin, Y., Stadtherr, M.A.: Advances in interval methods for deterministic global optimization in chemical engineering. J. Glob. Optim. **29**(3), 281–296 (2004)
10. Neumaier, A.: New techniques for the analysis of linear interval equations. Linear Algebra Appl. **58**, 273–325 (1984)
11. Neumaier, A.: Interval Methods for Systems of Equations. Cambridge University Press, Cambridge (1990)
12. Popova, E.: Quality of the solution sets of parameter-dependent interval linear systems. ZAMM, Z. Angew. Math. Mech. **82**(10), 723–727 (2002)
13. Popova, E.D.: On the solution of parametrised linear systems. In: Krämer, W., von Gudenberg, J.W. (eds.) Scientific Computing, Validated Numerics, Interval Methods, pp. 127–138. Kluwer (2001)
14. Popova, E.D., Hladík, M.: Outer enclosures to the parametric AE solution set. Soft Comput. **17**(8), 1403–1414 (2013)
15. Popova, E.D., Krämer, W.: Visualizing parametric solution sets. BIT **48**(1), 95–115 (2008)
16. Rump, S.M.: INTLAB - INTerval LABoratory. In: Csendes, T. (ed.) Developments in Reliable Computing, pp. 77–104. Kluwer Academic Publishers, Dordrecht (1999)
17. Rump, S.M.: Verification methods: Rigorous results using floating-point arithmetic. Acta Numer. **19**, 287–449 (2010)

Reproducible and Accurate Matrix Multiplication

Roman Iakymchuk[1,2,3]([✉]), David Defour[4], Caroline Collange[5],
and Stef Graillat[1,2]

[1] Sorbonne Universités UPMC Univ Paris 06, UMR 7606, LIP6, 75005 Paris, France
{roman.iakymchuk,stef.graillat}@lip6.fr
[2] CNRS, UMR 7606, LIP6, 75005 Paris, France
[3] Sorbonne Universités UPMC Univ Paris 06, ICS, 75005 Paris, France
[4] DALI–LIRMM, Université de Perpignan,
52 Avenue Paul Alduy, 66860 Perpignan, France
david.defour@univ-perp.fr
[5] INRIA – Centre de Recherche Rennes – Bretagne Atlantique,
Campus de Beaulieu, 35042 Rennes Cedex, France
caroline.collange@inria.fr

Abstract. Due to non-associativity of floating-point operations and
dynamic scheduling on parallel architectures, getting a bit-wise reproducible floating-point result for multiple executions of the same code
on different or even similar parallel architectures is challenging. In this
paper, we address the problem of reproducibility in the context of matrix
multiplication and propose an algorithm that yields both reproducible
and accurate results. This algorithm is composed of two main stages: a
filtering stage that uses fast vectorized floating-point expansions in conjunction with error-free transformations; an accumulation stage based
on Kulisch long accumulators in a high-radix carry-save representation.
Finally, we provide implementations and performance results in parallel
environments like GPUs.

Keywords: Matrix multiplication · Reproducibility · Accuracy ·
Kulisch long accumulator · Error-free transformation · Floating-point
expansion · Rounding-to-nearest · GPUs

1 Introduction

In many fields of science and engineering, the process of finding the solution for a
specific problem requires solving a system of linear equations, or a least squares
problem, or eigenvalue problem. A common approach is to develop solvers for
each specific task and then spend a tremendous amount of time on tuning them.
However, best practice suggests to use already optimized solver-routines contained in linear algebra libraries.

The development of linear algebra libraries began in the early 1970s. Since
that time many libraries have been released. With the influence of common HPC

© Springer International Publishing Switzerland 2016
M. Nehmeier et al. (Eds.): SCAN 2014, LNCS 9553, pp. 126–137, 2016.
DOI: 10.1007/978-3-319-31769-4_11

computers, which were based on vector processors, in 1979 a first set of Basic Linear Algebra Subprograms (BLAS-1) was designed as a set of *basic vector operations*. In 1988 the idea of BLAS was developed further, yielding a second set of routines for *matrix-vector operations* (BLAS-2). For those routines the amount of data required and floating-point operations (Flops) performed have quadratic complexity.

When architectures with multiple layers of cache memory appeared, the performance of both BLAS-1 and BLAS-2 operations became an issue: for these routines the ratio between the numbers of Flops and memory accesses is only $O(1)$. In order to attain high performance on architectures with a hierarchical memory system, in 1990 the third level of BLAS (BLAS-3) with *matrix-matrix operations* was defined. These routines perform $O(n^3)$ Flops over $O(n^2)$ data, giving the opportunity to hide memory latency and offer performance close to the achievable peak.

A generic implementation of the BLAS specification is provided since the announcement of the library in 1979. This reference implementation is equipped with the complete functionality, but it is not optimized for any architecture. Thus, processor manufacturers as well as scientists developed tuned implementations of the BLAS for each architecture. Prominent examples of these implementations are Intel MKL, AMD ACML, IBM ESSL, ATLAS, and GotoBLAS (now OpenBLAS). ATLAS [1] is based on an auto-tuned empirical approach while GotoBLAS [2] is a hand-tuned machine-specific implementation of the BLAS. Due to the raising popularity of GPUs for high-performance computing, NVIDIA provided a GPU-version of the BLAS (cuBLAS).

The core of the BLAS library is xGEMM[1], which is a BLAS-3 routine, that computes the matrix-matrix products as

$$C := \alpha op(A)op(B) + \beta C, \tag{1}$$

where α and β are scalars; $op(A), op(B)$, and C are general matrices with $op(A)$ a $m \times k$ matrix, $op(B)$ a $k \times n$ matrix, and C a $m \times n$ matrix; $op(X)$ represents either a non-transposed X or a transposed X^T matrix. xGEMM performs $2mnk$ floating-point operations over $mk + kn + mn$ data. All the other BLAS-3 routines can be expressed in terms of xGEMM. Moreover, when different implementations of BLAS are compared, the first criteria used for this comparison is the performance of xGEMM.

The profitable ratio between the computation and the memory references of the BLAS-3 routines has a strong impact on the design and automatic generation of linear algebra algorithms. For instance, in order to exploit the optimized BLAS implementations, the Linear Algebra PACKage (LAPACK) builds its blocked algorithms on top of the BLAS-3 operations. Furthermore, scientists either try to generate algorithms relying more on the BLAS-3 routines, in particular xGEMM, or try to rewrite their algorithms in order to benefit from the performance provided by the BLAS-3 routines [3].

---

[1] In general, x stands for four different formats, but in the scope of this article we consider x to correspond to single (S) or double (D) precision.

In general, matrix-matrix products relies on optimized version of parallel reduction and dot-product involving floating-point additions and multiplications which are non-associative operations. Hence, as the order of operations may vary from one parallel machine to another or even from one run to another [4], reproducibility of results is not guaranteed. These discrepancies worsen on heterogeneous architectures – such as clusters composed of standard CPUs in conjunction with GPUs and/or accelerators like Intel Xeon Phi – which combine together different programming environments that may obey various floating-point models and offer different intermediate precision or different operators [5,6]. In some cases, such non-reproducibility of floating-point computations on parallel machines causes validation and debugging issues, and may even lead to deadlocks [7].

By reproducibility, we mean getting a bit-wise identical floating-point result from multiple runs of the same code on the same data. Numerical reproducibility can be addressed by targeting either the order of operations or the error resulting from finite arithmetic. One solution consists in providing the deterministic control over rounding errors by, for example, enforcing the execution order for each operation. However, these approach is not portable and/or does not scale well with the number of processing cores. The other solution aims at avoiding cancellation and rounding errors by using, for instance, a long accumulator such as the one proposed by Kulisch [8]. This solution increases the accuracy at the price of more operations and memory transfers per output data. Because of that, for a long time, it was considered too expensive for the little benefit it was providing.

To enhance reproducibility, Intel proposed a "Conditional Numerical Reproducibility" (CNR) in its Math Kernel Library (MKL). Although, CNR guarantees reproducibility, it does not ensure correct rounding, meaning the accuracy is arguing. Additionally, the cost of obtaining reproducible results with CNR is high. For instance, for large arrays the MKL's summation with CNR is $85-93\%$ slower than both the regular MKL's and our reproducible summation; the later two deliver comparable performance. The performance gap between the MKL's reproducible matrix multiplication and its classic implementation is even higher and is roughly $3-4$ times.

Demmel and Nguyen introduced a family of algorithms for reproducible summation in floating-point arithmetic [9]. They have extended this concept to reproducible BLAS routines (covering, mainly, the BLAS-1 routines) that are distributed in the ReproBLAS library[2].

Recently, we introduced in [10] an approach to compute deterministic sums of floating-point numbers. Our approach is based on a multi-level algorithm that combines efficiently floating-point expansions and long accumulators. The proposed implementations on recent Intel desktop and server processors, on Intel Xeon Phi co-processors, and on both AMD and NVIDIA GPUs, showed that the numerical reproducibility and bit-perfect accuracy can be achieved at no additional cost for large sums that have dynamic ranges of up to 90 orders of

---

[2] http://bebop.cs.berkeley.edu/reproblas/.

magnitude. This speed-up is possible thanks to arithmetic units that are left underused by the standard reduction algorithms.

In this article, we propose an approach to ensure both reproducibility and accuracy (rounding-to-nearest) of the product of two matrices composed of floating-point numbers. The derived algorithm is based on the standard non-deterministic xGEMM and our deterministic summation algorithm. Moreover, we provide implementations of this algorithm on GPU accelerators. To our knowledge, this is the first work on reproducible matrix-matrix multiplication.

The paper is organized as follows. Section 2 reviews related aspects of floating-point arithmetic and highlights floating-point expansions and long accumulators. Section 3 presents our approach to derive exact, meaning both reproducible and accurate, matrix-matrix product. In Sect. 4, we expose implementations and performance results on GPU accelerators. Finally, we draw conclusions in Sect. 5.

2   Background

Without loss of generality, in the rest of this article, we will consider double precision format (`binary64`) from the IEEE-754 standard [11]. Floating-point representation of numbers allows to cover a wide *dynamic range*. Dynamic range refers to the absolute ratio between the number with the largest magnitude and the number with the smallest non-zero magnitude in a set. For instance, `binary64` can represent positive numbers from 4.9×10^{-324} to 1.8×10^{308}, so it covers a dynamic range of 3.7×10^{631}.

Non-associativity of floating-point addition implies that the result depends on the order of the operations. For example, in double precision $(-1 \oplus 1) \oplus 2^{-100}$ is different from $-1 \oplus (1 \oplus 2^{-100})$ where $\oplus$ denotes the result of a floating-point addition. Thus, the accuracy of a floating-point summation depends on the order of evaluation. More details about this phenomenon can be found in the main references [12,13].

Two approaches exist to execute one floating-point addition without introducing rounding error. The first solution aims at computing the error which occurred during rounding using floating-point expansions in conjunction with error-free transformations, see Sect. 2.1. The second solution exploits the finite range of representable floating-point numbers by storing every bit in a very long vector of bits, see Sect. 2.2.

2.1   Floating-Point Expansion

Floating-point expansions represent the result as an unevaluated sum of floating-point numbers, whose components are ordered in magnitude with minimal overlap to cover a wide range of exponents. Floating-point expansions of sizes 2 and 4 are described in [14,15], accordingly. They are based on error-free transformation. Indeed, when working with rounding-to-nearest, the rounding error in addition or multiplication can be represented as a floating-point number and can also be computed in floating-point arithmetic. The traditional error-free transformation for addition is TwoSum [16], see Algorithm 1, and for multiplication is TwoProduct, see Algorithm 2. For TwoSum, it means that $r + s = a + b$ with $r = a \oplus b$ and s, which is a floating-point number that corresponds to the rounding error. For TwoProduct, we use the fused multiply and add (FMA) instruction that is widely available on modern architectures. $\mathrm{FMA}(a, b, c)$ makes it possible to compute $a \times b + c$ with only one rounding. Thus, we have $r + s = a \times b$ with $r = a \otimes b$ and $s = \mathrm{FMA}(a, b, -r)$, where $\otimes$ stands for floating-point multiplication.

Algorithm 1. Error-free transformation for sum of two floating-point numbers.

Function $[r, s] = \mathrm{TwoSum}(a, b)$
  $r \leftarrow a + b$
  $z \leftarrow r - a$
  $s \leftarrow (a - (r - z)) + (b - z)$

Algorithm 2. Error-free transformation for product of two floating-point numbers.

Function $[r, s] = \mathrm{TwoProduct}(a, b)$
  $r \leftarrow a \times b$
  $s \leftarrow \mathrm{FMA}(a, b, -r)$

Adding one floating-point number to an expansion is an iterative operation. The floating-point number is first added to the head of the expansion and the rounding error is recovered as a floating-point number using an error-free transformation such as TwoSum. The error is then recursively accumulated to the remainder of the expansion.

With expansions of size p – that correspond to the unevaluated sum of p floating-point numbers – it is possible to accumulate floating-point numbers without losing accuracy as long as every intermediate result can be represented exactly as a sum of p floating-point numbers. This situation occurs when the dynamic range of the sum is lower than $2^{53 \cdot p}$ in case of binary64.

The main advantage of expansions is that they can be placed in registers during the whole computation. However, the accuracy is insufficient for the summation of numerous floating-point numbers or sums with large dynamic ranges. Moreover, the complexity of this algorithm grows linearly with the size of expansion.

2.2 Long Accumulator

An alternative algorithm to floating-point expansions uses very long fixed-point accumulators. This accumulator can be viewed as a projection of the set of floating-point numbers from minimum (emin) to maximum (emax) exponents into a long register, where each spot covers numbers with a certain exponent range. The length of the accumulator is selected in such a way that it represents

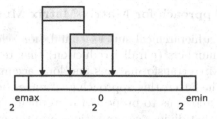

Fig. 1. Kulisch long accumulator.

every bit of information of the input format, e.g. `binary64`; this covers the range from the smallest representable floating-point value to the largest one, independently of the sign. For instance, Kulisch [8] proposed to use an accumulator of 4288 bits to handle the dot product of two vectors composed of `binary64` values. The summation is performed without loss of information by accumulating every floating-point input numbers in the long accumulator, see Fig. 1. The long accumulator is the perfect solution to produce the exact result of a very large amount of floating-point numbers of arbitrary magnitude. However, for a long period this approach was considered impractical as it induces a very large memory overhead. Furthermore, without dedicated hardware support, its performance is limited by indirect memory accesses that makes vectorization challenging.

3   Exact Matrix-Matrix Multiplication

In order to achieve best performance for linear algebra kernels, machine-specific hand tuning of those kernels is often applied; a good example is the Goto's implementation of xGEMM. Scientists aim at optimizing this process for existing and upcoming architectures through the automatic generation of linear algebra kernels. As the matrix-matrix multiplication is the core of the BLAS library, in several works [1,17] the problem of optimizing this routine for a given architecture was tackled by applying the automatic generation approach. For instance, the ATLAS project [1] provides a very good implementation of BLAS by tuning routines for various architectures; those are centralized around a highly tuned matrix-matrix product that is automatically optimized for different levels of memory hierarchy. The idea of auto-tuning was extended to GPUs architectures applying different programming models such as CUDA and OpenCL. Apart from both code generation and heuristic search in conjunction with OpenCL, Matsumoto et. al. [17] proposed to store data in memory not only in a standard row-/column-major order, but also in a block-major order. We revise these ideas and employ some of them in our implementations of exact xGEMM, which is described in Sect. 4.1. Therefore, we combine together auto-tuning for standard non-deterministic xGEMM and machine-specific hand tuning for our reproducible approach.

3.1   Hierarchical Approach for Matrix-Matrix Multiplication

We introduced in [10] a hierarchical superaccumulation scheme for the summation of floating-point numbers (parallel reduction) that relies on floating-point expansions with error-free transformations and long accumulators as described in Sect. 2. Thanks to the latter, this approach guarantees both reproducible and accurate results. This allows us to propose a reproducible and accurate matrix multiplication scheme that divides computations into three stages: filtering, private superaccumulation, and rounding. This decomposition is suitable for the nested parallelism of modern architectures and it makes a full use of SIMD and multi-threads.

In the first stage, each partial product is computed using error-free transformation. In order to ensure accuracy, this steps generates two floating-point numbers, see Algorithm 2: the result and the rounding error. Both resulting floating-point numbers are accumulated using Algorithm 1 into an expansion of size p $(p \geq 3)$ that is stored in registers or private memory for each threads. This step benefits from vectorization and pipelining by maintaining one expansion per GPU thread.

In case the accuracy provided by floating-point expansions for product and/or summation is not enough, meaning a non-zero residue x still remains after filtering, each residue x is added to a long accumulator. We also propose an optimized version of floating-point expansions of size p that relies on the stopping criteria $(x \equiv 0)$ in the accumulation loop. This technique is called *early-exit* and exhibits performance which depends on the distribution of input numbers and the ability of the architecture to handle irregular branches.

A trade-off between speed and usage of the hardware resources lies in the proper choice of the size p of the floating-point expansion. A small value of p will lead to numerous transfers from the expansion towards the long accumulators, which will slow down the computation. A large value of p will lead to the overuse of registers and eventually to the register spilling.

Once all the input number are accumulated, each floating-point expansion is flushed to a long accumulator, independently of the parameter p. Hence, the second stage is based on superaccumulation, meaning summation to long accumulators, and it is involved either when the accuracy provided by expansions is not enough or at the end of the computation. Depending on the amount of memory available, long accumulators are stored in either fast local memory, e.g. cache or shared memory, or global memory.

In the third stage, rounding of private long accumulators back to the desired floating-point format is performed in order to obtain reproducible and correctly rounded results.

4   Implementations and Experimental Results

This section presents our implementations of the multi-level reproducible matrix multiplication and their evaluation on both NVIDIA and AMD GPUs, see Table 1 for the detailed description of these GPU architectures. We compared

Table 1. Hardware platforms used for the experiments.

| A | NVIDIA Tesla K20c | 13 SMs × 192 CUDA cores | 0.705 GHz |
| B | AMD Radeon HD 7970 | 32 CUs × 64 units | 0.925 GHz |

the accuracy of our implementations with results produced by the multiple precision library MPFR on CPUs; the MPFR library is not multi-threaded and does not support GPUs. In case of `binary64`, we used 4196 bits ($2 \times$ (emin + emax + mantissa) = 2 × (1022 + 1023 + 53)) within MPFR in order to guarantee the bit-wise reproducibility as well as the accuracy of the results independently of rounding errors and dynamic ranges.

4.1 Implementations

We follow the strategy proposed by Matsumoto et al. [17] regarding their matrix partitioning technique in order to exploit multi-level memory hierarchies on GPU architectures, see Fig. 2. An adequate matrix partitioning improves significantly the reuse of data and keeps the computational units busy while performing memory transfers.

Our solution is different from Matsumoto's one, as we divide memory space among matrices, floating-point expansions, and long accumulators. The latter may require 76 (76 × 64 bits is the size of each long accumulator) times more storage, because the matrix C is entirely composed of long accumulators in the non-optimized case, meaning when long accumulators are not reused. Thus, we use two levels of blocking in our matrix multiplication algorithms to amortize the cost of data accesses to the three levels of memory on GPUs, namely private (registers), local or data caches, and global. The first level of blocking focuses on enhancing the access latency between the global and local memories for each group of threads (or warp or work-group on GPUs). We assume that $m_l, n_l,$ and k_l are three block sizes multiple of m, n, and k, respectively. Figure 2a represents the partitioning of the matrices $C, A,$ and B into blocks of sizes $m_l \times n_l, m_l \times k_l,$ and $k_l \times n_l$, accordingly. Each $m_l \times n_l$ block of C is computed by a work-group that involves an $m_l \times k$ panel of A and a $k \times n_l$ panel of B.

This panel-panel multiplication iterates k/k_l times in the outermost loop of our xGEMM algorithm using the block-block multiplication. Thus, on each iteration the work-group updates each resulting $m_l \times n_l$ block of C with the product of an $m_l \times k_l$ block of A by a $k_l \times n_l$ block of B. This second level of blocking optimizes the use of private memory for each thread (work-item on GPUs). Figure 2b shows further partitioning of matrices within their blocks in such a way that each work-item in the work-group is responsible for updating an $m_s \times n_s$ sub-block of C through the multiplication of an $m_s \times k_l$ sub-panel of A by a $k_l \times n_s$ sub-panel of B.

In order to ensure both reproducibility and accuracy of xGEMM, we use one floating-point expansion with error-free transformation per thread.

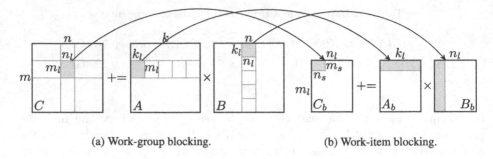

(a) Work-group blocking.   (b) Work-item blocking.

Fig. 2. Partitioning of matrix-matrix multiplication.

When the accuracy provided by these expansions is not enough, we switch to long accumulators that are allocated for each thread of a given work-group. However, this induces pressure on the memory hierarchy due to the required storage. So, we reuse both floating-point expansions and long accumulators for computing multiple elements of the resulting matrix.

Our implementations attempt to get the maximum performance by using all resources of the considered GPU architectures: SIMD instructions, fused multiply-add, private and local memory as well as atomic instructions. We developed both unique and hand-tuned OpenCL implementations for NVIDIA and AMD GPUs.

We use a long accumulator of finite length that represents the whole range of double precision floating-point numbers (4196 bits in case of `binary64`). We use such a long accumulator to avoid partial over/underflow that may occurs while accumulating partial product of the same sign. For instance, for matrices of size $n \times n$, only n partial-products need to be summed per resulting element, which leads to only $log_2(n)$ carry bits. With matrix size of $2^{20} \times 2^{20}$ that requires 8 Terabytes, only 20 extra bits are necessary to ensure that this phenomena will not occur.

4.2 Performance Results

As a baseline we consider the vectorized and parallelized non-deterministic double precision matrix multiplication. We prefer our tuned implementation to the one from cuBLAS and base our ExGEMM on it, because cuGEMM squeezes every percent of the architecture performance and does not leave a room for our approach. Figures 3a and b present the measured time achieved by the matrix multiplication algorithms as a function of the matrix size n on two GPUs, see Table 1. Apart from "Parallel DGEMM", all implementations are ours: "Superacc" corresponds to our matrix multiplication algorithm that is solely based on long accumulators and it is the slowest due to its extensive memory usage; "FPEp + Superacc" stands for algorithm with floating-point expansions of size p ($p = 3 : 8$) in conjunction with error-free transformations and long accumulators; "FPE4EE + Superacc" represents an optimized version of the expansion

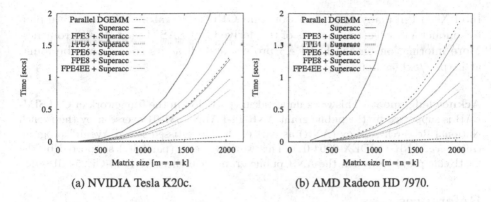

(a) NVIDIA Tesla K20c.

(b) AMD Radeon HD 7970.

Fig. 3. The matrix-matrix multiplication performance results on GPUs, see Table 1.

of size 4 with the early-exit technique. The implementations with expansions obtain better performance than with long accumulators only. However, due to switching to the long accumulator at the final stage of computing each element of the resulting matrix C as well as when the accuracy of expansions is not enough, the performance of implementations with expansions is bounded and it is at most 12 and 16 times off the DGEMM's performance on NVIDIA and AMD GPUs, respectively. We believe that there is a possibility to tune these preliminary implementations in order to be within 10 times slower. Nevertheless, our matrix multiplication algorithm delivers constantly reproducible and accurate results.

5   Conclusions and Future Work

xGEMM is the core of the BLAS library and all the other BLAS-3 routines are virtually built on top of it. Furthermore, the development and automatic generation of linear algebra algorithms are driven by the goal of achieving best performance on various architectures. One step towards this goal is made by using blocked versions of algorithms that are capable to obtain much higher performance compared to non-blocked algorithmic variants. This is achieved thanks to the usage of BLAS-3 routines, in particular xGEMM. Understanding such importance of the matrix multiplication routine, we targeted xGEMM and for the first time delivered a multi-level reproducible approach along with implementations. Even though the performance, which corresponds to roughly 5 % of the efficiency, can be argued (we think that a 10 times overhead at most for reproducible compute-bound algorithms is reasonable), the output of ExGEMM is consistently bit-wise reproducible and accurate, in terms of rounding-to-nearest, independently of threads scheduling, instruction set, and data partitioning.

Our ultimate goal is to apply the multi-level approach to derive a reproducible, accurate, and fast library for fundamental linear algebra operations – like those included in the BLAS library – on new parallel architectures such as

Intel Xeon Phi many-core processors and GPU accelerators. Moreover, we plan to conduct a priori error analysis of the derived ExBLAS (Exact BLAS) routines. More information on the ExBLAS project as well as its sources can be found at https://exblas.lip6.fr.

Acknowledgement. This work undertaken (partially) in the framework of CALSIM-LAB is supported by the public grant ANR-11-LABX-0037-01 overseen by the French National Research Agency (ANR) as part of the "Investissements d'Avenir" program (reference: ANR-11-IDEX-0004-02). This work was also (partially) supported by the FastRelax project through the ANR public grant (reference: ANR-14-CE25-0018-01).

References

1. Whaley, R.C., Dongarra, J.J.: Automatically tuned linear algebra software. In: Proceedings of the 1998 ACM/IEEE Conference on Supercomputing (CDROM). Supercomputing 1998, 1–27. IEEE Computer Society (1998)
2. Goto, K., van de Geijn, R.A.: High-performance implementation of the level-3 BLAS. ACM Trans. Math. Softw. **35**(1), 1–14 (2008)
3. Fabregat-Traver, D., Bientinesi, P.: Computing petaflops over terabytes of data: the case of genome-wide association studies. ACM Trans. Math. Softw. **40**(4), 27:1–27:22 (2014)
4. Bergman, K., al.: Exascale computing study: technology challenges in achieving exascale systems. DARPA report, September 2008
5. Whitehead, N., Fit-Florea, A.: Precision & performance: Floating point and IEEE 754 compliance for NVIDIA GPUs. Technical report, NVIDIA (2011)
6. Corden, M.: Differences in floating-point arithmetic between Intel Xeon processors and the Intel Xeon Phi^{TM} coprocessor. Technical report, Intel (2013)
7. Doertel, K.: Best known method: Avoid heterogeneous precision in control flow calculations. Technical report, Intel (2013)
8. Kulisch, U., Snyder, V.: The exact dot product as basic tool for long interval arithmetic. Computing **91**(3), 307–313 (2011)
9. Demmel, J., Nguyen, H.D.: Fast reproducible floating-point summation. In: Proceedings of the 21st IEEE Symposium on Computer Arithmetic, Austin, Texas, USA, pp. 163–172 (2013)
10. Collange, C., Defour, D., Graillat, S., Iakymchuk, R.: Full-Speed Deterministic Bit-Accurate Parallel Floating-Point Summation on Multi- and Many-Core Architectures. Technical report HAL: hal-00949355, INRIA, DALI-LIRMM, LIP6, ICS, February 2014
11. IEEE Computer Society: IEEE Standard for Floating-Point Arithmetic. IEEE Standard 754-2008, August 2008
12. Higham, N.J.: Accuracy and stability of numerical algorithms, 2nd edn. Society for Industrial and Applied Mathematics (SIAM), Philadelphia, PA (2002)
13. Muller, J.M., Brisebarre, N., de Dinechin, F., Jeannerod, C.P., Lefèvre, V., Melquiond, G., Revol, N., Stehlé, D., Torres, S.: Handbook of Floating-Point Arithmetic. Birkhäuser, Boston (2010)
14. Li, X.S., Demmel, J.W., Bailey, D.H., Henry, G., Hida, Y., Iskandar, J., Kahan, W., Kang, S.Y., Kapur, A., Martin, M.C., Thompson, B.J., Tung, T., Yoo, D.J.: Design, implementation and testing of extended and mixed precision BLAS. ACM Trans. Math. Softw. **28**(2), 152–205 (2002)

15. Hida, Y., Li, X.S., Bailey, D.H.: Algorithms for quad-double precision floating point arithmetic. In: Proceedings of the 15th IEEE Symposium on Computer Arithmetic, CA, USA, 155–162. IEEE Computer Society Press, Los Alamitos (2001)
16. Knuth, D.E.: The Art of Computer Programming. Seminumerical Algorithms, vol. 2, 3rd edn. Addison-Wesley, Boston (1997)
17. Matsumoto, K., Nakasato, N., Sakai, T., Yahagi, H., Sedukhin, S.G.: Multi-level optimization of matrix multiplication for gpu-equipped systems. In: ICCS. Procedia Computer Science, vol. 4, pp. 342–351. Elsevier (2011)

Outer Bounds for the Parametric Controllable Solution Set with Linear Shape

Evgenija D. Popova[✉]

Institute of Mathematics and Informatics, Bulgarian Academy of Sciences,
Acad. G. Bonchev Street, Block 8, 1113 Sofia, Bulgaria
epopova@bio.bas.bg

Abstract. We consider linear algebraic equations, where the elements of the matrix and of the right-hand side vector are linear functions of interval parameters, and their parametric AE-solution sets, which are defined by universal and existential quantifiers for the parameters. We present how some sufficient conditions for a parametric AE-solution set to have linear boundary can be exploited for obtaining sharp outer bounds of that parametric AE-solution set. For a parametric controllable solution set having linear boundary we present a numerical method for outer interval enclosure of the solution set. The new method has better properties than some other methods available so far.

Keywords: Interval linear systems · Parameter dependencies · AE-solution set · Controllable solution set · Solution enclosure · Iteration method

1 Introduction

Consider linear algebraic systems involving linear dependencies between a number of interval parameters $p = (p_1, \ldots, p_K)^\top \in \mathbf{p} = (\mathbf{p}_1, \ldots, \mathbf{p}_K)^\top$

$$A(p)x = b(p)$$

$$A(p) := A_0 + \sum_{k=1}^{K} p_k A_k, \qquad b(p) := b_0 + \sum_{k=1}^{K} p_k b_k, \tag{1}$$

where $A_k \in \mathbb{R}^{m \times n}$, $b_k \in \mathbb{R}^m$, $k = 0, \ldots, K$, and $\mathbb{R}^{m \times n}$ is the set of real $m \times n$ matrices, $\mathbb{R}^m := \mathbb{R}^{m \times 1}$ denotes the set of real vectors with m components. A real compact interval is $\mathbf{a} = [a_1, a_2] := \{a \in \mathbb{R} \mid a_1 \le a \le a_2; a_1, a_2 \in \mathbb{R}\}$. By $\mathbb{IR}^m$, $\mathbb{IR}^{m \times n}$ we denote the sets of interval m-vectors and interval $m \times n$ matrices, respectively.

We consider the parametric AE-solution sets of the system (1)

$$\Sigma_{AE}^p := \left\{ x \in \mathbb{R}^n \mid (\forall p_\mathcal{A} \in \mathbf{p}_\mathcal{A})(\exists p_\mathcal{E} \in \mathbf{p}_\mathcal{E})(A(p)x = b(p)) \right\}, \tag{2}$$

This work is dedicated to the memory of Prof. Dr. Walter Krämer (1952-2014), Univ. of Wuppertal, Germany.

M. Nehmeier et al. (Eds.): SCAN 2014, LNCS 9553, pp. 138–147, 2016.
DOI: 10.1007/978-3-319-31769-4_12

for the sets of indexes $\mathcal{A}, \mathcal{E}$ such that $\mathcal{A} \cup \mathcal{E} = \{1, \ldots, K\}$, $\mathcal{A} \cap \mathcal{E} = \emptyset$. For a given index set $\Pi = \{\pi_1, \ldots, \pi_k\}$, p_Π denotes $(p_{\pi_1}, \ldots, p_{\pi_k})$. Among the AE-solution sets most studied and of particular practical interest are: the (parametric) united solution set

$$\Sigma_{\mathrm{uni}}(A(p), b(p), \mathbf{p}) := \{x \in \mathbb{R}^n \mid (\exists p \in \mathbf{p})(A(p)x = b(p))\},$$

the (parametric) tolerable solution set

$$\Sigma(A(p_\mathcal{A}), b(p_\mathcal{E}), \mathbf{p}) := \{x \in \mathbb{R}^n \mid (\forall p_\mathcal{A} \in \mathbf{p}_\mathcal{A})(\exists p_\mathcal{E} \in \mathbf{p}_\mathcal{E})(A(p_\mathcal{A})x = b(p_\mathcal{E}))\}$$

and the (parametric) controllable solution set

$$\Sigma(A(p_\mathcal{E}), b(p_\mathcal{A}), \mathbf{p}) := \{x \in \mathbb{R}^n \mid (\forall p_\mathcal{A} \in \mathbf{p}_\mathcal{A})(\exists p_\mathcal{E} \in \mathbf{p}_\mathcal{E})(A(p_\mathcal{E})x = b(p_\mathcal{A}))\}.$$

A parametric solution set is usually smaller [4, Theorem 5.6] than the solution set of the corresponding nonparametric interval linear system. Therefore, the former describes more precisely a physical phenomenon, whose model is a linear algebraic system involving dependencies between interval model parameters; for practical examples see [3,5] and the references given therein.

A nonempty parametric AE-solution set, in general, has a complicated structure, see [4]. Its boundary is defined by parts of polynomials that may have arbitrary high degree. It is proven in [4] that the universally quantified parameters contribute linearly to the boundary of a parametric AE-solution set and therefore only the existentially quantified parameters determine the shape (boundary) of a nonempty parametric AE-solution set. On the other hand, the existentially quantified parameters can be classified in two groups: parameters which contribute linearly to the boundary of a solution set and parameters which determine the nonlinear boundary of a parametric AE-solution set. Recently, in [5], some sufficient conditions for an existentially quantified parameter to contribute linearly to the boundary of a parametric united solution set were proven. Based on these conditions, the scope of applicability of an efficient interval method [3] finding outer bounds for the parametric united solution set with linear shape was greatly expanded.

The goal of the present work is two-fold:

(a) to present the applicability of the above sufficient conditions for obtaining sharp outer bounds of a parametric AE-solution set;
(b) to generalize the interval method, proposed in [3], for parametric controllable solution sets.

The paper is organized as follows. Section 2 introduces some notions that will be used. Section 3 discusses the parametric AE-solution sets with linear shape and the goal (a). A new interval method for outer enclosure of the parametric controllable solution set with linear shape is presented in Sect. 4 together with some illustrative examples. The paper ends by some conclusions.

2   Theoretical Background

For $\mathbf{a} = [a_1, a_2]$, define mid-point $\breve{a} := (a_1 + a_2)/2$, radius $\hat{a} := (a_2 - a_1)/2$, width (diameter) $\omega(\mathbf{a}) := 2\hat{a}$, magnitude (absolute value) $|\mathbf{a}|$ and mignitude $\langle \mathbf{a} \rangle$ by

$$|\mathbf{a}| := \max\{|a_1|, |a_2|\}$$
$$\langle \mathbf{a} \rangle := \min\{|a_1|, |a_2|\} \text{ if } 0 \notin \mathbf{a}, \ \langle \mathbf{a} \rangle := 0 \text{ otherwise.}$$

These functions are applied to interval vectors and matrices componentwise. Without loss of generality and in order to have a unique representation of the parameter dependencies, we assume that $\hat{p}_k > 0$ for all $1 \leq k \leq K$. For a bounded $\Sigma_{AE}^p \neq \emptyset$, $\Box \Sigma_{AE}^p := \bigcap \{\mathbf{x} \in \mathbb{IR}^n \mid \mathbf{x} \supseteq \Sigma_{AE}^p\}$.

In order to simplify the presentation, in Sect. 4 we use the arithmetic on proper and improper intervals [1,2], called Kaucher complete arithmetic or modal interval arithmetic, and its properties. The set of proper intervals $\mathbb{IR}$ is extended in [2] by the set $\overline{\mathbb{IR}} := \{[a_1, a_2] \mid a_1, a_2 \in \mathbb{R}, a_1 \geq a_2\}$ of *improper* intervals obtaining thus the set $\mathbb{IR} \bigcup \overline{\mathbb{IR}} = \{[a_1, a_2] \mid a_1, a_2 \in \mathbb{R}\}$ of all ordered couples of real numbers called also generalized intervals. The conventional interval arithmetic and lattice operations, order relations and other interval functions are isomorphically extended onto the whole set $\mathbb{IR} \bigcup \overline{\mathbb{IR}}$, [2]. *Modal interval analysis* imposes a logical-semantic background on generalized intervals (considered there as modal intervals) and allows giving a logical meaning to the interval results, see [1] for more details.

An element-to-element symmetry between proper and improper intervals is expressed by the "Dual" operator, $\text{Dual}(\mathbf{a}) := [a_2, a_1]$ for $\mathbf{a} = [a_1, a_2] \in \mathbb{IR} \bigcup \overline{\mathbb{IR}}$. Dual is applied componentwise to vectors and matrices. For $\mathbf{a}, \mathbf{b} \in \mathbb{IR} \bigcup \overline{\mathbb{IR}}$

$$\text{Dual}(\text{Dual}(\mathbf{a})) = \mathbf{a}, \quad \text{Dual}(\mathbf{a} \circ \mathbf{b}) = \text{Dual}(\mathbf{a}) \circ \text{Dual}(\mathbf{b}), \quad \circ \in \{+, -, \times, /\},$$
$$\mathbf{a} \subseteq \mathbf{b} \Leftrightarrow \text{Dual}(\mathbf{a}) \supseteq \text{Dual}(\mathbf{b}). \tag{3}$$

The generalized interval arithmetic structure possesses group properties with respect to the operations addition and multiplication of intervals that do not involve zero. For $\mathbf{a}, \mathbf{b} \in \mathbb{IR} \bigcup \overline{\mathbb{IR}}$, $0 \notin \mathbf{b}$

$$\mathbf{a} - \text{Dual}(\mathbf{a}) = 0, \quad \mathbf{b}/\text{Dual}(\mathbf{b}) = 1. \tag{4}$$

Lattice operations are closed with respect to the inclusion relation; handling of norm and metric are very similar to norm and metric in linear spaces [2]. Mid-point, radius, absolute value and mignitude are extended on generalized intervals by the same formulae. For $\mathbf{a} \in \overline{\mathbb{IR}}$, $\omega(\mathbf{a}) := 2|\hat{a}|$. For

$$\mathbf{b} \in \mathbb{IR} \bigcup \overline{\mathbb{IR}}, \quad \mathbf{b} \neq 0, \quad 0 \notin \text{interior of} \begin{cases} \mathbf{b} & \text{if } \mathbf{b} \in \mathbb{IR} \\ \text{Dual}(\mathbf{b}) & \text{if } \mathbf{b} \in \overline{\mathbb{IR}}, \end{cases} \tag{5}$$

$\text{sgn}(\mathbf{b}) := \{1 \text{ if } b_1, b_2 \geq 0, -1 \text{ if } b_1, b_2 \leq 0\}$. For $\mathbf{a} \in \mathbb{IR}$, $0 \in \mathbf{a}$, and $\mathbf{b} \in \overline{\mathbb{IR}}$ satisfying (5),

$$\mathbf{a} * \mathbf{b} = \text{sgn}(\mathbf{b}) \langle \mathbf{b} \rangle \mathbf{a}. \tag{6}$$

3   Parametric AE-solution Sets with Linear Shape

Since the universally quantified parameters contribute linearly to the boundary of a parametric AE-solution set, cf. [4], the theory about parametric united solution sets with linear shape can be generalized to parametric AE-solution sets with linear shape; the latter are also called polyhedral AE-solution sets, cf. [9]. Theorem 2 and Lemma 1 from [5] imply the following theorem.

Theorem 1. *A parameter p_k, $k \in \mathcal{E}$, contributes linearly to the boundary of a parametric AE-solution set if some of the following three equivalent conditions holds true*

(i) the nonzero elements of $A_k x - b_k$ are linearly dependent
(ii) if $A_k = 0$ or the polynomial greatest common divisor (GCD) of the elements of $A_k x - b_k$ is a nonconstant linear polynomial of $x_1, \ldots, x_n$
(iii) $\mathrm{rank}((A_k | b_k)) = 1$, where $(A_k | b_k) \in \mathbb{R}^{m \times (n+1)}$ is the matrix obtained by augmenting the columns of A_k with the vector b_k.

The following theorem follows from the property of the universally quantified parameters mentioned above and [5, Theorem 3].

Theorem 2. *Let $\mathcal{E} = \mathcal{E}_1 \cup \mathcal{E}_2$, $\mathcal{E}_1 \cap \mathcal{E}_2 = \emptyset$ be such that $A_k \neq 0$ for $k \in \mathcal{E}_1$. Denote $k_1 := \mathrm{Card}(\mathcal{E}_1)$, $k_2 := \mathrm{Card}(\mathcal{E}_2)$. Denote by $g_k(x)$ the GCD of the elements of $A_k x$ and let $g_k(x)$ be a nonconstant linear polynomial for every $k \in \mathcal{E}_1$. Define*

$$L := (l_1 | \ldots | l_{k_1}) \in \mathbb{R}^{m \times k_1}, \quad \text{where } l_k := A_k x / g_k(x) \in \mathbb{R}^m$$

$$R := (r_1 | \ldots | r_{k_1})^\top \in \mathbb{R}^{k_1 \times n}, \quad \text{where } r_k := (\frac{\partial g_k(x)}{\partial x_1}, \ldots, \frac{\partial g_k(x)}{\partial x_n})^\top \in \mathbb{R}^n.$$

If there exists $t_k \in \mathbb{R}$ such that $t_k l_k = b_k := \partial b(p)/\partial p_k$ for every $k \in \mathcal{E}_1$, then

$$A_0 x - b_0 + \sum_{k \in \mathcal{E} \cup \mathcal{A}} p_k (A_k x - b_k) = LDRx - LDt - F(p_1, \ldots, p_{k_2})^\top +$$

$$A_0 x - b_0 + \sum_{k \in \mathcal{A}} p_k (A_k x - b_k),$$

where $F := (b_1 | \ldots | b_{k_2}) \in \mathbb{R}^{m \times k_2}$, $t = (t_1, \ldots, t_{k_1})^\top$ and $D = Diag(p_1, \ldots, p_{k_1})$.

Theorem 2 contains Theorem 3 of [5] as a special case. The following corollary is important for finding sharp outer bounds of parametric AE-solution sets.

Corollary 1 ([5], Corollary 4). *For a bounded $\Sigma^p_{AE} \neq \emptyset$,*

$$\inf\{\Box \Sigma^p_{AE}\}_i \quad \text{and} \quad \sup\{\Box \Sigma^p_{AE}\}_i, \qquad i = 1, \ldots, n,$$

are attained at particular end-points of the intervals for the parameters that contribute linearly to the boundary of the solution set.

For a given index set $\mathcal{A}$, define the set $\mathcal{B}_{\mathcal{A}}$ of all end-points (vertices) of $\mathbf{p}_{\mathcal{A}}$.

Proposition 1 ([6], **Corollary 1**). *For a bounded parametric AE-solution set* $\Sigma_{AE}^{p} \neq \emptyset$ *and a set* $\mathcal{B}_{A}'$, *such that* $\mathcal{B}_{A}' \subseteq \mathcal{B}_{A}$ *and* $\Sigma(A(\tilde{p}_{A}, p_{\mathcal{E}}), b(\tilde{p}_{A}, p_{\mathcal{E}}), \mathbf{p}_{\mathcal{E}})$ *is bounded for every* $\tilde{p}_{A} \in \mathcal{B}_{A}'$, *we have*

$$\square\Sigma_{AE}^{p} \subseteq \bigcap_{\tilde{p}_{A} \in \mathcal{B}_{A}'} \square\Sigma(A(\tilde{p}_{A}, p_{\mathcal{E}}), b(\tilde{p}_{A}, p_{\mathcal{E}}), \mathbf{p}_{\mathcal{E}}).$$

Proposition 1 shows that outer bounds of a parametric AE-solution set can be found by bounding only parametric united solution sets. Then, the methodology for finding sharp bounds of parametric united solution sets with linear shape applies to parametric AE-solution set via Proposition 1.

Let $\mathcal{E} = \mathcal{E}_1 \cup \mathcal{E}_2$, $\mathcal{E}_1 \cap \mathcal{E}_2 = \emptyset$ be such that p_k, $k \in \mathcal{E}_1$, satisfy Theorem 1. Then,

$$\square\Sigma_{AE}^{p} \subseteq \bigcap_{\tilde{p}_{A} \in \mathcal{B}_{A}'} \square \left(\bigcup_{\tilde{p}_{\mathcal{E}_1} \in \mathcal{B}_{\mathcal{E}_1}} \square\Sigma(A(\tilde{p}_{A}, \tilde{p}_{\mathcal{E}_1}, p_{\mathcal{E}_2}), b(\tilde{p}_{A}, \tilde{p}_{\mathcal{E}_1}, p_{\mathcal{E}_2}), \mathbf{p}_{\mathcal{E}_2}) \right). \quad (7)$$

Note, that the above methodology can be applied to parametric AE-solution sets such that not all existentially quantified parameters satisfy Theorem 1 or contribute linearly to the boundary of the parametric AE-solution set. A larger discussion is contained in [5].

4   Enclosure of $\Sigma_{ctrl}(A(p), b(q), \mathbf{p}, \mathbf{q})$ with Linear Shape

Consider a parametric interval algebraic system

$$A(p)x = b(q), \qquad p \in \mathbf{p} \in \mathbb{IR}^{K_p}, q \in \mathbf{q} \in \mathbb{IR}^{K_q},$$

$$A(p) := A_0 + \sum_{k=1}^{K_p} p_k A_k, \quad b(q) := b_0 + \sum_{k=1}^{K_q} q_k b_k. \quad (8)$$

We assume that the structure of the dependencies between the parameters p is such that the conditions defined in Theorem 1 hold true for all the parameters and, therefore, the system (8) has the equivalent representation

$$(A_0 + L\mathrm{Diag}(p)R)\, x = b_0 + Fq, \quad p \in \mathbf{p} \in \mathbb{IR}^{K_p}, q \in \mathbf{q} \in \mathbb{IR}^{K_q}, \quad (9)$$

with suitable numerical matrices L, R, F found by Theorem 2.

We search for an outer interval enclosure of the parametric controllable solution set

$$\begin{aligned} \Sigma_{ctrl}^{p} &= \Sigma_{ctrl}(A(p), b(q), \mathbf{p}, \mathbf{q}) \\ &:= \{x \in \mathbb{R}^n \mid (\forall q \in \mathbf{q})(\exists p \in \mathbf{p})(A(p)x = b(q))\}. \end{aligned}$$

4.1   Iteration Method

Theorem 3. *If $x \in \Sigma_{ctrl}^{p} \neq \emptyset$ for the system (9), $D_0 \subset \mathrm{Diag}(\mathbf{p})$ and $A_0 + LD_0R$ is invertible, then $x \in (A_0 + LD_0R)^{-1}(b_0 + F\mathrm{Dual}(\mathbf{q}) + L\mathbf{d})$, wherein*

$$\mathbf{d} := (D_0 - \mathrm{Diag}(\mathbf{p}))Rx, \tag{10}$$

and $b_0 + F\mathrm{Dual}(\mathbf{q}) + L\mathbf{d}$ is a proper interval vector in Kaucher interval arithmetic.

Proof. If $x \in \Sigma_{ctrl}^{p} \neq \emptyset$, according to [7, Theorem 3.2] we have

$$(A_0 + L\mathrm{Diag}(\mathbf{p})R)\,x \supseteq b_0 + F\mathbf{q}. \tag{11}$$

We apply the Dual operator to the above inclusion relation, the relation (3), and the distributivity of multiplication by a point vector x. Then, we add $-L\mathrm{Diag}(\mathbf{p})Rx$ to both sides of the obtained inclusion. Due to (4) we obtain

$$A_0 x \subseteq b_0 + F\mathrm{Dual}(\mathbf{q}) - L\mathrm{Diag}(\mathbf{p})Rx. \tag{12}$$

The relation (10) implies $-\mathrm{Diag}(\mathbf{p})Rx = \mathbf{d} - D_0Rx$ which we substitute in (12) and obtain

$$(A_0 + LD_0R)x \subseteq b_0 + F\mathrm{Dual}(\mathbf{q}) + L\mathbf{d}.$$

The inclusion (11), which holds true for $x \in \Sigma_{ctrl}^{p} \neq \emptyset$, implies $\omega(F\mathbf{q}) \leq \omega(L\mathbf{d})$. The latter implies that $b_0 + F\mathrm{Dual}(\mathbf{q}) + L\mathbf{d}$ is a proper interval vector since $L\mathbf{d}$ is a proper interval vector and $F\mathrm{Dual}(\mathbf{q})$ is an improper one. Due to invertibility of $A_0 + LD_0R$, we obtain $x \in (A_0 + LD_0R)^{-1}(b_0 + F\mathrm{Dual}(\mathbf{q}) + L\mathbf{d})$.

Theorem 4. *Let $D_0 \in \mathbb{R}^{K_p \times K_p}$, $D_0 \in \mathrm{Diag}(\mathbf{p}) = \mathbf{D}$ be such that $A_0 + LD_0R$ is invertible and put*

$$C := (A_0 + LD_0R)^{-1}.$$

Define

$$w' := w - |D_0 - \mathbf{D}|\,|RCL|w, \tag{13}$$

$$w'' := |D_0 - \mathbf{D}|\langle RCb_0 + (RCF)\mathrm{Dual}(\mathbf{q})\rangle, \tag{14}$$

for some vector $w \geq 0$, and

$$\mathbf{u} := [-\alpha w, \alpha w], \qquad \alpha = \max_i \frac{w_i''}{w_i'}. \tag{15}$$

(i) $x = x(p,q) \in \Sigma_{ctrl}^{p} \neq \emptyset$ *for (9) is related to* $y = Rx(p,q)$ *by the inclusions*

$$x \in Cb_0 + (CF)\mathrm{Dual}(\mathbf{q}) + CL\mathbf{d}, \tag{16}$$

$$y \in RCb_0 + (RCF)\mathrm{Dual}(\mathbf{q}) + (RCL)\mathbf{d}, \tag{17}$$

where

$$\mathbf{d} = (D_0 - \mathbf{D})y. \tag{18}$$

(ii) If $w' > 0$ and $0 \notin b_0 + F\mathbf{q}$, then $\mathbf{d} \subseteq \mathbf{u}$.

(iii) If $\mathbf{x} := Cb_0 + (CF)\mathrm{Dual}(\mathbf{q}) + (CL)\mathbf{u}$ is a proper interval vector, then every $x \in \Sigma_{ctrl}^{p}$ satisfies $x \in \mathbf{x}$.

Proof. (i) follows from Theorem 3.

(ii) Since $w' > 0$ we put

$$\beta = \max_i |\mathbf{d}_i|/w_i$$

and note that $|\mathbf{d}| \leq \beta w$, with equality in some component i. The definition of α and $0 \notin b_0 + F\mathbf{q}$ imply $0 \leq w'' \leq \alpha w'$. From (16)–(18) and a subdistributive law in Kaucher arithmetic we have

$$\mathbf{d} \subseteq (D_0 - \mathbf{D})(RCb_0 + (RCF)\text{Dual}(\mathbf{q}) + (RCL)\mathbf{d})$$
$$\subseteq (D_0 - \mathbf{D})(RCb_0 + (RCF)\text{Dual}(\mathbf{q})) + (D_0 - \mathbf{D})(RCL)\mathbf{d}.$$

Then, formula (6) implies

$$|\mathbf{d}| \quad \leq \quad |D_0 - \mathbf{D}|\langle RCb_0 + (RCF)\text{Dual}(\mathbf{q})\rangle + |D_0 - \mathbf{D}||RCL|\beta w$$
$$\underset{(14),(13)}{\leq} \quad w'' + \beta(w - w') \leq \alpha w' + \beta(w - w').$$

Thus $\beta w_i = |\mathbf{d}_i| \leq \alpha w_i' + \beta(w_i - w_i')$, hence $\beta w_i' \leq \alpha w_i'$. As $w' > 0$, we conclude that $\beta \leq \alpha$, and $\mathbf{d} \subseteq \mathbf{u}$ follows.

(iii) follows by (ii) and Theorem 3.

In the computations we take D_0 as the midpoint of $\mathbf{D}$, and w, e.g., as the vector with all entries one. In order to provide guaranteed enclosures, w' should be rounded downward, w'' and α should be rounded upward. If $w' \leq 0$ in (13), we may apply the approach proposed in [3] to compute the largest eigenvalue ϱ (= the spectral radius) of the nonnegative matrix

$$M := |D_0 - \mathbf{D}||RCL|.$$

If $\varrho < 1$, any $w > 0$ sufficiently close to an associated eigenvector makes $w' > 0$. In practice, one could run a Lanczos iteration and stop as soon as an intermediate eigenvector approximation $w > 0$ satisfies $Mw < w$, [3].

The computed initial interval enclosure $\mathbf{u}$ of $\mathbf{d}$ can be further improved by iterating and intersecting with the previously computed enclosures. It is sufficient to iterate the enclosures for y and d, and compute the enclosures for x when the intersected results no longer improve significantly. Thus we iterate

$$\mathbf{y} = \{(RCb_0) + (RCF)\text{Dual}(\mathbf{q}) + (RCL)\mathbf{u})\} \cap \mathbf{y},$$
$$\mathbf{u} = \{(D_0 - \mathbf{D})\mathbf{y}\} \cap \mathbf{u}$$

until some stopping test holds, and then get the enclosure

$$\mathbf{x} := (Cb_0) + (CF)\text{Dual}(\mathbf{q}) + (CL)\mathbf{u}$$

for all x that belong to Σ_{ctrl}^p of (9). In the implementation of the method we used the stopping criterion proposed in [3], namely, the iteration stops when the sum of widths of the components of $\mathbf{u}$ does not improve by a factor of 0.999 or after at most 10 iterations.

The method presented here can be considered as an extension of the so-called formal (algebraic) approach, cf. [10], for enclosing nonparametric AE-solution sets to parametric controllable solution sets.

4.2   Numerical Examples

Here we illustrate the advantages of the above parametric iterative method by some numerical examples and compare this method to the only discussed by now methods for outer enclosure of the parametric controllable solution set presented in [6]. The implementations and the numerical computations are done in the environment of *Mathematica*® using the package `directed.m` [8]. The latter package supports the arithmetic of proper and improper intervals and provides compatibility with the conventional interval arithmetic supported by the *Mathematica*® kernel. The numerical computations are done exactly if all input data are represented exactly (e.g., by rational numbers) or by appropriate directed rounding in floating-point if some input data are in floating point arithmetic. This software environment and the implementation provide obtaining numerical interval vectors which are guaranteed to contain the considered parametric solution set.

The iteration method, proposed in Sect. 4.1, can be implemented in any software environment which does not support the arithmetic of proper and improper intervals if the lower and upper bounds of the corresponding intervals are computed separately applying the corresponding formulae for the arithmetic operations and applying correct directed rounding in floating-point arithmetic. The iteration method from [3] is implemented in the environment of Matlab, see [3], and in C-XSC, see [11].

Explicit representation of any parametric controllable solution set, considered below, is obtained by methods from [4] as a system of real inequalities in the coordinate variables, which is then solved by suitable *Mathematica*® functions.

Example 1. Find an enclosure of the parametric controllable solution set to the system

$$\begin{pmatrix} p_1 + p_2, p_1 - p_2 \\ p_1 - p_2, p_1 + p_2 \end{pmatrix} x = \begin{pmatrix} 3/2 + q \\ q \end{pmatrix},$$

where $p_1, p_2 \in [1/2, 3/2]$, $q \in [-1/10, 1/10]$. According to Theorem 1 the parametric controllable solution set has linear shape and the application of Theorem 4 with an iterative refinement gives the following (rounded outward) interval enclosure

$$\begin{pmatrix} [0.099951, 1.400049] \\ [-0.650049, 0.650049] \end{pmatrix}.$$

Since the parametric matrix is not strongly regular, the parametric Bauer-Skeel method from [6, Corollary 6] fails. Due to the same reason Proposition 1 cannot be applied together with the parametric Bauer-Skeel method while it can be applied together with Theorem 4.

Example 2. Find an enclosure of the parametric controllable solution set to

$$\begin{pmatrix} p_1 + p_2, & p_1 - 2p_2 \\ p_1 - p_2/2, & p_1 + p_2 \end{pmatrix} x = \begin{pmatrix} 3/2 + q/3 \\ q/2 \end{pmatrix},$$

where $p_1 \in [1, 3/2]$, $p_2 \in [-1, -1/2]$, $q \in [9/10, 11/10]$. According to Theorem 1 the parametric controllable solution set has linear shape and the application of Theorem 4 with an iterative refinement gives the interval enclosure

$$([-0.17779, 0.39507], \ [0.39875, 0.89507])^\top,$$

while the parametric Bauer-Skeel method from [6] gives the enclosure

$$([-0.32840, 0.54568], \ [0.29135, 0.1.00247])^\top.$$

The former interval enclosure overestimates the exact interval hull

$$([-\frac{59}{405}, \frac{3}{10}], [\frac{421}{810}, \frac{239}{270}])^\top$$

of the parametric controllable solution set by $(22.2, 26.4)^\top \%$, while the latter enclosure overestimates the hull by $(49, 48.6)^\top \%$. In general, the percentage of overestimation depends on the particular problem (parameter dependencies), the problem size, the number of the parameters, and the width of the parameter intervals. The parametric Bauer-Skeel method can be applied to the above system, where the parameter intervals $\mathbf{p}_1, \mathbf{p}_2$ are with enlarged radius from $r = 1/4$ (the intervals considered above) to a radius $r = 0.481$ which still provides strong regularity of the parametric matrix. In the latter case the overestimation of the corresponding exact interval hull is $(99.81, 99.79)^\top \%$. The method from Theorem 4 is applicable to the above system where the parameter intervals $\mathbf{p}_1, \mathbf{p}_2$ have radius $r = 0.749$ still providing regularity of the parametric matrix. The overestimation of the corresponding exact interval hull for these intervals is $(35.86, 37.94)^\top \%$.

Further improvement of an enclosure obtained by some applicable enclosure method (or by some method for obtaining the exact interval hull of Σ^p_{uni} with linear shape) could be obtained by Proposition 1, respectively (7), at the expense of a bigger computational effort. The application of (7) to Example 2 with the initial parameter intervals gives an interval enclosure which overestimates the exact interval hull of the parametric controllable solution set by $(0, 5.7)^\top \%$.

In the following example, we consider the behaviour of the proposed method on parametric controllable solution sets that are empty sets or unbounded.

Example 3. Consider the parametric linear system

$$\begin{pmatrix} p_1 - p_2, & p_2 \\ p_1 + p_2, & -p_2 \end{pmatrix} x = \begin{pmatrix} 2q \\ 2q \end{pmatrix}.$$

(a) For $p_1 \in [3/4, 5/4]$, $p_2 \in [0, 1]$, $q \in [1, 2]$, $\Sigma^p_{ctrl} = \emptyset$. The application of the method considered above yields $w' < 0$ and the largest eigenvalue ($=$ the spectral radius) of the matrix $|D_0 - \mathbf{D}||RCL|$ equal to 1.

(b) With the same data as in (a) but twice bigger radius of p_1, $p_1 \in [1/2, 3/2]$, the corresponding Σ^p_{ctrl} is unbounded and defined by $8/3 \leq x_1 \leq 4$, $x_2 \in \mathbb{R}$. The application of the method proposed in this paper yields the same output as in (a).

The method, proposed in this paper, for bounding parametric controllable solution sets with linear shape possesses the same scalability property as the methods from [3,5] for bounding parametric united solution sets. Examples of large parametric linear systems with over 5000 variables and over 10 000 parameters which appear in finite element analysis of uncertain truss structures can be found in [3], while [5] presents some examples coming from modeling of electrical circuits and models in biology.

5 Conclusion

We presented a new interval method for outer enclosure of a class of nonempty and bounded parametric controllable solution sets with linear shape. Contrary to other available so far interval methods for bounding general parametric controllable solution sets, which require strong regularity of the parametric matrix, the new method does not have such a restriction. Furthermore, the method is applicable to parametric linear systems of high dimensions that involve many parameters, see [3], and when the parameter intervals are large, see the end of Example 2. Further improvement of the solution enclosure may be achieved by methods presented in Sect. 3 and [5, Sect. 2].

References

1. Sainz, M.A., Armengol, J., Calm, R., Herrero, P., Jorba, L., Vehi, J.: Modal Interval Analysis: New Tools for Numerical Information. Lecture Notes in Mathematics, vol. 2091. Springer, Switzerland (2014)
2. Kaucher, E.: Interval analysis in the extended interval space IR. Computing, Suppl. 2, 33–49 (1980). http://www.math.bas.bg/~epopova/Kaucher-80-CS_33-49.pdf
3. Neumaier, A., Pownuk, A.: Linear systems with large uncertainties, with applications to truss structures. Reliable Comput. **13**, 149–172 (2007)
4. Popova, E.D.: Explicit description of AE-solution sets for parametric linear systems. SIAM J. Matrix Anal. Appl. **33**, 1172–1189 (2012)
5. Popova, E.D.: Improved enclosure for some parametric solution sets with linear shape. Comput. Math. Appl. **68**(9), 994–1005 (2014)
6. Popova, E.D., Hladík, M.: Outer enclosures to the parametric AE-solution set. Soft Comput. **17**, 1403–1414 (2013)
7. Popova, E.D., Krämer, W.: Characterization of AE-solution sets to a class of parametric linear systems. C. R. Acad. Bulgare Sci. **64**(3), 325–332 (2011)
8. Popova, E.D., Ullrich, C.: Directed interval arithmetic in Mathematica: implementation and applications. Technical report 96-3, Universität Basel (1996). http://www.math.bas.bg/~epopova/papers/tr96-3.pdf
9. Sharaya, I.A.: Boundary intervals method for visualization of polyhedral solution sets. Comput. Technol. **20**(1), 75–103 (2015). (in Russian)
10. Shary, S.P.: A new technique in systems analysis under interval uncertainty and ambiguity. Reliable Comput. **8**(5), 321–418 (2002)
11. Zimmer, M.: Software zur hocheffizienten Loesung von Intervallgleichungssystemen mit C-XSC, Ph.D. thesis, Bergische Universität Wuppertal (2013)

Reserve of Characteristic Inclusion as Recognizing Functional for Interval Linear Systems

Irene A. Sharaya and Sergey P. Shary[✉]

Institute of Computational Technologies, 6, Lavrentiev ave., Novosibirsk, Russia
shary@ict.nsc.ru
http://www.nsc.ru/interval/sharaya

Abstract. The paper considers the interval linear inclusion $Cx \subseteq d$ in the Kaucher interval arithmetic. We introduce a quantitative measure of its fulfillment, called "reserve", and investigate its properties and application. We show that the reserve proves useful in the study of AE-solutions and quantifier solutions to interval linear problems. In particular, using the reserve can help to recognize position of a point with respect to the solution set, emptiness of the solution set and of its interior, etc.

Keywords: Interval linear system · AE-solutions · Quantifier solutions · Solution set · Characteristic inclusion · Reserve · Recognizing functional

1 Introduction

Let $\mathbb{KR} = \{[\underline{v}, \overline{v}] \mid \underline{v}, \overline{v} \in \mathbb{R}\}$ be the set of Kaucher intervals, and $\mathbb{K}\overline{\mathbb{R}} = \{[\underline{v}, \overline{v}] \mid \underline{v}, \overline{v} \in \overline{\mathbb{R}}\}$ be the set of Kaucher intervals over the extended real axis $\overline{\mathbb{R}} = \mathbb{R} \cup \{-\infty, \infty\}$ (see [5,6]). It makes sense to remind that the Kaucher complete interval arithmetic $\mathbb{KR}$, apart from usual (proper) intervals, also includes improper intervals $[\underline{z}, \overline{z}]$ with $\underline{z} > \overline{z}$.

Our object under study is the interval linear inclusion

$$Cx \subseteq d, \tag{1}$$

where $x \in \mathbb{R}^n$, C is an $m \times n$-matrix with the elements from $\mathbb{KR}$, d is an m-vector made up of the extended intervals from $\mathbb{K}\overline{\mathbb{R}}$. We introduce and investigate a quantitative measure, called "reserve", of the fulfillment of inclusion (1). Then an answer to the question 'What can the reserve serve for?' is given.

The set of (formal) solutions to inclusion (1) is naturally defined as

$$\Xi = \{ x \in \mathbb{R}^n \mid Cx \subseteq d \}$$
$$= \Big\{ x \in \mathbb{R}^n \;\Big|\; \sum_j C_{ij}x_j \geq \underline{d}_i, \; \overline{\sum_j C_{ij}x_j} \leq \overline{d}_i, \; i = 1, 2, \ldots, m \Big\}.$$

The inclusion $Cx \subseteq d$ in the Kaucher arithmetic and its solutions prove useful for many purposes. First of all, they provide a tool for unified treatment

© Springer International Publishing Switzerland 2016
M. Nehmeier et al. (Eds.): SCAN 2014, LNCS 9553, pp. 148–167, 2016.
DOI: 10.1007/978-3-319-31769-4_13

of so-called AE-solutions and quantifier solutions to interval linear problems. In the next section, we remind briefly the corresponding concepts and main results (see details in [13, 15, 16]).

Our notation follows mainly the informal standard [7].

2 Quantifier Solutions and AE-solutions

In practice, we usually consider intervals

- from the set of proper intervals $\mathbb{IR} = \{ v = [\underline{v}, \overline{v}] \mid \underline{v}, \overline{v} \in \mathbb{R}, \ \underline{v} \le \overline{v} \}$ and
- in connection with a certain property $P(v)$ that can be either fulfilled or not fulfilled for the point members v of the intervals.

For instance, the property P may have the form "to be a solution to an equation", "to be a solution to a problem" with some parameters that can take values from prescribed intervals, and so on. Then the following different situations may occur:

(1) either the property $P(v)$ holds for *all* members v from the given interval v,
(2) or the property $P(v)$ holds only for *some* members v from the interval v, not necessarily all, or even for a single value from v.

Formally, the above distinction can be expressed by logical quantifiers:

- In the first case, we write "$(\forall v \in v)\, P(v)$"
  and speak of *interval A-uncertainty*,

- In the second case, we write "$(\exists v \in v)\, P(v)$"
  and speak of *interval E-uncertainty*.

$$(2)$$

We thus have to distinguish between the two above types of interval uncertainty. The quantifier solutions and AE-solutions to interval system of relations are the solutions that take into account the difference between the A-type and E-type of uncertainty in the input interval parameters [13, 15, 16].

Let us consider an interval system of relations

$$F(a, x) \,\sigma\, b, \tag{3}$$

where $F = \big(F_1(a, x), F_2(a, x), \ldots, F_m(a, x) \big)^{\top}$ is a vector-function with some mappings $F_i : \mathbb{R}^l \times \mathbb{R}^n \to \mathbb{R}$ as components, $a \in \mathbb{IR}^l$, $x \in \mathbb{R}^n$, $\sigma \in \{=, \le, \ge\}^m$, $b \in \mathbb{IR}^m$. Taking into account our conclusion about uncertainty types, we can assume that the property determining "solutions" to system (3) should look like

$$(Q_1 v_{\pi_1} \in v_{\pi_1})(Q_2 v_{\pi_2} \in v_{\pi_2}) \cdots (Q_{l+m} v_{\pi_{l+m}} \in v_{\pi_{l+m}})\big(F(a, x) \,\sigma\, b \big), \tag{4}$$

where

$Q_1, Q_2, \ldots, Q_{l+m}$ are the logical quantifiers "$\forall$" or "$\exists$",

$(v_1, v_2, \ldots, v_{l+m}) := (a_1, a_2, \ldots, a_l, b_1, b_2, \ldots, b_m) \in \mathbb{R}^{l+m}$
   is an aggregated parameter vector,

$(\boldsymbol{v}_1, \boldsymbol{v}_2, \ldots, \boldsymbol{v}_{l+m}) := (\boldsymbol{a}_1, \boldsymbol{a}_2, \ldots, \boldsymbol{a}_l, \boldsymbol{b}_1, \boldsymbol{b}_2, \ldots, \boldsymbol{b}_m) \in \mathbb{IR}^{l+m}$
   is an aggregated interval vector of their possible values,

$(\pi_1, \pi_2, \ldots, \pi_{l+m})$ is a permutation of the positive integer
   numbers $1, 2, \ldots, l + m$.

We will call the logical formula (4) *selecting predicate* of the solutions to (3).

A vector y will be referred to as *quantifier solution* to the interval system of relations $F(\boldsymbol{a}, x)\,\sigma\,\boldsymbol{b}$ if the selecting predicate (4) is true for $x = y$. This is a very general construction that can describe a great variety of specific solutions (their total number far exceeding 2^{l+m}). Sometimes, it makes sense to somehow restrict the generality by specializing the form of the selecting predicate.

A quantifier solution to the interval system of relations for which, in the selecting predicate, all occurrences of the universal quantifier "$\forall$" precede those of the existential quantifier "$\exists$" will be referred to as *AE-solution*. The AE-solutions are thus a particular case of the quantifier solutions obtained by fixing a certain order of the logical quantifiers in the logical formula (selecting predicate) that determines the solution.

For the last two decades, since their introduction in [15], the AE-solutions have been the area of active research. They have a clear practical interpretation as solutions to single-step decision making processes with intervally defined data when we have to choose a compromise between perturbations (expressed by interval parameters with A-uncertainty) and our controls (expressed by interval parameters with E-uncertainty) [16]. A generalization of the concept of AE-solutions has penetrated into fuzzy sets theory and its applications (see [2]). The interested reader can find a lot of further results on AE-solutions, e.g., in the works [3,4,8,9,11,12] and others.

The inclusion $Cx \subseteq d$ arises in connection with quantifier solutions and AE-solutions to the systems of interval linear relations of the form

$$Ax\,\sigma\,b,$$

where $\sigma \in \{=, \leq, \geq\}^m$, $A \in \mathbb{IR}^{m \times n}$, $x \in \mathbb{R}^n$, and $b \in \mathbb{IR}^m$. If the quantifier matrix $\mathcal{A} \in \{\forall, \exists\}^{m \times n}$ and the quantifier vector $\beta \in \{\forall, \exists\}^m$ specify the uncertainty types of the separate interval parameters A_{ij}, b_i for all i and j, we introduce the auxiliary interval matrices $A^{\forall}$, $A^{\exists}$ and vectors $b^{\forall}$, $b^{\exists}$ as follows:

$$A_{ij}^{\forall} := \begin{cases} A_{ij}, & \text{if } \mathcal{A}_{ij} = \forall, \\ 0, & \text{if } \mathcal{A}_{ij} = \exists, \end{cases} \qquad b_i^{\forall} := \begin{cases} b_i, & \text{if } \beta_i = \forall, \\ 0, & \text{if } \beta_i = \exists, \end{cases}$$

$$A_{ij}^{\exists} := \begin{cases} A_{ij}, & \text{if } \mathcal{A}_{ij} = \exists, \\ 0, & \text{if } \mathcal{A}_{ij} = \forall, \end{cases} \qquad b_i^{\exists} := \begin{cases} b_i, & \text{if } \beta_i = \exists, \\ 0, & \text{if } \beta_i = \forall. \end{cases} \tag{5}$$

Then the AE-solution set to the interval linear system of equations $Ax = b$ can be alternatively defined as the set

$$\Xi_{A\beta}(A, b) = \{\, x \in \mathbb{R}^n \mid$$

$$(\forall A' \in A^\forall)\,(\forall b' \in b^\forall)\,(\exists A'' \in A^\exists)\,(\exists b'' \in b^\exists)$$

$$((A' + A'')\,x = b' + b'')\,\}.$$

Besides, the following equivalent characterization is valid in the Kaucher interval arithmetic [16]:

$$x \in \Xi_{A\beta}(A, b) \quad \Longleftrightarrow \quad (A^\forall + \operatorname{dual} A^\exists)\,x \subseteq \operatorname{dual} b^\forall + b^\exists, \qquad (6)$$

where "dual" means dualization operator dual $: \mathbb{KR} \to \mathbb{KR}$ reverting the interval endpoints, i.e., such that $\operatorname{dual}[\underline{z}, \overline{z}] = [\overline{z}, \underline{z}]$. Inclusion (6), which coincides with (1) in form, is called *characteristic inclusion* for the AE-solution set determined by the uncertainty distribution (5) over the interval elements of the system $Ax = b$. See examples in Sect. 3.

The next important particular case is quantifier solutions to the systems of interval linear inequalities

$$Ax \le b \qquad \text{or} \qquad Ax \ge b, \qquad (7)$$

where $A \in \mathbb{IR}^{m \times n}$, $x \in \mathbb{R}^n$, and $b \in \mathbb{IR}^m$. A remarkable fact about the interval linear systems of inequalities is that the order of logical quantifiers in the selecting predicate does not matter, and general quantifier solutions coincide with AE-solutions for interval linear inequalities [13]. Specifically, if $Q(A, b, \mathcal{A}, \beta)$ is a quantifier prefix of the selecting predicate, made up of the quantifier terms that correspond to individual interval parameters, then

$$Q(A, b, \mathcal{A}, \beta)(Ax \ge b) \quad \Longleftrightarrow \quad (A^\forall + \operatorname{dual} A^\exists)\,x \subseteq [\,\overline{b}^\forall + \underline{b}^\exists, +\infty\,],$$

$$Q(A, b, \mathcal{A}, \beta)(Ax \le b) \quad \Longleftrightarrow \quad (A^\forall + \operatorname{dual} A^\exists)\,x \subseteq [\,-\infty, \underline{b}^\forall + \overline{b}^\exists\,];$$

see details in [13].

As a particular case of the inclusion $Cx \subseteq d$, it is worth noting some special quantifier solutions to the interval linear system of relations $Ax\,\sigma\,b$. Let the selecting predicate of the solutions be such that, for the interval parameters from the equality relations, all the quantifiers "$\forall$" precede all the quantifiers "$\exists$". If $Q^\sigma(A, b, \mathcal{A}, \beta)$ is the quantifier prefix of the above selecting predicate, then there holds (see [13]):

$$Q^\sigma(A, b, \mathcal{A}, \beta)(Ax\,\sigma\,b) \quad \Longleftrightarrow \quad (A^\forall + \operatorname{dual} A^\exists)\,x \subseteq \operatorname{dual} b^\forall + b^\exists + w,$$

where the interval m-vector w is defined as

$$w_i := \begin{cases} 0, & \text{if } \sigma_i \text{ is "=",} \\ [0, \infty], & \text{if } \sigma_i \text{ is "$\ge$",} \\ [-\infty, 0], & \text{if } \sigma_i \text{ is "$\le$".} \end{cases}$$

To sum up, considering the inclusion $Cx \subseteq d$ in the Kaucher complete interval arithmetic enables us to study all the particular cases of the quantifier solutions to interval linear inequality systems and of the AE-solutions to interval linear systems of relations

- ▶ simultaneously and in a uniform way,
- ▶ by interval methods.

3   Definition and Main Properties of Reserve

By *reserve of the inclusion* $Cx \subseteq d$ (or just *reserve*), we call the maximal number $\mathrm{Rsv} \in \mathbb{R}$ such that

$$Cx + [-\mathrm{Rsv}, \mathrm{Rsv}]\,e \subseteq d$$

for m-vector $e = (1, 1, \ldots, 1)^{\top}$. Notice that, if $\mathrm{Rsv} < 0$, then $[-\mathrm{Rsv}, \mathrm{Rsv}]$ is an improper interval, that is, all the arithmetic operations and relations are understood in the Kaucher complete interval arithmetic.

From the above definition, one can easily deduce the following representation for Rsv:

$$\begin{aligned}
\mathrm{Rsv} &= \min_{1 \le i \le m}\; \min\left\{ \boldsymbol{C}_{i:}x - \boldsymbol{d}_i,\; -\overline{C}_{i:}x + \overline{d}_i \right\} \\
&= \min_{1 \le i \le m}\; \min\left\{ \boldsymbol{C}_{i:}x^+ - \overline{C}_{i:}x^- - \boldsymbol{d}_i,\; -\overline{C}_{i:}x^+ + \boldsymbol{C}_{i:}x^- + \overline{d}_i \right\} \\
&= \min_{1 \le i \le m}\; \min\left\{ \sum_{j=1}^{n} C_{ij}^{-\mathrm{sgn}\,x_j} x_j - \boldsymbol{d}_i,\; -\sum_{j=1}^{n} C_{ij}^{\mathrm{sgn}\,x_j} x_j + \overline{d}_i \right\},
\end{aligned} \tag{8}$$

where $x^+, x^- \in \mathbb{R}^n_+$, $x^+ = \max\{0, x\}$, $x^- = \max\{0, -x\}$ are positive and negative parts of the vector x respectively, and for every i

$$C_{ij}^{-\mathrm{sgn}\,x_j} = \begin{cases} \boldsymbol{C}_{ij}, & \text{if } x_j \ge 0, \\ \overline{C}_{ij}, & \text{otherwise}, \end{cases} \qquad C_{ij}^{\mathrm{sgn}\,x_j} = \begin{cases} \overline{C}_{ij}, & \text{if } x_j \ge 0, \\ \boldsymbol{C}_{ij}, & \text{otherwise}. \end{cases}$$

For fixed C and d, we can consider the reserve as a functional of x, that is, as a function $\mathrm{Rsv}\,(x) : \mathbb{R}^n \to \mathbb{R}$. Then it characterizes properties of the point x with respect to the interval inclusion $Cx \subseteq d$ and interval linear systems of relations described by this inclusion (see Sect. 2). From (8), it follows that the functional $\mathrm{Rsv}\,(x)$ is defined on the entire $\mathbb{R}^n$. Also, it is evidently continuous and even Lipschitz continuous.

We are reminded that the set of points $x = (x_1, x_2, \ldots, x_n)^{\top} \in \mathbb{R}^n$ having a definite sign of each their component x_j, supplemented with its boundary, is called *orthant* of the space $\mathbb{R}^n$.

Proposition 1. *The function* $\mathrm{Rsv}\,(x)$ *is concave in each orthant of* $\mathbb{R}^n$.

Proof. In every fixed orthant of the space $\mathbb{R}^n$, the values of $C_{ij}^{-\mathrm{sgn}\,x_j}$ and $C_{ij}^{\mathrm{sgn}\,x_j}$ are constant. Therefore, the last formula of (8) implies that the functional $\mathrm{Rsv}\,(x)$ is the minimum of $2m$ linear functions

$$\left(\sum_{j=1}^{n} C_{ij}^{-\mathrm{sgn}\,x_j} x_j - \underline{d}_i \right) \text{ and } \left(-\sum_{j=1}^{n} C_{ij}^{\mathrm{sgn}\,x_j} x_j + \overline{d}_i \right), \qquad i = 1, 2, \ldots, m,$$

within any fixed orthant.

Proposition 2. *The function* $\mathrm{Rsv}\,(x)$ *is piecewise-linear.*

Proof. It is almost obvious from representations (8) and the proof of the preceding proposition.

As follows from Sect. 2, the reserve of inclusion (1) is a very general construction that covers many particular instances of interval linear systems of relations and a lot of their solution sets. Still, some special cases of the reserve have been successfully applied in the earlier works on interval systems of equations, although under the name of "recognizing functionals".

Historically, the first recognizing functional, $\mathrm{Tol}\,(x)$, was proposed for the tolerable solution set to interval linear systems of equations [14]. Given an interval linear system $\boldsymbol{A}x = \boldsymbol{b}$, its *tolerable solution set* is defined as the set

$$\Xi_{tol} = \{\, x \in \mathbb{R}^n \mid (\forall A \in \boldsymbol{A})(\exists b \in \boldsymbol{b})(Ax = b)\}, \tag{9}$$

formed by all such x's that the product Ax falls into $\boldsymbol{b}$ for any possible $A \in \boldsymbol{A}$. The functional

$$\mathrm{Tol}(x) \;=\; \min_i \left\{\, \mathrm{rad}\,\boldsymbol{b}_i - \left| \mathrm{mid}\,\boldsymbol{b}_i - \boldsymbol{A}_{i:}x \right| \right\} \tag{10}$$

is the reserve of the corresponding characteristic inclusion $\boldsymbol{A}x \subseteq \boldsymbol{b}$ (see [14] for further details and history of the problem).

For the interval system of linear equations $\boldsymbol{A}x = \boldsymbol{b}$, the *united solution set* is known to be defined as

$$\Xi_{uni} = \{\, x \in \mathbb{R}^n \mid (\exists A \in \boldsymbol{A})(\exists b \in \boldsymbol{b})(Ax = b)\}, \tag{11}$$

being the solution set for the collection of all the systems $Ax = b$ with $A \in \boldsymbol{A}$ and $b \in \boldsymbol{b}$. So, the selecting predicate is $(\exists A \in \boldsymbol{A})(\exists b \in \boldsymbol{b})(Ax = b)$, and the corresponding characteristic inclusion is $(\mathrm{dual}\,\boldsymbol{A})\,x \subseteq \boldsymbol{b}$. Its reserve as a function of x coincides with the recognizing functional

$$\mathrm{Uss}(x) \;=\; \min_i \left\{\, \mathrm{rad}\,\boldsymbol{b}_i + (\mathrm{rad}\,\boldsymbol{A}_{i:})\,|x| - \left| \mathrm{mid}\,\boldsymbol{b}_i - (\mathrm{mid}\,\boldsymbol{A}_{i:})\,x \right| \right\}$$

introduced in [19, 20] (see also [18]).

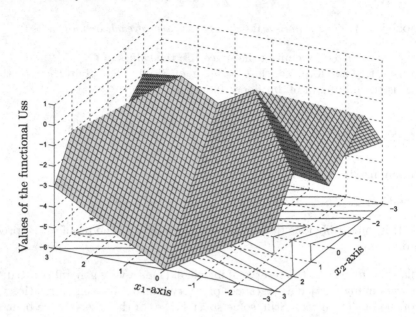

Fig. 1. Graph of the reserve (recognizing functional Uss) for system (12)

Example 1. For the united solution set to the interval linear equation system

$$\begin{pmatrix} [2,4] & [-1,1] \\ [-1,1] & [2,4] \end{pmatrix} x = \begin{pmatrix} [-3,3] \\ 0 \end{pmatrix},$$ (12)

the functional $\mathrm{Rsv}\,(x)$ (i. e., Uss defined by (11)) has the graph depicted at Fig. 1 (see also [18]).

For what purpose can one use the reserve? We will show that, using the reserve as a functional of x, one can get extensive information on

> position of a point with respect to the solution set,
>
> whether the solution set $\varXi$ is empty or not,
>
> whether the interior of $\varXi$ is empty or not,
>
> the 'best' points for the inclusion $Cx \subseteq d$.

4   Properties of the Solution Set

Prior to formulating the main results of our paper, we need to revise some geometric and topological properties of the solution set $\varXi$ to the interval linear inclusion $Cx \subseteq d$.

The intersection of the solution set Ξ with each orthant of the space $\mathbb{R}^n$ is a convex polyhedral set. This fact is well-known for particular solution sets to interval linear relations (see, e.g., [1,14,16]), but it does not hurt to outline its substantiation for the general inclusion $Cx \subseteq d$.

The membership of a vector y in an orthant of $\mathbb{R}^n$ is determined by fixing the signs of its components y_i, $i = 1, 2, \ldots, n$. Also, for any interval $m \times n$-matrix C, the components of the product $Cy = ((Cy)_1, (Cy)_2, \ldots, (Cy)_m)^\top$ can be represented as

$$
(Cy)_i = \sum_{j=1}^{n} C_{ij} y_j = \left[\sum_{j=1}^{n} \underline{C_{ij}} y_j , \sum_{j=1}^{n} \overline{C_{ij}} y_j \right]
$$

$$
= \left[\sum_{j=1}^{n} C'_{ij} y_j , \sum_{j=1}^{n} C''_{ij} y_j \right], \tag{13}
$$

where C'_{ij} and C''_{ij} are numbers from the endpoint set $\{ \underline{C}_{ij}, \overline{C}_{ij} \}$, and they are fixed for any separate orthant containing y. Next, writing out the inclusion $Cy \subseteq d$ in a componentwise manner and changing, on the basis of (13), each one-dimensional inclusion to two inequalities between the interval endpoints, we arrive at a system of $3n$ linear inequalities

$$
\begin{cases}
C'y \geq \underline{d}, \\
C''y \leq \overline{d}, \\
\text{conditions on the signs of } y_j, \; j = 1, 2, \ldots, n,
\end{cases} \tag{14}
$$

where C', C'' are point matrices formed by endpoints of the entries of C. Each non-strict inequality of the system (14) determines a closed half-space of $\mathbb{R}^n$, and the solution set to the entire system is the intersection of the half-spaces, that is, a convex polyhedral set in the space $\mathbb{R}^n$.

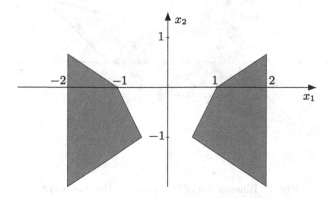

Fig. 2. Disconnected solution set to the interval linear inclusion (15)

Example 2. For the interval linear inclusion system

$$\begin{pmatrix} 1 & 0 \\ [2,-2] & [-3,-1] \end{pmatrix} x \subseteq \begin{pmatrix} [-2,2] \\ 2 \end{pmatrix}, \tag{15}$$

the solution set is depicted at Fig. 2. It is polyhedral within each orthant, but disconnected as a whole.

Proposition 3. *A point belongs to the interior of the solution set $\varXi$ if and only if every sufficiently small perturbation along each coordinate axis do not cause the point to leave the solution set. To be more specific,*

$$y \in \text{int}\, \varXi$$

$$\Updownarrow \tag{16}$$

$$y \in \varXi \ \ and \ \ (\exists \varDelta > 0)(\forall j \in \{1,2,\dots,n\})(\forall |\varepsilon| < \varDelta)\big(x + \varepsilon e^j \in \varXi\big),$$

where $e^j = (0,\dots,0,1,0,\dots,0)^\top$ is the vector with the only nonzero component at the j-th place, i.e., the unit vector of the coordinate axis $0x_j$.

Proof. In equivalence (16), the direct (downward) implication is obvious, and, in fact, we have to prove only the upward implication. It will follows from the convexity of the solution set $\varXi$ within every orthant of the space $\mathbb{R}^n$.

Our substantiation of the upward implication in (16) is constructive. Assuming that the point y complies with the "coordinate-wise stability", i.e., there exists such $\varDelta > 0$ that $y \pm \varepsilon e^j \in \varXi$ for every ε satisfying $|\varepsilon| < \varDelta$ and each $j = 1,2,\dots,n$, we explicitly produce a neighborhood of y that is entirely contained in the solution set $\varXi$.

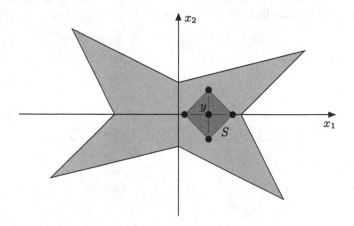

Fig. 3. Illustration of the proof of Proposition 3

Without loss in generality, we suppose that the value of Δ is taken so that the signs of nonzero components y_j are preserved in all the perturbations $y \pm \varepsilon e^j$, $|\varepsilon| < \Delta$. Otherwise, we can always decrease the positive Δ to meet the above requirement.

We take the convex hull S of the $2n + 1$ points y, $y \pm \varepsilon e^j$, $j = 1, 2, \ldots, n$ (see Fig. 3). Since the unit vectors e^j are n linearly independent vectors, then the convex set S also has the dimension n. Hence, S is a bodily convex set in $\mathbb{R}^n$, having nonempty interior [10]. Then S must contain an open ball centered at y. As a consequence, the proof of Proposition 3 is completed if we show that S is included in the solution set Ξ. The latter will be proved by demonstration that the intersection of S with each orthant is included in Ξ.

Since the points $y \pm \varepsilon e^j$, $j = 1, 2, \ldots, n$, represent perturbations of y directed along the coordinate axis, we can claim that the intersection of S with an orthant $\mathcal{O}$ is the convex hull of the points from the set $\{y, y \pm \varepsilon e^j, j = 1, 2, \ldots, n\}$ that belong to the orthant $\mathcal{O}$ itself. Let us denote this convex hull as $S_\mathcal{O}$. In general, $S_\mathcal{O}$ may be a proper subset of the intersection $S \cap \mathcal{O}$, but in our specific case the structure of the point set $\{y, y \pm \varepsilon e^j, j = 1, 2, \ldots, n\}$ is "in conformity" with the partition of $\mathbb{R}^n$ to orthants, since all e^j are the unit coordinate vectors.

It only remains to note that $S_\mathcal{O}$, being the convex hull of points from $\mathcal{O}$, is included in the solution set Ξ, because Ξ is convex within the orthant $\mathcal{O}$.

5   Position of a Point with Respect to the Solution Set

From the definition of the reserve and continuity of the functional $\mathrm{Rsv}(x)$, it is obvious that

$$\mathrm{Rsv}(y) \geq 0 \iff y \in \Xi, \tag{17}$$

$$\mathrm{Rsv}(y) > 0 \implies y \in \mathrm{int}\,\Xi, \tag{18}$$

$$\mathrm{Rsv}(y) = 0 \impliedby y \in \partial\Xi, \tag{19}$$

where $\mathrm{int}\,\Xi$ is the topological interior of the solution set Ξ, and $\partial\Xi$ is the boundary of the solution set Ξ. A natural question is whether we can reverse the logical implications in the second and third cases, thus getting equivalences. That would allow us to completely investigate the position of a point with respect to the interior and boundary of solution sets. Localizing the position of a point within the solution set has practical significance. In particular, if the point is in the interior of the solution set, it is stable under data perturbations and, moreover, we can construct an inner estimating box around the point as a center (see, e.g., [14,17]).

Simple examples show, however, that additional requirements should be imposed on the system under study as well as on the point y in order to make the two-sided implications in (18)–(19) possible.

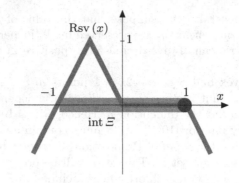

Fig. 4. Graph of the reserve for system (20) in Example 3: the points -1 and 1 is the boundary $\partial \Xi$ of the solution set Ξ

Example 3. Let us consider the following 2×1-system of interval inclusions

$$\begin{cases} [0, -1]\, x \subseteq [0, 1], \\ [-2, 2]\, x \subseteq [-2, 2]. \end{cases} \tag{20}$$

The graph of its reserve is depicted at Fig. 4. We can see that $\mathrm{Rsv}\,(0.5) = 0$, while 0.5 is in the interior of the solution set $\Xi = [-1, 1]$.

Given an inclusion of the form $\boldsymbol{C}x \subseteq \boldsymbol{d}$ and a point $y \in \mathbb{R}^n$, let us divide the index sets of the matrix elements to the following index subsets:

for the row index $i \in \{1, 2, \ldots, m\}$,

$$L := \big\{\, i \mid \underline{\boldsymbol{C}}_{i:}y = \underline{\boldsymbol{d}}_i \big\}, \qquad\qquad R := \big\{\, i \mid \overline{\boldsymbol{C}}_{i:}y = \overline{\boldsymbol{d}}_i \big\},$$

$$\neg L := \big\{\, i \mid \underline{\boldsymbol{C}}_{i:}y \neq \underline{\boldsymbol{d}}_i \big\}, \qquad\quad \neg R := \big\{\, i \mid \overline{\boldsymbol{C}}_{i:}y \neq \overline{\boldsymbol{d}}_i \big\};$$

for the column index $j \in \{1, 2, \ldots, n\}$,

$$P := \{\, j \mid y_j > 0 \,\},$$
$$N := \{\, j \mid y_j < 0 \,\},$$
$$E := \{\, j \mid y_j = 0 \,\}.$$

Overall, we have

$$L \cup \neg L = \{1, 2, \ldots, m\},$$
$$R \cup \neg R = \{1, 2, \ldots, m\},$$
$$P \cup N \cup E = \{1, 2, \ldots, n\}.$$

The special condition on $\boldsymbol{C}$, $\boldsymbol{d}$ and the point y, which we denote $\mathrm{SpeC}\,(y)$, is formulated as follows:

$$\mathrm{SpeC}\,(y) := \begin{cases} \underline{\boldsymbol{C}}_{LP} = 0, & \overline{\boldsymbol{C}}_{LN} = 0, \\[4pt] \overline{\boldsymbol{C}}_{RP} = 0, & \underline{\boldsymbol{C}}_{RN} = 0, \end{cases} \quad \boldsymbol{C}_{(L \cup R)E} \subseteq 0, \tag{21}$$

where, for example, $\underline{C}_{LP}$ means a submatrix within the matrix $\underline{C}$ formed by all the elements having their index pair (i, j) in the set $L \times P$. $(L \cup R)$ means the union of the index subsets L and R, and so on.

Proposition 4. *Let $\varXi$ be the solution set to an interval inclusion $Cx \subseteq d$ with $C \in \mathbb{KR}^{m \times n}$, $d \in \mathbb{KR}^m$. For any $y \in \mathbb{R}^n$, there holds*

$$y \in \text{int}\, \varXi \quad \Longleftrightarrow \quad (\,\text{Rsv}\,(y) > 0\,) \text{ or } (\,\text{Rsv}\,(y) = 0 \,\&\, \text{SpeC}\,(y)\,),$$
$$y \in \partial\varXi \quad \Longleftrightarrow \quad (\,\text{Rsv}\,(y) = 0 \,\&\, \neg\,\text{SpeC}\,(y)\,),$$

where "$\neg$" is the logical negation.

Proof. Our intention is to prove the first equivalence of Proposition 4 using the result of Proposition 3. The second equivalence of Proposition 4 is, in fact, a logical consequence of the first one.

We take a point $y \in \mathbb{R}^n$ and fix an index $j \in P$ (providing that $P \neq \varnothing$), i.e., such that $y_j > 0$. If $\varepsilon > 0$, then

$$C(y + \varepsilon e^j) = \sum_{k \neq j} C_{:k} y_k + C_{:j}(y_j + \varepsilon) \tag{22}$$

$$= \sum_{k \neq j} C_{:k} y_k + [\underline{C}_{:j}(y_j + \varepsilon), \overline{C}_{:j}(y_j + \varepsilon)] \tag{23}$$

$$\text{since } y_j + \varepsilon > 0$$

$$= \sum_{k \neq j} C_{:k} y_k + [\underline{C}_{:j} y_j + \underline{C}_{:j}\varepsilon, \overline{C}_{:j} y_j + \overline{C}_{:j}\varepsilon]$$

$$\text{since } \underline{C}_{:j} \text{ and } \overline{C}_{:j} \text{ are point (noninterval)}$$

$$= \sum_{k \neq j} C_{:k} y_k + [\underline{C}_{:j} y_j, \overline{C}_{:j} y_j] + [\underline{C}_{:j}\varepsilon, \overline{C}_{:j}\varepsilon]$$

$$= \sum_{k=1}^{n} C_{:k} y_k + [\underline{C}_{:j}\varepsilon, \overline{C}_{:j}\varepsilon] \quad \text{since } y_j > 0$$

$$= [\,\underline{C}y + \underline{C}_{:j}\varepsilon, \; \overline{C}y + \overline{C}_{:j}\varepsilon\,]. \tag{24}$$

The membership $y + \varepsilon e^j \in \varXi$ is equivalent to $C(y + \varepsilon e^j) \subseteq d$, which means, due to (24), that

$$\underline{C}y + \underline{C}_{:j}\,\varepsilon \; \geq \; \underline{d} \quad \text{and} \quad \overline{C}y + \overline{C}_{:j}\,\varepsilon \; \leq \; \overline{d}. \tag{25}$$

If $y \in \varXi$, then, from the definition of the index subset L, we get, first,

$$C_{L:}\,y \;=\; \underline{d}_L,$$

and, second,

$$C_{\neg L:}\,y \;>\; \underline{d}_{\neg L}.$$

The latter strict inequality remains true for sufficiently small perturbations ε, while the former equality has nontrivial consequences.

The requirement that $y + \varepsilon e^j \in \varXi$ entails

$$\underline{C}_{L:}\, y + \underline{C}_{Lj}\, \varepsilon \geq \underline{d}_L,$$

and the inequality is satisfied only for $\underline{C}_{Lj}\, \varepsilon \geq 0$, which means, in view of $\varepsilon > 0$, that $\underline{C}_{Lj} \geq 0$. Also, similar arguments applied to the second inequality from (25) and the index subset R imply that $\overline{C}_{Rj} \leq 0$.

On the other hand, we can take the point $(y - \varepsilon e^j)$ instead of $(y + \varepsilon e^j)$ in our above reasoning, starting from (22). The only reservation is that ε should be chosen sufficiently small to keep the inequality $y_j + \varepsilon > 0$ so as we could pass from (22) to (23). This leads to the conclusion that $\underline{C}_{Lj} \leq 0$ and $\overline{C}_{Rj} \geq 0$, which, combined with our previous results on $\underline{C}_{Lj}$ and $\overline{C}_{Rj}$, yields $\underline{C}_{Lj} = 0$ and $\overline{C}_{Rj} = 0$ for every $j \in P$. To put it another way, $\underline{C}_{LP} = 0$ and $\overline{C}_{RP} = 0$.

In exactly the same manner, after fixing an index $j \in N$ (providing that $N \neq \varnothing$), we can prove that $\overline{C}_{LN} = 0$ and $\underline{C}_{RN} = 0$; we omit the expanded reasoning for brevity.

To prove that $C_{(L \cup R)E} \subseteq 0$, we fix an index $j \in E$ among the components of the point y (providing that $E \neq \varnothing$). As the result of considering ε-perturbations of y along the j-th axis, similar to what has been done in the several preceding paragraphs, we get the inequalities $\underline{C}_{(L \cup R)j} \geq 0$ and $\overline{C}_{(L \cup R)j} \leq 0$. Hence, $C_{(L \cup R)E} \subseteq [0,0]$ in the Kaucher complete interval arithmetic, as is required. The details are again omitted.

Summing up, we get

$$y \in \operatorname{int} \varXi \iff y \in \varXi \ \&$$
$$\underline{C}_{LP} = 0 \ \& \ \overline{C}_{RP} = 0 \ \& \ \overline{C}_{LN} = 0 \ \& \ \underline{C}_{RN} = 0 \ \& \ C_{(L \cup E)E} \subseteq 0.$$

In the right-hand side of the above equivalence, the condition at the second line is nothing but SpeC (y). We can further transform this result taking into account that the membership $y \in \varXi$ means Rsv $(y) \geq 0$:

$$y \in \operatorname{int} \varXi \iff \operatorname{Rsv}(y) \geq 0 \ \& \ \operatorname{SpeC}(y)$$
$$\iff \big(\operatorname{Rsv}(y) > 0 \ \& \ \operatorname{SpeC}(y)\big) \vee \big(\operatorname{Rsv}(y) = 0 \ \& \ \operatorname{SpeC}(y)\big)$$
$$\iff \operatorname{Rsv}(y) > 0 \vee \big(\operatorname{Rsv}(y) = 0 \ \& \ \operatorname{SpeC}(y)\big),$$

since for Rsv $(y) > 0$ we have $L = \varnothing$ and $R = \varnothing$, and then the condition SpeC holds true. The last logical formula is exactly what stands in the right-hand side of the first equivalence of Proposition 4.

Finally, we have to prove the second equivalence of Proposition 4. In fact, it is the negation of the first equivalence we have already substantiated, taken under the condition that $y \in \varXi$. Since every point of the solution set $\varXi$ is either interior or boundary, then the negation of $x \in \text{int}\,\varXi$ is the membership $x \in \partial\varXi$. Next, negate the logical formula in the right-hand side of the first equivalence:

$$\neg\Big(\big(\text{Rsv}\,(y) > 0 \big) \vee \big(\text{Rsv}\,(y) = 0 \ \& \ \text{SpeC}\,(y) \big) \Big)$$

$$\updownarrow \qquad \text{by de Morgan's law}$$

$$\neg\big(\text{Rsv}\,(y) > 0 \big) \ \& \ \neg\big(\text{Rsv}\,(y) = 0 \ \& \ \text{SpeC}\,(y) \big)$$

$$\updownarrow \qquad \text{by de Morgan's law}$$

$$\big(\text{Rsv}\,(y) = 0 \big) \ \& \ \big(\text{Rsv}\,(y) > 0 \ \vee \ \neg\,\text{SpeC}\,(y) \big)$$

$$\updownarrow \qquad \text{by distributivity of "\&" and "$\vee$"}$$

$$\big(\text{Rsv}\,(y) = 0 \ \& \ \text{Rsv}\,(y) > 0 \big) \ \vee \ \big(\text{Rsv}\,(y) = 0 \ \& \ \neg\,\text{SpeC}\,(y) \big)$$

$$\updownarrow$$

$$\text{Rsv}\,(y) = 0 \ \& \ \neg\,\text{SpeC}\,(y),$$

since $\big(\text{Rsv}\,(y) = 0 \ \& \ \text{Rsv}\,(y) > 0 \big)$ is always false. The logical formula we have obtained coincides with the right-hand side of the second equivalence in Proposition 4.

The special condition $\text{SpeC}\,(y)$ can be reduced to a more convenient, although less general, form. To give its formulation, we need the concept of *vertex* of an interval vector $\boldsymbol{u} \in \mathbb{K}\overline{\mathbb{R}}^{l}$: it is any such $u \in \overline{\mathbb{R}}^{l}$ that $u_k \in \{\underline{\boldsymbol{u}}_k, \overline{\boldsymbol{u}}_k\}$, $k = 1, 2, \ldots, l$.

Proposition 5. *Assume that*

(i) at least one of the following conditions is true:
 - *y does not lie on a coordinate hyperplane,*
 - *the matrix $\boldsymbol{C}$ is proper;*

(ii) the augmented matrix $(\boldsymbol{C}, \boldsymbol{d})$ does not have rows with zero vertices.

Then $\quad y \in \text{int}\,\varXi \iff \text{Rsv}\,(y) > 0,$
$$y \in \partial\varXi \iff \text{Rsv}\,(y) = 0.$$

Proof. Due to Proposition 4,

$$y \in \text{int}\,\varXi \iff \big(\text{Rsv}\,(y) > 0 \big) \vee \big(\text{Rsv}\,(y) = 0 \ \& \ \text{SpeC}\,(y) \big).$$

So, to substantiate Proposition 5, it suffices to show that the second term of the right-hand side disjunction, i.e., the condition $\big(\text{Rsv}\,(y) = 0 \ \& \ \text{SpeC}\,(y) \big)$ is incompatible with the premise of Proposition 5.

To put the above plan into practice, we are going to demonstrate that if both $\big(\text{Rsv}\,(y) = 0 \ \& \ \text{SpeC}\,(y) \big)$ and the condition (i) hold true, then the condition (ii), i.e.,

$$\forall\, i \in \{1, 2, \ldots, m\} \quad \big(0 \notin \text{vert}\,(\boldsymbol{C}_{i:}, \boldsymbol{d}_i) \big),$$

is violated. As far as

$$\underline{d}_L = \underline{C}_{LP}\,x_P + \overline{C}_{LN}\,x_N + C_{LE}\cdot 0,$$
$$\overline{d}_R = \overline{C}_{RP}\,x_P + \underline{C}_{RN}\,x_N + C_{RE}\cdot 0,$$

we have

$$\text{SpeC}\,(y) \quad \Longrightarrow \quad \underline{d}_L = 0 \ \&\ \overline{d}_R = 0. \tag{26}$$

At the same time, it is obvious that

$$\text{Rsv}\,(y) = 0 \quad \Longrightarrow \quad L \neq \varnothing \ \lor \ R \neq \varnothing. \tag{27}$$

Implications (26) and (27) entail that, under $\big(\text{Rsv}\,(y) = 0 \ \&\ \text{SpeC}\,(y)\big)$, the following is true:

$$(\exists l)\big(\underline{C}_{lP} = 0 \ \&\ \overline{C}_{lN} = 0 \ \&\ C_{lE} \subseteq 0 \ \&\ \underline{d}_l = 0\big)$$
$$\text{or}\ \ (\exists r)\big(\overline{C}_{rP} = 0 \ \&\ \underline{C}_{rN} = 0 \ \&\ C_{rE} \subseteq 0 \ \&\ \overline{d}_r = 0\big). \tag{28}$$

If the point y does not lie on a coordinate hyperplane, then $E = \varnothing$ and (28) implies that there exists such i that $0 \in \text{vert}\,(C_{i:}, d_i)$, which runs contrary to the premise (ii).

If the matrix C is proper, then $C_{lE} \subseteq 0$ is equivalent to $C_{lE} = 0$, and $C_{rE} \subseteq 0$ is equivalent to $C_{rE} = 0$. Again, (28) implies that the premise (ii) is violated.

6   Solvability of the Inclusion $Cx \subseteq d$

In this section, we study the solvability of the inclusion $Cx \subseteq d$, that is, answer the question whether its solution set is empty or not. In the sequel, we denote for brevity

$$\max \text{Rsv} := \max_{x \in \mathbb{R}^n} \text{Rsv}\,(x).$$

If $\text{Rsv}\,(x)$ is unbounded from above, we assign $\max \text{Rsv} = \infty$.

From the equivalence $\big(\text{Rsv}\,(y) \geq 0 \iff y \in \varXi\big)$, it follows that

$$\varXi \neq \varnothing \quad \iff \quad \max \text{Rsv} \geq 0.$$

Therefore, examination of solvability of the inclusion $Cx \subseteq d$ (and of the related interval linear problems as well) amounts to the solution of the unconstrained optimization problem

$$\text{find} \quad \max_{x \in \mathbb{R}^n} \text{Rsv}\,(x).$$

Then one has to inquire into the sign of the maximum.

Finally, we can consider the question on whether the topological interior of $\varXi$ is empty or not. In other words, do there exist solutions to the inclusion $Cx \subseteq d$ that are stable under small perturbations in their position? In the general case, it follows from the implication $\big(\text{Rsv}\,(y) > 0 \implies y \in \text{int}\,\varXi\big)$ that

$$\max \text{Rsv} > 0 \implies \text{int}\,\varXi \neq \varnothing.$$

The following counterexample shows that the reverse implication may prove false.

Example 4 ($\not\Longleftarrow$). For the inclusion $[0,1]\,x \subseteq [0,1]$, we have $\operatorname{int} \varXi =]0,1[\neq \varnothing$, but $\max \operatorname{Rsv} = 0$. The graph of the reserve is depicted at Fig. 5.

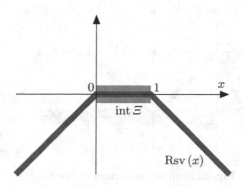

Fig. 5. Zero maximum of the reserve in Example 4

Nevertheless, after imposing special conditions on C and d, we can draw conclusions in the opposite direction too.

Proposition 6. *If the augmented matrix (C, d) does not have rows with zero vertices, then*

$$\operatorname{int} \varXi \neq \varnothing \iff \max \operatorname{Rsv} > 0.$$

Proof. If $\max \operatorname{Rsv} > 0$, there exists such $y \in \mathbb{R}^n$ that $\operatorname{Rsv}(y) > 0$. Therefore, $y \in \operatorname{int} \varXi$ in view of (18), and so $\operatorname{int} \varXi \neq \varnothing$.

Conversely, let $\operatorname{int} \varXi \neq \varnothing$. Then $\operatorname{int} \varXi$ contains an open ball B. Within B, we can take a point y that does not belong to the coordinate planes. Additionally, the augmented matrix (C, d) does not have rows with zero vertices, and this is why we can apply Proposition 5 concluding that $\operatorname{Rsv}(y) > 0$. Hence, $\max \operatorname{Rsv} \geq \operatorname{Rsv}(y) > 0$, as required.

7   The 'best' Points for the Inclusion $Cx \subseteq d$

It follows from the above constructions that the function $\operatorname{Rsv}(x)$ provides us with a quantitative measure of 'how strong' (how 'good') the inclusion $Cx \subseteq d$ is fulfilled. The points where the maximum of the function $\operatorname{Rsv}(x)$ is attained are of special importance, since they satisfy the inclusion $Cx \subseteq d$ in the largest possible amount. Such points are usually the 'best' points (in a certain sense) from the solution set or points that comply with some additional optimality conditions.

Below, we briefly describe the corresponding results, using the notation

$$\operatorname{Arg\,max} := \{ y \in \mathbb{R}^n \mid \operatorname{Rsv}(y) = \max \operatorname{Rsv} \}.$$

We distinguish three cases, when max Rsv is positive, zero, and negative respectively.

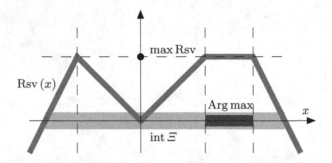

Fig. 6. Positive maximum of the reserve

Case max Rsv > 0 (see Fig. 6):

- Arg max consists of all such points for which $Cx \subseteq d$
  holds with maximum positive reserve.
- Arg max $\subseteq$ int $\varXi$.

We can regard the points from Arg max as 'the most stable' under data perturbations, i. e., under variations in C and d.

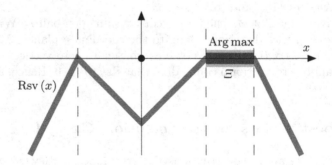

Fig. 7. Zero maximum of the reserve

Case max Rsv $= 0$ (see Fig. 7):

- Arg max consists of all such points for which $Cx \subseteq d$
  holds with maximum reserve, although this reserve is zero.
- Arg max $= \varXi$.

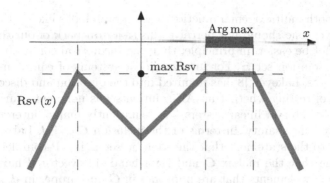

Fig. 8. Negative maximum of the reserve

Case $\max \mathrm{Rsv} < 0$ (see Fig. 8):

- Arg max consists of all such points for which $Cx \subseteq d$ is violated in the minimum amount.
- $\Xi = \varnothing$. Arg max is the solution set to the inclusion $Cx \subseteq d + e\,[\max \mathrm{Rsv}, -\max \mathrm{Rsv}]$.

Although the solution set is empty, the points from the set Arg max can be taken as 'pseudosolutions' to the corresponding system of interval relations, since such points minimize the discrepancy in the characteristic inclusion $Cx \subseteq d$. In particular, the points from Arg max are the first points that appear in nonempty solution set after uniform widening of the right-hand side d by $\max \mathrm{Rsv}$. This follows from (8).

It is worth noting that a particular case of the above construction has been implemented in [18,19] as a promising approach to data fitting problem under interval uncertainty that works even for inconsistent input data.

8 Computational Complexity

We conclude the paper with a discussion on the complexity of the developed technique that relies on the use of the reserve function. Let us consider the first formula of (8):

$$\mathrm{Rsv}\,(x) \;=\; \min_{1 \le i \le m}\ \min\big\{\, \underline{C_{i:}x} - \underline{d_i},\ -\overline{C_{i:}x} + \overline{d_i} \,\big\}.$$

We can see that computation of the value $\mathrm{Rsv}(x)$ requires $2n$ multiplications and 2 additions for each of the two subexpressions under inner minimum. The overall expression for Rsv thus takes $m(4n + 4)$ arithmetic operations, which is quite cheap. Since testing condition SpeC (21) and its derivatives is also inexpensive, we can assert that examining position of a point with respect to AE-solution sets based on the reserve function (Sect. 5) is computationally efficient.

Solvability issues considered in Sect. 6 and the constructions related to the "best points" of the solution sets from Sect. 7 require unconstrained maximization of the reserve function. This is a hard problem since in general $\mathrm{Rsv}\,(x)$ may be a non-smooth multiextremal function whose graph looks like that depicted at Fig. 1. However, one should not perceive this as a drawback of our technique as far as it cannot be easier, in principle, than the theoretical complexity of recognition of the solution set Ξ. For interval linear systems of equations and their AE-solution sets, Lakeyev [8] has inquired into the question and discovered that the problem of testing whether an AE-solution set is nonempty turns out NP-complete if the interval linear system has "sufficiently many" interval elements with E-type of uncertainty. In terms of the inclusion $Cx \subseteq d$, Lakeyev's result is equivalent to the statement that the solution set Ξ is NP-complete to recognize providing that the matrix C and right-hand side vector d have, in total, sufficiently many elements that are improper in C and proper in d.

On the other hand, if an interval linear system of relations has "few" interval elements with E-type of uncertainty, then its solution set can be recognized by polynomial time algorithms. Then the reserve function $\text{Rsv}(x)$ is not hard for maximization too. Such is, for instance, the situation with the tolerable solution set (9) and its recognizing functional (10) which can be efficiently maximized for polynomial time by modern non-smooth optimization procedures.

Anyway, reducing examination of the solvability of the inclusion $Cx \subseteq d$ to the unconstrained maximization problem with the objective function $\text{Rsv}(x)$ provides flexibility in choosing our instruments and in further actions. In particular, for a specific problem, we can take this or that optimization procedure depending on our needs, convenience, capability and available resources.

References

1. Beeck, H.: Über die Struktur und Abschätzungen der Lösungsmenge von linearen Gleichungssystemen mit Intervallkoeffizienten. Computing **10**(3), 231–244 (1972). doi:10.1007/BF02316910
2. Dymova, L.: Soft Computing in Economics and Finance. Springer, Heidelberg (2011). doi:10.1007/978-3-642-17719-4
3. Goldsztejn, A., Chabert, G.: On the approximation of linear AE-solution sets. Proceedings of the 12th GAMM-IMACS International Symposium on Scientific Computing, Computer Arithmetic and Validated Numerics SCAN 2006, Duisburg, Germany, 26–29 September 2006, 18 p (2006). doi:10.1109/SCAN.2006.33
4. Hladík, M.: AE-solutions and AE-solvability to general interval linear systems. Linear Algebra Appl. **465**, 221–238 (2015). doi:10.1016/j.laa.2014.09.030
5. Kaucher, E.: Algebraische Erweiterungen der Intervallrechnung unter Erhaltung der Ordnungs- und Verbandsstrukturen. Comput. Suppl. **1**, 65–79 (1977)
6. Kaucher, E.: Interval analysis in the extended interval space $\mathbb{IR}$. Comput. Suppl. **2**, 33–49 (1980)
7. Kearfott, R.B., Nakao, M., Neumaier, A., Rump, S., Shary, S.P., van Hentenryck, P.: Standardized notation in interval analysis. Comput. Technol. **15**(1), 7–13 (2010)
8. Lakeyev, A.V.: Computational complexity of estimation of generalized solution sets for interval linear systems. Comput. Technol. **8**(1), 12–23 (2003). http://www.nsc.ru/interval/lakeyev/publications/03ct.pdf
9. Popova, E.: Explicit description of AE-solution sets for parametric linear systems. SIAM Journal on Matrix Analysis and Applications **33**(4), 1172–1189 (2012). doi:10.1137/120870359

10. Rockafellar, R.T.: Convex Analysis. Princeton University Press, Princeton (1997). (Reprint of the 1979 Princeton Mathematical Series 28th edn.)
11. Rohn, J.: A Handbook of Results on Interval Linear Problems. Electronic book, Institute of Computer Science, Academy of Sciences of the Czech Republic, Prague (2012). http://uivtx.cs.cas.cz/~rohn/publist/!aahandbook.pdf
12. Sainz, M.Á., Gardeñes, E., Jorba, L.: Interval estimations of solution sets to real-valued systems of linear or non-linear equations. Reliable Comput. 8, 283–305 (2002). doi:10.1023/A:1016385132064
13. Sharaya, I.A.: Quantifier-free descriptions of interval-quantifier linear systems. Proc. Inst. Math. Mech. UB RAS (Trudy Instituta Matematiki i Mekhaniki UrO RAN) 20(2), 311–323 (2014). (in Russian) http://www.nsc.ru/interval/sharaya/Papers/trIMM14.pdf
14. Shary, S.P.: Solving the linear interval tolerance problem. Math. Comput. Simul. 39, 53–85 (1995)
15. Shary, S.P.: Algebraic solutions to interval linear equations and their applications. In: Alefeld, G., Herzberger, J. (eds.) Numerical Methods and Error Bounds, Proceedings of IMACS-GAMM International Symposium on Numerical Methods and Error Bounds, Oldenburg, Germany, July 9–12, 1995. Mathematical Research, vol. 89, pp. 224–233. Akademie Verlag, Berlin (1996). http://www.nsc.ru/interval/shary/Papers/Herz.pdf
16. Shary, S.P.: A new technique in systems analysis under interval uncertainty and ambiguity. Reliable Comput. 8(5), 321–418 (2002). http://www.nsc.ru/interval/shary/Papers/ANewTech.pdf
17. Shary, S.P.: A new method for inner estimation of solution sets to interval linear systems. In: Rauh, A., Auer, E. (eds.) Modeling, Design, and Simulation of Systems with Uncertainties. Mathematical Engineering, vol. 3, pp. 21–42 (2011). doi:10.1007/978-3-642-15956-5_2
18. Shary, S.P.: Maximum consistency method for data fitting under interval uncertainty. J. Glob. Optim. 1–16 (2015). doi:10.1007/s10898-015-0340-1
19. Shary, S.P., Sharaya, I.A.: Recognizing solvability of interval equations and its application to data analysis. Comput. Technol. 18(3), 80–109 (2013). (in Russian)
20. Shary, S.P., Sharaya, I.A.: On solvability recognition for interval linear systems of equations. Optim. Lett. 10(2), 247–260 (2015). doi:10.1007/s11590-015-0891-6

Global Optimisation

Convergence and Inclusion Isotonicity of the Tensorial Rational Bernstein Form

Jürgen Garloff[1]([⊠]) and Tareq Hamadneh[2]

[1] University of Applied Sciences/HTWG Konstanz, Konstanz, Germany
garloff@htwg-konstanz.de
[2] University of Konstanz, Konstanz, Germany
tareq.hamadneh@uni-konstanz.de

Abstract. A method is investigated by which tight bounds on the range of a multivariate rational function over a box can be computed. The approach relies on the expansion of the numerator and denominator polynomials in Bernstein polynomials. Convergence of the bounds to the range with respect to degree elevation of the Bernstein expansion, to the width of the box and to subdivision are proven and the inclusion isotonicity of the related enclosure function is shown.

Keywords: Bernstein polynomial · Rational function · Range bounding

1 Introduction

The expansion of a given (multivariate) polynomial p into Bernstein polynomials provides bounds on the range of p over a box. This is now a well-established tool as documented in [6]. In [8] the approach is extended to rational functions, however, without any proof of the convergence of the bounds to the range. In this paper we aim at filling this gap. Furthermore, we show that the related rational Bernstein form is inclusion isotone, a property which is of fundamental importance in interval computations, see, e.g., [9, Sect. 1.4]. The organization of our paper is as follows. In Sects. 2 and 3 we recall the polynomial and the rational Bernstein forms. In Sect. 4 we present our main results. Related results for the *simplicial* Bernstein form which relies on the expansion of a polynomial into Bernstein polynomials over a simplex are given in [13]. The Bernstein form considered in this paper is also called the *tensorial* Bernstein form. But for simplicity we use here only the term 'Bernstein form'.

2 The Polynomial Bernstein Form

In this section we briefly recall the most important properties of the Bernstein expansion, which will be used in the following sections. Let $\mathbb{I}(\mathbb{R})$ be the set of the compact, non-empty real intervals. We denote the distance q between two intervals $A = [\underline{a}, \overline{a}]$, $B = [\underline{b}, \overline{b}]$ by

$$q([\underline{a}, \overline{a}], [\underline{b}, \overline{b}]) := \max\{|\underline{a} - \underline{b}|, |\overline{a} - \overline{b}|\}.$$

© Springer International Publishing Switzerland 2016
M. Nehmeier et al. (Eds.): SCAN 2014, LNCS 9553, pp. 171–179, 2016.
DOI: 10.1007/978-3-319-31769-4_14

Without loss of generality we may consider the unit box $I := [0,1]^n$ since any compact non-empty box in $\mathbb{R}^n$ can be mapped thereupon by an affine transformation.

Comparisons and arithmetic operations on multiindices $i = (i_1, \ldots, i_n)^T$ are defined componentwise. For $x \in \mathbb{R}^n$ its monomials are $x^i := x_1^{i_1} \ldots x_n^{i_n}$. Using the compact notation $\sum_{i=0}^k := \sum_{i_1=0}^{k_1} \cdots \sum_{i_n=0}^{k_n}$, $\binom{k}{i} := \prod_{\mu=1}^n \binom{k_\mu}{i_\mu}$, an n-variate polynomial p, $p(x) = \sum_{i=0}^l a_i x^i$, can be represented as

$$p(x) = \sum_{i=0}^k b_i^{(k)}(p) \, B_i^{(k)}(x), \ x \in I, \tag{1}$$

where

$$B_i^{(k)}(x) = \binom{k}{i} x^i (1-x)^{k-i} \tag{2}$$

is the ith Bernstein polynomial of degree $k \geq l$, and the so-called *Bernstein coefficients* $b_i^{(k)}(p)$ are given by

$$b_i^{(k)}(p) = \sum_{j=0}^i \frac{\binom{i}{j}}{\binom{k}{j}} a_j, \ 0 \leq i \leq k, \text{ where } a_j := 0 \text{ for } l \leq j, j \neq l. \tag{3}$$

In particular, we have the *endpoint interpolation property*

$$b_i^{(k)}(p) = p(\frac{i}{k}), \text{ for all } i, \ 0 \leq i \leq k, \tag{4a}$$

$$\text{with } i_\mu \in \{0, k_\mu\}. \tag{4b}$$

A fundamental property for our approach is the *convex hull property*, which states that the graph of p over I is contained within the convex hull of the control points derived from the Bernstein coefficients, i.e.,

$$\left\{ \begin{pmatrix} x \\ p(x) \end{pmatrix} : x \in I \right\} \subseteq conv \left\{ \begin{pmatrix} \frac{i}{k} \\ b_i^{(k)}(p) \end{pmatrix} : 0 \leq i \leq k \right\}, \tag{5}$$

where *conv* denotes the convex hull. This implies the *interval enclosing property* [1]

$$\min_{0 \leq i \leq k} b_i^{(k)}(p) \leq p(x) \leq \max_{0 \leq i \leq k} b_i^{(k)}(p), \text{ for all } x \in I. \tag{6}$$

Equality holds on the left or right hand side of (6), if the minimum or maximum, respectively, is attained at an index i satisfying (4b). This condition is called the *vertex condition*. For an efficient computation of the Bernstein coefficients, see [4].

A disadvantage of the direct use of (3) is that the number of the Bernstein coefficients to be computed explicity grows exponentially with the number of variables n. Therefore, it is advantageous to use a method [11] by which the number of coefficients which are needed for the enclosure only grows approximately linearly with the number of the terms of the polynomial.

In many cases it is desired to calculate the Bernstein expansion of p over a general n-dimensional box X in the $\mathbb{I}(\mathbb{R})^n$,

$$X = [\underline{x}_1, \overline{x}_1] \times \cdots \times [\underline{x}_n, \overline{x}_n]$$

with

$$\underline{x}_\mu < \overline{x}_\mu, \ \mu = 1, \ldots, n.$$

The *width* of X is denoted by $w(X)$,

$$w(X) := \overline{x} - \underline{x}.$$

It is possible to firstly apply the affine transformation which maps X on the unit box I and to apply (3) using the coefficients of the transformed polynomial. However, in Sect. 4 it will be useful to consider the direct computation. Here, the ith Bernstein polynomial of degree $k \geq l$ is given by

$$B_i^{(k)}(x) = \binom{k}{i}(x - \underline{x})^i(\overline{x} - x)^{k-i} w(X)^{-k}, \ 0 \leq i \leq k. \tag{7}$$

The Bernstein coefficients $b_i^{(k)}$ of p of degree k over X are given by

$$b_i^{(k)}(p) = \sum_{j=0}^{i} \frac{\binom{i}{j}}{\binom{k}{j}} c_j, \ 0 \leq i \leq k, \tag{8}$$

$$\text{where} \quad c_j = w(X)^j \sum_{\tau=j}^{k} \binom{\tau}{j} a_\tau \underline{x}^{\tau-j} \tag{9}$$

with the convention $a_j := 0$ for $l \leq j, \ l \neq j$.

The interval

$$B^{(k)}(p, X) := [\min_{0 \leq i \leq k} b_i^{(k)}, \max_{0 \leq i \leq k} b_i^{(k)}]$$

encloses the range of p over X and is called the *polynomial Bernstein form* of p.

If the degree of the Bernstein expansion is elevated, the Bernstein coefficients of order $k+1$ can easily be computed as convex combinations of the coefficients of order k, e.g., [2, formula (13)], [4, formula (3.11)]. It follows that

$$B^{(k+1)}(p, X) \subseteq B^{(k)}(p, X). \tag{10}$$

The following theorem, see [10, formula (16)] for the univariate case and [4, Theorem 3] for its multivariate extension, will be used to derive our main results.

Theorem 1. For $l \leq k$, the following bound holds for the overestimation of the range $p(X)$ of p over X by the Bernstein form

$$q(p(X), B^{(k)}(p, X)) \leq \sum_{i=0}^{l} \sum_{\mu=1}^{n} \frac{[\max(0, i_\mu - 1)]^2}{k_\mu} |c_i|, \tag{11}$$

where the coefficients c_i are given by (9).

Remark 1. If $2 \leq k_\mu$ the bound on the right hand side of (11) can be improved slightly, see [10, formula (17)]. For later use we note an extension of [10, Theorem 4]. Let $x_i^{(k)}$ be the grid point the μth component of which is given by

$$x_{i,\mu}^{(k)} = \underline{x}_\mu + \frac{i_\mu}{k_\mu}(\overline{x}_\mu - \underline{x}_\mu), \ \mu = 1, \ldots, n. \tag{12}$$

Then by [4, p. 42] the difference $|(p(x_i^{(k)}) - b_i^{(k)}|$ can be bounded from above for all i, $0 \leq i \leq k$, by the right-hand side of (11).

3   The Rational Bernstein Form

Let p and q be polynomials in variables $x_1, \ldots, x_n$ with Bernstein coefficients $b_i^{(k)}(p)$ and $b_i^{(k)}(q)$, $0 \leq i \leq k$, over a box X, respectively. We consider the rational function $f := p/q$. We may assume that both p and q have the same degree l since otherwise we can elevate the degree of the Bernstein expansion of either polynomial by component where necessary to ensure that their Bernstein coefficients are of the same order $k \geq l$. We call

$$b_i^{(k)}(f) := \frac{b_i^{(k)}(p)}{b_i^{(k)}(q)}, \ 0 \leq i \leq k,$$

the *rational Bernstein coefficients* of f.

Theorem 2. [8, Theorem 3.1] Assume that all Bernstein coefficients $b_i^{(k)}(q)$ have the same sign and are non-zero (this implies that $q(x) \neq 0$, for all $x \in X$). Then the following enclosure for the range of f over X holds:

$$\underline{m}^{(k)} := \min_{0 \leq i \leq k} b_i^{(k)}(f) \leq f(x) \leq \max_{0 \leq i \leq k} b_i^{(k)}(f) =: \overline{m}^{(k)}, \ \text{for all} \ x \in X. \tag{13}$$

The interval spanned by the left and right hand sides of (13) constitutes the *rational Bernstein form* $B(f, X)$,

$$B^{(k)}(f, X) := [\underline{m}^{(k)}, \ \overline{m}^{(k)}].$$

Remark 2. The convex hull property (5) does not in general carry over to rational functions and control points formed from the rational Bernstein coefficients even in the univariate case ($n = 1$). For a counterexample see [8].

4   Main Results

Let throughout $f = p/q$ be a rational function, where p and q are polynomials of degree l and let the range of f over X be $f(X) = [\underline{f}, \overline{f}]$. Without loss of generality we assume that

$$0 < b_i^{(l)}(q), \quad \text{for all } i, \ 0 \le i \le l, \tag{14}$$

and prove the statements only for the upper bounds since the proofs for the lower bounds are entirely analogous. The polynomial r,

$$r := p - \overline{m}^{(k)} q, \tag{15}$$

will serve as a vehicle to convey the results from the polynomial to the rational case. Note that the Bernstein coefficients of a polynomial are linear, hence

$$b_i^{(k)}(r) = b_i^{(k)}(p) - \overline{m}^{(k)} b_i^{(k)}(q). \tag{16}$$

First we show that the vertex condition remains in force.

Proposition 3. It holds that $\underline{m}^{(k)} = \underline{f}$ ($\overline{m}^{(k)} = \overline{f}$) if and only if $\underline{m}^{(k)}$ ($\overline{m}^{(k)}$) $= b_i^{(k)}(f)$ with i satisfying (4b).

Proof. By (4a), $b_i^{(k)}(f)$ with i satisfying (4b) is a value of f at a vertex of X. If follows that $\overline{m}^{(k)}$ is sharp if it is attained at such a Bernstein coefficient. Conversely, assume that $\overline{m}^{(k)} = \overline{f}$,

$$\overline{m}^{(k)} = b_{i_0}^{(k)}(f), \quad \text{for some } i_0, \ 0 \le i_0 \le k, \tag{17}$$

and $\overline{f} = f(\hat{x})$ for some $\hat{x} \in X$. Then we can conclude that

$$\frac{r(\hat{x})}{q(\hat{x})} = f(\hat{x}) - \overline{m}^{(k)} = 0,$$

hence $r(\hat{x}) = 0$. Since r is nonpositive on X it attains its maximum at $\hat{x}$. On the other hand, we have by (16)

$$b_i^{(k)}(r) \le 0, \quad \text{for all } i, \ 0 \le i \le k, \tag{18}$$

and by (17) $b_{i_0}^{(k)}(r) = 0$. So we can conclude that

$$\max_{x \in X} r(x) = b_{i_0}^{(k)}(r). \tag{19}$$

By the polynomial vertex condition if follows that the index i_0 satisfies (4b). $\square$

4.1   Linear Convergence with Respect to Degree Elevation

We start with the observation that the monotonicity property (10) carries over to the rational case.

Proposition 4. For $l \leq k$ it holds that $B^{(k+1)}(f, X) \subseteq B^{(k)}(f, X)$.

Proof. By application of (10) to polynomial r (15) and noting (16) we obtain for all j, $0 \leq j \leq k+1$,

$$b_j^{(k+1)}(p) - \overline{m}^{(k)} b_j^{(k+1)}(q) \leq \max_{0 \leq i \leq k+1} \{ b_i^{(k+1)}(p) - \overline{m}^{(k)} b_i^{(k+1)}(q) \}$$

$$\leq \max_{0 \leq i \leq k} \{ b_i^{(k)}(p) - \overline{m}^{(k)} b_i^{(k)}(q) \} \leq 0,$$

hence $b_j^{(k+1)}(f) \leq \overline{m}^{(k)}$.   □

Theorem 5. For $l \leq k$ it holds that

$$q(f(X), B^{(k)}(f, X)) \leq \frac{\beta}{k}, \tag{20}$$

where β is a constant not depending on k.

Proof. Without loss of generality we consider only the case $0 \leq \overline{m}^{(k)}$. We assume again that (17) holds and use the corresponding grid point $x_{i_0}^{(k)}$, see (12). By (10) we may estimate for $l \leq k$

$$\overline{m}^{(k)} \leq \overline{m}^{(l)} \leq \frac{\max b_i^{(l)}(p)}{\min b_i^{(l)}(q)} =: \beta'. \tag{21}$$

We can conclude from (17) that

$$\overline{m}^{(k)} - \overline{f} \leq \overline{m}^{(k)} - f(x_{i_0}^{(k)})$$

$$= \frac{\overline{m}^{(k)} \cdot q(x_{i_0}^{(k)}) - p(x_{i_0}^{(k)}) + b_{i_0}^{(k)}(p) - \overline{m}^{(k)} \cdot b_{i_0}^{(k)}(q)}{q(x_{i_0}^{(k)})}$$

$$= \frac{\overline{m}^{(k)} (q(x_{i_0}^{(k)}) - b_{i_0}^{(k)}(q)) + b_{i_0}^{(k)}(p) - p(x_{i_0}^{(k)})}{q(x_{i_0}^{(k)})}.$$

Taking absolute values and using Remark 1 and (10) we can estimate

$$\overline{m}^{(k)} - \overline{f} \leq \frac{\beta' \frac{\beta_1}{k} + \frac{\beta_2}{k}}{\min b_i^{(l)}(q)}, \tag{22}$$

where β_1, β_2 are constants not depending on k, which completes the proof.   □

4.2   Quadratic Convergence with Respect to the Width of an Interval

Inspection of (22) shows that we can extract the square of $\max_{\mu=1}^{n}(\overline{x}_\mu - \underline{x}_\mu)$ from the constant β in (20), cf. (9), (11). Therefore, we obtain the following extension of [12, Corollary 3.4.16].

Theorem 6. Let $A \in \mathbb{I}(\mathbb{R})^n$ be fixed. Then for all $X \in \mathbb{I}(\mathbb{R})^n$, $X \subseteq A$, and $l \leq k$ it holds that

$$q(f(X), B^{(k)}(f, X)) \leq \gamma \, \|w(X)\|_\infty^2, \tag{23}$$

where γ is a constant not depending on X.

4.3   Quadratic Convergence with Respect to Subdivision

Since the convergence with respect to degree elevation is only linear we will choose $k = l$ in the sequel and reserve in this subsection the upper index of the Bernstein coefficients for the subdivision level. For simplicity we consider the unit box I. Repeated bisection of $I^{(0,1)} := I$ in all n coordinate directions results at subdivision level $1 \leq h$ in subboxes $I^{(h,\nu)}$ of edge length 2^{-h}, $\nu = 1,\ldots,2^{nh}$. Denote the Bernstein coefficients of f over $I^{(h,\nu)}$ by $b_i^{(h,\nu)}(f)$. For their computation see [4,14]. Put

$$B^{(h)}(f) := [\min_{\substack{0 \leq i \leq l, \\ 1 \leq \nu \leq 2^{nh}}} b_i^{(h,\nu)}(f), \ \max_{\substack{0 \leq i \leq l, \\ 1 \leq \nu \leq 2^{nh}}} b_i^{(h,\nu)}(f)].$$

We obtain the following extension of [3, formula (23)].

Theorem 7. For each $1 \leq h$ it holds

$$q(f(X), B^{(h)}(f)) \leq \delta(2^{-h})^2, \tag{24}$$

where δ is a constant not depending on h.

Proof. Assume that

$$\max_{\substack{0 \leq i \leq l, \\ 1 \leq \nu \leq 2^{nh}}} b_i^{(h,\nu)} = \max_{0 \leq i \leq l} b_i^{(h,\nu_0)}, \quad \text{for some } \nu_0, \ 0 \leq \nu_0 \leq 2^{nh}.$$

Then it follows by Theorem 6

$$\max_{\substack{0 \leq i \leq l, \\ 1 \leq \nu \leq 2^{nh}}} b_i^{(h,\nu)} - \max_{x \in I} f(x) \leq \max_{0 \leq i \leq l} b_i^{(h,\nu_0)} - \max_{x \in I^{(h,\nu_0)}} f(x)$$

$$\leq \delta \|w(I^{(h,\nu_0)})\|_\infty^2 = \delta \, (2^{-h})^2. \qquad \square$$

Remark 3. Note that by (9), (11) the constants β, γ and δ in (20), (23), and (24) can be given explicitly.

4.4   Inclusion Isotonicity

We continue with choosing $k = l$ and suppress therefore the upper index for the Bernstein coefficients. An interval function $F : \mathbb{I}(\mathbb{R})^n \longrightarrow \mathbb{I}(\mathbb{R})$ is called *inclusion isotone,* if, for all $X, Y \in \mathbb{I}(\mathbb{R})^n$, $X \subseteq Y$ implies $F(X) \subseteq F(Y)$.

In [7] it was shown by a lengthy proof that the polynomial Bernstein form is inclusion isotone. In [5] a brief proof of this property and an extension to the multivariate case are presented. We show that the inclusion isotonicity carries over to rational functions.

Theorem 8. The rational Bernstein form is inclusion isotone.

Proof. We consider without loss of generality the unit box I and denote the Bernstein coefficients of the rational function f over I by $b_i(f)$, $0 \le i \le l$. It suffices to show that the inclusion isotonicity holds if we shrink only one edge of I and this is done in turn separately at its left and right endpoint. Without loss of generality we consider only the first case and the first component interval of I and denote by $b_i^*(f)$, $0 \le i \le l$, the Bernstein coefficients of f over $[\epsilon, 1] \times [0, 1]^{n-1}, 0 < \epsilon < 1$. Put

$$\overline{m}^* := \max_{0 \le i \le l} b_i^*(f).$$

We proceed by contradiction and assume that

$$\overline{m}^* = b_{i_0}^*(f), \text{ for some } i_0, \ 0 \le i_0 \le l, \tag{25}$$

and

$$\overline{m} := \max_{0 \le i \le l} b_i(f) < \overline{m}^*. \tag{26}$$

Since the Bernstein form of the polynomial $p - \overline{m}^* q$ is inclusion isotone we obtain from (26) that

$$b_{i_0}^*(p) - \overline{m}^* b_{i_0}^*(q) \le \max_{0 \le i \le l} \{b_i^*(p) - \overline{m}^* b_i^*(q)\}$$

$$\le \max_{0 \le i \le l} \{b_i(p) - \overline{m}^* b_i(q)\}$$

$$< \max_{0 \le i \le l} \{b_i(p) - \overline{m}\, b_i(q)\} \le 0$$

from which we get a contradiction to (25).   □

Acknowledgements. The authors gratefully acknowledge support from the University of Applied Sciences/HTWG Konstanz through the SRP program.

References

1. Cargo, G.T., Shisha, O.: The Bernstein form of a polynomial. J. Res. Nat. Bur. Stand. Sect. B **70B**, 79–81 (1966)
2. Farouki, R.T.: The Bernstein polynomial basis: a centennial retrospective. Comput. Aided Geom. Des. **29**, 379–419 (2012)
3. Fischer, H.C.: Range computation and applications. In: Ulrich, C. (ed.) Contributions to Computer Arithmetic and Self-Validating Numerical Methods, pp. 197–211. Balzer, Amsterdam (1990)
4. Garloff, J.: Convergent bounds for the range of multivariate polynomials. In: Nickel, K. (ed.) Interval Mathematics 1985. LNCS, vol. 212, pp. 37–56. Springer, Heidelberg (1986)
5. Garloff, J., Jansson, C., Smith, A.P.: Inclusion isotonicity of convex-concave extensions for polynomials based on Bernstein expansion. Computing **70**, 111–119 (2003)
6. Garloff, J., Smith, A.P.: Special issue on the use of Bernstein polynomials in reliable computing: a centennial anniversary. Reliab. Comput. **17** (2012)
7. Hong, H., Stahl, V.: Bernstein form is inclusion monotone. Computing **55**, 43–53 (1995)
8. Narkawicz, A., Garloff, J., Smith, A.P., Muñoz, C.A.: Bounding the range of a rational function over a box. Reliab. Comput. **17**, 34–39 (2012)
9. Neumaier, A.: Interval Methods for Systems of Equations. Encyclopedia of Mathematics and Its Applications, vol. 37. Cambridge University Press, Cambridge (1990)
10. Rivlin, T.: Bounds on a polynomial. J. Res. Nat. Bur. Stand. Sect. B **74B**, 47–54 (1970)
11. Smith, A.P.: Fast construction of constant bound functions for sparse polynomials. J. Global Optim. **43**, 445–458 (2009)
12. Stahl, V.: Interval methods for bounding the range of polynomials and solving systems of nonlinear equations. Dissertation, Johannes Kepler University, Linz (1995)
13. Titi, J., Hamadneh, T., Garloff, J.: Convergence of the simplicial rational Bernstein form. In: Le Thi, H.A., Pham, D.T., Nguyen, N.T. (eds.) MCO 2015 - Part I. AISC, vol. 359, pp. 433–441. Springer, Heidelberg (2015)
14. Zettler, M., Garloff, J.: Robustness analysis of polynomials with polynomial parameter dependency using Bernstein expansion. IEEE Trans. Automat. Control **43**, 425–431 (1998)

The Bernstein Branch-and-Bound Unconstrained Global Optimization Algorithm for MINLP Problems

Bhagyesh V. Patil[1]($\boxtimes$) and P.S.V. Nataraj[2]

[1] Université de Nantes, 2 rue de la Houssiniére, 44322 Nantes Cedex 3, France
bhagyesh.patil@gmail.com
[2] Systems and Control Engineering, IIT Bombay, Powai, Mumbai 400 076, India
nataraj@sc.iitb.ac.in

Abstract. In this work a Bernstein global optimization algorithm to solve unconstrained polynomial mixed-integer nonlinear programming (MINLP) problems is proposed. The proposed algorithm use a branch-and-bound framework and possesses several new features, such as a modified subdivision procedure, the Bernstein box consistency and the Bernstein hull consistency procedures to prune the solution search space. The performance of the proposed algorithm is numerically investigated and compared with previously reported Bernstein global optimization algorithm on a set of 10 test problems. The findings of the tests establishes the efficacy of the proposed algorithm over the previously reported Bernstein algorithm in terms of the chosen performance metrics.

1 Introduction

Global optimization of MINLP problems is a promising research area and has been a point of attraction to many researchers from academia as well as industry. In this work, we attempt to solve such MINLP problems which are of the following form:

$$\min_{x} \quad f(x)$$
$$x_k \in \mathbf{x} \subseteq \mathbb{R}, \quad k = 1, 2, \ldots, l_d \tag{1}$$
$$x_k \in \mathbb{Z}, \quad k = l_d + 1, \ldots, l,$$

where $f : \mathbb{R}^l \mapsto \mathbb{R}$ is the (possibly polynomial nonlinear) objective function, $\mathbf{x} := [\underline{x}, \overline{x}]$ is an interval in $\mathbb{R}$, x_k ($k = 1, 2, \ldots, l_d$) are continuous decision variables, and the rest of x_k ($k = l_d + 1, \ldots, l$) are integer decision variables.

A widely used strategy for solving MINLP problems of the form (1) is to use a branch-and-bound (BB) framework [5]. Specifically, a relaxed nonlinear programming (NLP) problem is solved at each node of the branch-and-bound tree. Different variants of the BB approach have been reported in the literature and are widely adapted by several state-of-the-art MINLP solvers (cf. Bonmin [2], SBB [6], BARON [14]). Recently, a new technique based on the separable

© Springer International Publishing Switzerland 2016
M. Nehmeier et al. (Eds.): SCAN 2014, LNCS 9553, pp. 180–198, 2016.
DOI: 10.1007/978-3-319-31769-4_15

underestimators for box-constrained MINLP problems has been proposed in [3], and found to be well competent with state-of-the-art MINLP solvers. We direct the interested readers to [4] and references therein for more details about state-of-the-art available tools for solving MINLP problems. However, despite of the widespread enjoyed interest by the BB approach in the field of MINLPs, we note that sometimes the type of NLP solver used has found to limit its performance in practice. This seems to be true since most of the NLP solvers assume generalized *convexity*. To solve *polynomial* NLP problems, an alternative approach is provided by the Bernstein global optimization algorithms which use the Bernstein form of a polynomial function. A notable feature of the Bernstein form is that the Bernstein range enclosure of a function is sharper than those obtained with most interval forms [13]. Hence, algorithms based on the Bernstein form in practice are found to be more effective than existing interval algorithms. Several variants of such Bernstein algorithms to solve unconstrained and constrained polynomial NLPs have been reported in literature (see, for instance, [9,11]). However, we note that no work has yet been reported in the literature for global optimization of *bound constrained polynomial* MINLP problems using the Bernstein form. This motivates us to investigate and dig more findings with this elegant Bernstein form approach in the direction of bound constrained MINLP problems.

In this work, we propose a Bernstein algorithm for bound constrained global optimization of MINLPs of the form (1). The proposed algorithm is similar in philosophy to interval branch-and-bound procedures for the global optimization of bound constrained NLPs and extends the work proposed for solving unconstrained NLP problems by [9]. The proposed algorithm use combination of the several enhanced tools, such as the monotonicity and the concavity tests, a modified subdivision procedure, and the Bernstein box consistency and the Bernstein hull consistency procedures to prune the solution search space. It may be noted that the Bernstein algorithm in [9] lack such solution search space pruning procedures and hence may be computationally expensive. Further, the performance of the proposed algorithm is evaluated on a collection of 10 test problems taken from NLP literature. These problems are appropriately modified as MINLPs and the test results are compared with the Bernstein algorithm in [9][1].

The rest of the paper is organized as follows. In Sect. 2, we introduce the reader with some background of the Bernstein form. In Sect. 3, we suggest some improvements in the Bernstein algorithm used to solve NLPs. We also present consistency techniques based on the Bernstein form to prune the solution search space. Finally, we present our main proposed unconstrained global optimization algorithm for the MINLP problems. In Sect. 4, numerical experiments are reported along with their findings. Lastly, we draw some conclusions from the present work in the Sect. 5.

---

[1] Albeit, the Bernstein global optimization algorithm in [9] is for NLP problems, we modify it at appropriate places to handle integer decision variables.

2   Background

In this section, we introduce few notions about the Bernstein form. We would like to direct the interested reader to reference [11] for more details about the topic.

Let $l \in \mathbb{N}$ be the number of variables and $x = (x_1, x_2, ..., x_l) \in \mathbb{R}^l$. A multi-index I is defined as $I = (i_1, i_2, ..., i_l) \in \mathbb{N}^l$ and the multi-power x^I is defined as $x^I = (x_1^{i_1}, x_2^{i_2}, ..., x_l^{i_l})$. A multi-index N is defined as $N = (n_1, n_2, ..., n_l)$. Inequalities $I \leq N$ for multi-indices are meant component-wise. With $I = (i_1, ..., i_{r-1}, i_r, i_{r+1}, ..., i_l)$ we associate the index $I_{r,k}$ given by $I_{r,k} = (i_1, ..., i_{r-1}, i_r + k, i_{r+1}, ..., i_l)$, where $0 \leq i_r + k \leq n_r$. Also we write $\binom{N}{I}$ for $\binom{n_1}{i_1} \cdots \binom{n_l}{i_l}$ and (N/I) for $(n_1/i_1, n_2/i_2, ..., n_l/i_l)$ provided that $0 < i_k$, $k = 1, 2, ..., l$.

A real bounded and closed interval $\mathbf{x}$ is defined as

$$\mathbf{x} = [\underline{x}, \overline{x}] := [\inf \mathbf{x}, \sup \mathbf{x}] \in \mathbb{IR},$$

where $\mathbb{IR}$ denotes the set of compact intervals. Let $w(\mathbf{x})$ denote the width of $\mathbf{x}$, that is $w(\mathbf{x}) := \overline{x} - \underline{x}$, and $m(\mathbf{x})$ denote the midpoint of $\mathbf{x}$, that is $m(\mathbf{x}) := (\overline{x} + \underline{x})/2$. Similarly, for an l-dimensional interval vector or box $\mathbf{x} = (\mathbf{x}_1, \mathbf{x}_2, ..., \mathbf{x}_l) \in \mathbb{IR}^l$, the width of $\mathbf{x}$ is $w(\mathbf{x}) := \max(w(\mathbf{x}_1), w(\mathbf{x}_2), ..., w(\mathbf{x}_l))$.

We can write an l-variate polynomial p in the form

$$p(x) = \sum_{I \leq N} a_I x^I, \quad x \in \mathbb{R}^l \tag{2}$$

with N being the degree of p. We expand a given multivariate polynomial into Bernstein polynomials to obtain bounds for its range over an l-dimensional box $\mathbf{x}$. The Ith Bernstein basis polynomial of degree N is defined as

$$B_I^N(x) = B_{i_1}^{n_1}(x_1) \cdots B_{i_l}^{n_l}(x_l), \quad x \in \mathbb{R}^l, \tag{3}$$

where, for $i_j = 0, 1, ..., n_j$, $j = 1, 2, ..., l$

$$B_{i_j}^{n_j}(x_j) = \binom{n_j}{i_j} \frac{(x_j - \underline{x}_j)^{i_j} (\overline{x}_j - x_j)^{n_j - i_j}}{(\overline{x}_j - \underline{x}_j)^{n_j}}. \tag{4}$$

The Bernstein coefficients $b_I(\mathbf{x})$ of p over the box $\mathbf{x}$ are given by

$$b_I(\mathbf{x}) = \sum_{J \leq I} \frac{\binom{I}{J}}{\binom{N}{J}} w(\mathbf{x})^J \sum_{K \leq J} \binom{K}{J} (\inf \mathbf{x})^{K-J} a_K, \; I \leq N. \tag{5}$$

The Bernstein form of a multivariate polynomial p is defined by

$$p(x) = \sum_{I \leq N} b_I(\mathbf{x}) B_I^N(x). \tag{6}$$

The Bernstein coefficients are collected in an array $(b_I(\mathbf{x}))_{I \in S}$, where $S = \{I : I \leq N\}$. We denote S_0 as a special subset of the index set S comprising indices of the vertices of this array, that is

$$S_0 := \{0, n_1\} \times \{0, n_2\} \times \cdots \times \{0, n_l\}.$$

Theorem 1. *(Range enclosure property) Let p be a polynomial of degree N, and let $\bar{p}(\mathbf{x})$ denote the range of p on a given box $\mathbf{x} \in \mathbb{IR}^l$. Then,*

$$\bar{p}(\mathbf{x}) \subseteq B(\mathbf{x}) := \left[\min \ (b_I(\mathbf{x}))_{I \in S}, \ \max \ (b_I(\mathbf{x}))_{I \in S}\right]. \tag{7}$$

Proof: See [7].

Remark 1. The above theorem says that the minimum and maximum coefficients of the array $(b_I(\mathbf{x}))_{I \in S}$ provide lower and upper bounds for the range. This forms the Bernstein range enclosure, defined by $B(\mathbf{x})$ in Eq. (7). The Bernstein range enclosure can successively be sharpened by the continuous domain subdivision procedure [7].

Lemma 2. *(Vertex property) [7] Consider the Bernstein form in Eq. (6) for a polynomial p of degree N, and let the range $\bar{p}(\mathbf{x}) = [a, b]$. Then*

$$a = \min_{0 \leq I \leq N} \ (b_I(\mathbf{x})) \ \text{if and only if} \ \min_{0 \leq I \leq N} \ (b_I(\mathbf{x})) = \min_{I \in S_0} \ (b_I(\mathbf{x}))$$

$$b = \max_{0 \leq I \leq N} \ (b_I(\mathbf{x})) \ \text{if and only if} \ \max_{0 \leq I \leq N} \ (b_I(\mathbf{x})) = \max_{I \in S_0} \ (b_I(\mathbf{x}))$$

Remark 3. The above Lemma says that the lower bound (respectively upper bound) is sharp if and only if $\min \ (b_I(\mathbf{x}))_{I \in S}$ (respectively $\max(b_I(\mathbf{x}))_{I \in S}$) is attained at a Bernstein coefficients of the array $(b_I(\mathbf{x}))$ with $I \in S_0$. This condition is known as the *vertex property*.

3   Proposed Algorithm

In this section, we present a modified subdivision procedure which is an extension of a classical subdivision procedure used by the Bernstein algorithms. We then introduce new constraints formulated based on the gradient and upper bound on the global minimum of the objective function (f). Further, the consistency techniques based on the Bernstein form are introduced. We shall use this Bernstein consistency techniques in combination with our new formulated constraints to isolate those stationary points from the solution search space that are not the global minimum. Finally, we shall combine all these enhancements and present our main proposed Bernstein algorithm to solve unconstrained MINLP problems.

3.1   Modified Subdivision Procedure

As explained in Remark 1, the range enclosure obtained using Bernstein coefficients can be improved by subdividing the domain of decision variables. A subdivision in the rth direction ($1 \leq r \leq l$) is a bisection perpendicular to this direction. Let

$$\mathbf{x} = [\underline{x_1}, \overline{x_1}] \times \cdots \times [\underline{x_r}, \overline{x_r}] \times \cdots \times [\underline{x_l}, \overline{x_l}], \tag{8}$$

be any subbox. Generally, $\mathbf{x}$ is bisected along the rth component direction (for NLPs having only continuous decision variables); resulting into two subboxes $\mathbf{x}_A$ and $\mathbf{x}_B$ as

$$\mathbf{x}_A = [\underline{\mathbf{x}_1}, \overline{\mathbf{x}_1}] \times \cdots \times [\underline{\mathbf{x}_r}, m(\mathbf{x}_r)] \times \cdots \times [\underline{\mathbf{x}_l}, \overline{\mathbf{x}_l}], \tag{9}$$

$$\mathbf{x}_B = [\underline{\mathbf{x}_1}, \overline{\mathbf{x}_1}] \times \cdots \times [m(\mathbf{x}_r), \overline{\mathbf{x}_r}] \times \cdots \times [\underline{\mathbf{x}_l}, \overline{\mathbf{x}_l}], \tag{10}$$

where $m(\mathbf{x}_r)$ denotes the midpoint of $[\underline{\mathbf{x}_r}, \overline{\mathbf{x}_r}]$.

Similar to the above, we suggest following modification in the subdivision procedure to cope with the *integer* decision variables in our proposed Bernstein algorithm. We bisect $\mathbf{x}$ along the rth component direction; such that, two subboxes $\mathbf{x}_A$ and $\mathbf{x}_B$ are formed as below

$$\mathbf{x}_A = [\underline{\mathbf{x}_1}, \overline{\mathbf{x}_1}] \times \cdots \times [\underline{\mathbf{x}_r}, \lfloor m(\mathbf{x}_r) \rfloor] \times \cdots \times [\underline{\mathbf{x}_l}, \overline{\mathbf{x}_l}], \tag{11}$$

$$\mathbf{x}_B = [\underline{\mathbf{x}_1}, \overline{\mathbf{x}_1}] \times \cdots \times [\lfloor m(\mathbf{x}_r) + 1 \rfloor, \overline{\mathbf{x}_r}] \times \cdots \times [\underline{\mathbf{x}_l}, \overline{\mathbf{x}_l}], \tag{12}$$

where $\lfloor m(\mathbf{x}_r) \rfloor$ denotes the *floor* of midpoint of $[\underline{\mathbf{x}_r}, \overline{\mathbf{x}_r}]$. We shall use the subdivision procedure in Eqs. (9) and (10), if the rth component direction is a continuous decision variable. Similarly, the subdivision procedure in Eqs. (11) and (12) will be used, if the rth component direction is a integer decision variable.

Similarly, other more sophisticated approaches to branch integer variables in a branch-and-bound tree are reported in a literature. For instance, [2,14] investigate use of a valid inequalities to discard fractional solutions at each node of a branch-and-bound tree. We direct the interested reader to [14], and references therein for more specific details about the topic.

3.2   Constraint Formulation

In the constrained global optimization algorithms, we determine the global minimum subjected to some (inequality and equality) feasibility constraints (see, for instance, work by Nataraj et al. [10]. We can apply the consistency techniques to these feasibility constraints to contract the bounds on the decision variables. However, in the bound constrained global optimization algorithms there are no such feasibility constraints, and this can defy the application of these consistency techniques. To alleviate this problem, we introduce the constraints based on the gradient and upper bound on the global minimum of the objective function (f), wherein we can apply these consistency techniques.

– Assume that the objective function (f) is continuously differentiable. Then, the gradient (∇f) of f is zero at local minima, at maxima, at saddle points, and at the global minima. Thus, we can find the zero(s) of the gradient at which f has a global minimum by discarding any that are not a global minimum of f. In practice the gradient will have l components and we can form l constraints corresponding these l components as

$$f'_{x_r} = 0, \quad r = 1, 2, \ldots, l, \tag{13}$$

where l being the number of variables in the objective function and $\nabla f = [f'_{x_1} \; f'_{x_2} \; \cdots \; f'_{x_l}]^T$.

- In the global optimization algorithm, at the outset we may compute an upper bound (say, $\widetilde{f}$) on the global minimum (f^*) on the box $\mathbf{x}$, that is

$$\widetilde{f} \geq f^*. \tag{14}$$

Thus, we can delete any point (or subbox) of $\mathbf{x}$ for which $f > \widetilde{f}$. This serves to delete a subbox that bounds a nonoptimal stationary point of f.

We shall apply the Bernstein box and Bernstein hull consistency techniques (refer to the Sect. 3.3) to Eqs. (13) and (14) to delete nonoptimal points from the box $\mathbf{x}$. Henceforth, we shall indicate the application of the Bernstein box consistency and the Bernstein hull consistency techniques by flags BCF and HCF, respectively to the Eqs. (13) and (14).

3.3   Consistency Techniques

We now present algorithms based on the consistency techniques that help pruning the solution search space. The pruning is achieved by assessing consistency of the algebraic equations (in our case inequality and equality constraints) over a given box, and thereby discarding regions of a box where no guarantee of global minimum lies. We note in the literature two types of consistency notions exist; box consistency which use a one-dimensional interval Newton method to compute a box consistent region for a given set of the algebraic equations, and hull consistency which use a constraint inversion procedure to compute a hull consistent region for the given set of the algebraic equations. The interested reader can refer [8] for more details. In sequel, we now present algorithms on the consistency ideas borrowed from [8], and expanded in context of the Bernstein form. Henceforth, we shall call these algorithms as Bernstein box consistency (BBC) and Bernstein hull consistency (BHC) algorithms. We shall use these Bernstein consistency algorithms for pruning purpose in our main proposed global optimization algorithm (see, algorithm IBBBU in Sect. 3.4).

Algorithm Bernstein box consistency: $\mathbf{x}' = \mathrm{BBC}((b_g(\mathbf{x})), \mathbf{x}, r, x_{status,r}, eq\_type)$

Inputs: The Bernstein coefficient array $(b_g(\mathbf{x}))$ of a given constraint function $g(x)$, the l-dimensional box $\mathbf{x}$, the direction r for which the bounds are to be contracted, flag $x_{status,r}$ to indicate whether r^{th} direction (variable) is continuous $(x_{status,r} = 0)$ or integer $(x_{status,r} = 1)$, and flag eq_type to indicate whether $g(x)$ is equality constraint $(eq\_type = 0)$ or inequality constraint $(eq\_type = 1)$.

Outputs: A box $\mathbf{x}'$ that is contracted using Bernstein box consistency technique for a given constraint function $g(x)$.

BEGIN Algorithm

1. Set $a = \inf \mathbf{x}_r$, $b = \sup \mathbf{x}_r$.
2. Compute the derivative enclosure $\mathbf{g}'_{x_r}$ in the direction x_r.[2]
3. (Consider left endpoint of $\mathbf{x}_r$). Obtain the Bernstein range enclosure $\mathbf{g}(a)$ as the minimum to maximum from the Bernstein coefficient array of $(b_g(\mathbf{x}))$ for $x_r = a$.
4. If $eq\_type = 1$, then modify $\mathbf{g}(a)$ as $\mathbf{g}(a) = [\min \mathbf{g}(a), \inf]$.
5. If $0 \in \mathbf{g}(a)$, then we cannot increase a. Go to step 8 and try from the right endpoint b of the interval $\mathbf{x}_r$.
6. Do one iteration of the univariate Bernstein Newton contractor

$$\mathbf{N}(\mathbf{x}_r) = a - (\mathbf{g}(a)/\mathbf{g}'_{x_r}).$$
$$\mathbf{x}'_{r_a} = \mathbf{x}_r \cap \mathbf{N}(\mathbf{x}_r).$$

7. If $\mathbf{x}'_{r_a} = \emptyset$, then there is no zero of $\mathbf{g}$ on entire interval $\mathbf{x}_r$ and hence the constraint g is infeasible over box $\mathbf{x}$. EXIT the algorithm in this case with $\mathbf{x}' = \emptyset$.
8. (Consider right endpoint of $\mathbf{x}_r$). Obtain the Bernstein range enclosure $\mathbf{g}(b)$ as the minimum to maximum from the Bernstein coefficient array of $(b(\mathbf{x}))$ for $x_r = b$.
9. If $eq\_type = 1$, then modify $\mathbf{g}(b)$ as $\mathbf{g}(b) = [\min \mathbf{g}(b), \inf]$.
10. If $0 \in \mathbf{g}(b)$, then we cannot decrease b. Go to step 13
11. Do one iteration of the univariate Bernstein Newton contractor

$$\mathbf{N}(\mathbf{x}_r) = b - (\mathbf{g}(b)/\mathbf{g}'_{x_r}).$$
$$\mathbf{x}'_{r_b} = \mathbf{x}_r \cap \mathbf{N}(\mathbf{x}_r).$$

12. If $\mathbf{x}'_{r_b} = \emptyset$, EXIT the algorithm with $\mathbf{x}' = \emptyset$.
13. Compute $\mathbf{x}'_r$ as follows:
    (a) $\mathbf{x}'_r = \mathbf{x}'_{r_a} \cap \mathbf{x}'_{r_b}$, if both $\mathbf{x}'_{r_a}$ and $\mathbf{x}'_{r_b}$ are computed.
    (b) $\mathbf{x}'_r = \mathbf{x}'_{r_a}$ or $\mathbf{x}'_{r_b}$, which ever is computed.
    (c) $\mathbf{x}'_r = \mathbf{x}_r$ (both $\mathbf{x}'_{r_a}$ and $\mathbf{x}'_{r_b}$ are not computed).
14. for $k = 1, 2$ if $x_{status,r} = 1$ then
    (a) if $\mathbf{x}(r, k)$ and $\mathbf{x}'_r(r, k)$ are equal then go to substep (e).
    (b) Set $t_a = \mathbf{x}(r, k)$ and $t_b = \mathbf{x}'_r(r, k)$.
    (c) if $t_a > t_b$ then set $\mathbf{x}'_r(r, k) = \lfloor \mathbf{x}'_r(r, k) \rfloor$.
    (d) if $t_a < t_b$ then set $\mathbf{x}'_r(r, k) = \lceil \mathbf{x}'_r(r, k) \rceil$.
    (e) end (of k-loop).
15. Return $\mathbf{x}' = \mathbf{x}'_r$.

END Algorithm

Algorithm Bernstein hull consistency: $\mathbf{x}' = \mathrm{BHC}((b_g(\mathbf{x})), a_I, I, \mathbf{x}, x_{status},$ $eq\_type)$

---

[2] The derivative of a polynomial function in a particular direction can be found from the Bernstein coefficients of the original polynomial function [13].

Inputs: The Bernstein coefficient array $(b_g(\mathbf{x}))$ of a given constraint function $g(x)$, coefficient a_l of the selected term t, power I of the each variable in term t, the l−dimensional box $\mathbf{x}$, a column vector x_{status} describing the status (continuous or integer) of the each variable x_i $(i = 1, 2, \ldots, l)$, and flag eq_type to indicate whether $g(x)$ is equality constraint $(eq\_type = 0)$ or inequality constraint$(eq\_type = 1)$.

Outputs: A box $\mathbf{x}'$, that is contracted using Bernstein hull consistency technique applied to a given constraint $g(x)$ and selected term t.

BEGIN Algorithm

1. Compute the Bernstein coefficient array of the selected term t as $(b_t(\mathbf{x}))$.
2. Obtain the Bernstein coefficients of the constraint inverse polynomial by subtracting $(b_g(\mathbf{x}))$ from $(b_t(\mathbf{x}))$, and then obtain its Bernstein range enclosure as the minimum to maximum of these Bernstein coefficients. Denote it as $\mathbf{h}'$.
3. if $eq\_type = 1$ then
    (a) Compute an interval $\mathbf{y}$ as $\mathbf{y} = [-\infty, 0] \cap [\min(b_g(x)), \max(b_g(x))]$.
    (b) if $\mathbf{y} = \emptyset$ then set $\mathbf{x}' = \emptyset$, and EXIT the algorithm. Else modify $\mathbf{h}'$ as $\mathbf{h}' = \mathbf{h}' + \mathbf{y}$.
4. (a) for $r = 1, 2, \ldots, l$ ($r :=$ number of variables)
    (b) Compute $\mathbf{x}'_r = \left(\dfrac{\mathbf{h}'}{a_I \prod \mathbf{x}_k^{i_k}} \right)^{1/i_r} \cap \mathbf{x}_r$
    (c) for $k = 1, 2$ if $x_{status}(r) = 1$ then
        (i) if $\mathbf{x}(r, k)$ and $\mathbf{x}'_r(r, k)$ are equal then go to substep (v).
        (ii) Set $t_a = \mathbf{x}(r, k)$ and $t_b = \mathbf{x}'_r(r, k)$.
        (iii) if $t_a > t_b$ then set $\mathbf{x}'_r(r, k) = \lfloor \mathbf{x}'_r(r, k) \rfloor$.
        (iv) if $t_a < t_b$ then set $\mathbf{x}'_r(r, k) = \lceil \mathbf{x}'_r(r, k) \rceil$.
        (v) end (of k−loop).
    (d) end (of r−loop).
5. Return $\mathbf{x}'$.

END Algorithm

3.4   Main Proposed Global Optimization Algorithm

We now propose an algorithm for bound constrained global optimization of multivariate MINLP problems, called as improved Bernstein branch-and-bound unconstrained (IBBBU) algorithm. The proposed algorithm use a modified subdivision procedure presented in the Sect. 3.1, the Bernstein box and hull consistency algorithms presented in the Sect. 3.3, and the accelerating devices, such as the cut-off test, the monotonicity test, and the concavity tests[3].

---

[3] Due to lack of space and time, we skip the presentation of the accelerating devices. However, the interested reader can refer to [8,10] for the exact details about the topic.

We next present our proposed algorithm.

Algorithm unconstrained optimization: $[\tilde{y}, \tilde{p}, U]$=IBBBU$(N, a_I, \mathbf{x}, x_{status},$
$\epsilon_p, \epsilon_x)$

Inputs: Degree N of the variables occurring in the objective function, the coefficients a_I of the objective function in the power form, the initial search domain $\mathbf{x}$, a column vector x_{status} describing the status (continuous or integer) of the each variable x_i ($i = 1, 2, \ldots, l$), the tolerance parameters ϵ_p and ϵ_x on the global minimum and global minimizer(s).

Outputs: A lower bound $\tilde{y}$ and an upper bound $\tilde{p}$ on the global minimum f^*, along with a set U containing all the global minimizer(s) $\mathbf{x}^{(i)}$.

BEGIN Algorithm

1. Set $\mathbf{y} := \mathbf{x}$ and $y_{status} := x_{status}$.
2. From a_I, compute the Bernstein coefficient array of the objective function on the box $\mathbf{y}$ as $(b_o(\mathbf{y}))$.
3. Set $\tilde{p} := \infty$ and $y := \min(b_o(\mathbf{y}))$.
4. Initialize list $\mathcal{L} := \{(\mathbf{y}, (b_o(\mathbf{y})), y)\}$, $\mathcal{L}^{sol} := \{\}$.
5. If $\mathcal{L}$ is empty then go to step 24. Otherwise, pick the first item $(\mathbf{y}, (b_o(\mathbf{y})), y)$ from $\mathcal{L}$, and delete its entry from $\mathcal{L}$.
6. Apply the Bernstein hull consistency algorithm to the relation $f(\mathbf{y}) \leq \tilde{p}$. If the result is empty, then delete item $(\mathbf{y}, (b_o(\mathbf{y})), y)$ and go to step 5.

$$\mathbf{y}' = \text{BHC}((b_o(\mathbf{y})), a_I, I, \mathbf{y}, y_{status}, 1),$$

7. Set $\mathbf{y} := \mathbf{y}'$ and compute the Bernstein coefficient array of the objective function on the box $\mathbf{y}$ as $(b_o(\mathbf{y}))$. Also set $y := \min(b_o(\mathbf{y}))$.
8. Apply the Bernstein box consistency algorithm to the $f(\mathbf{y}) \leq \tilde{p}$. If the result is empty, then delete item $(\mathbf{y}, (b_o(\mathbf{y})), y)$ and go to step 5.

$$\mathbf{y}' = \text{BBC}((b_o(\mathbf{y})), \mathbf{y}, r, y_{status,r}, 1) \quad r = 1, 2, \ldots, l,$$

where bound contraction will be applied in the rth direction.
9. Set $\mathbf{y} := \mathbf{y}'$ and compute the Bernstein coefficient array of the objective function on the box $\mathbf{y}$ as $(b_o(\mathbf{y}))$. Also set $y := \min(b_o(\mathbf{y}))$.
10. {Monotonicity test} If $0 \notin f'_r(\mathbf{y})$ for any $r \in \{1, 2, ..., l\}$, discard the item $(\mathbf{y}, (b_o(\mathbf{y})), y)$ and go to step 5.
11. {Concavity test} If $\overline{H}_{rr}(\mathbf{y}) < 0$ for some $r = 1, \ldots, l$, discard the item $(\mathbf{y}, (b_o(\mathbf{y})), y)$ and go to step 5.
12. Apply the Bernstein hull consistency algorithm to the relation $f'_{y_r} = 0 (r = 1, 2, \ldots, l)$, that is each component of the gradient of $f(\mathbf{y})$. If the result is empty, then delete item $(\mathbf{y}, (b_o(\mathbf{y})), y)$ and go to step 5.

$$\mathbf{y}' = \text{BHC}((b_r(\mathbf{y})), a_{I,r}, I, \mathbf{y}, y_{status}, 0) \quad r = 1, 2, \ldots, l,$$

where $(b_r(\mathbf{y}))$ is a Bernstein coefficient array of the ith component of gradient of the objective function $f(\mathbf{y})$, and $a_{I,r}$ is coefficient of the ith component of gradient of the objective function $f(\mathbf{y})$.

13. Set $\mathbf{y} := \mathbf{y}'$ and compute the Bernstein coefficient array of the objective function on the box $\mathbf{y}$ as $(b_o(\mathbf{y}))$. Also set $y := \min(b_o(\mathbf{y}))$.

14. Apply the Bernstein box consistency algorithm to the relation $f'_{y_r} = 0 (r = 1, 2, \ldots, l)$, that is each component of the gradient of $f(\mathbf{y})$. If the result is empty, then delete item $(\mathbf{y}, (b_o(\mathbf{y})), y)$ and go to step 5.

$$\mathbf{y}' = \text{BBC}((b_r(\mathbf{y})), \mathbf{y}, r, y_{status,r}, 0) \quad r = 1, 2, \ldots, l,$$

where $(b_r(\mathbf{y}))$ is a Bernstein coefficient array of the ith component of gradient of the objective function $f(\mathbf{y})$.

15. Set $\mathbf{y} := \mathbf{y}'$ and compute the Bernstein coefficient array of the objective function on the box $\mathbf{y}$ as $(b_o(\mathbf{y}))$. Also set $y := \min(b_o(\mathbf{y}))$.

16. Choose a coordinate direction λ parallel to which $\mathbf{y}_1 \times \cdots \times \mathbf{y}_l$ has an edge of maximum length, that is $\lambda \in \{i : w(\mathbf{y}) := w(\mathbf{y}_i), i = 1, 2, \ldots, l\}$.

17. Bisect $\mathbf{y}$ normal to direction λ, getting boxes $\mathbf{v}_1$, $\mathbf{v}_2$ such that $\mathbf{y} = \mathbf{v}_1 \cup \mathbf{v}_2$. We shall use the modified subdivision procedure given in Sect. 3.1.

18. for $k = 1, 2$
    (a) Find the Bernstein coefficient array and the corresponding Bernstein range enclosure of the objective function (f) over $\mathbf{v}_k$ as $(b_0(\mathbf{v}_k))$ and $B_0(\mathbf{v}_k)$, respectively.
    (b) Set $d_k := \min B_o(\mathbf{v}_k)$.
    (c) If $\widetilde{p} < d_k$, then go to substep (f).
    (d) Set $\widetilde{p} := \min(\widetilde{p}, \max B_o(\mathbf{v}_k))$.
    (e) Enter $(\mathbf{v}_k, (b_o(\mathbf{v}_k)), d_k)$ into the list $\mathcal{L}$ such that the third members of all items of the list do not decrease.
    (f) end (of $k-$loop).

19. {Cut-off test} Discard all items $(\mathbf{z}, (b_o(\mathbf{z})), z)$ in the list $\mathcal{L}$ that satisfy $\widetilde{p} < z$.

20. Denote the first item of the list by $(\mathbf{y}, (b_o(\mathbf{y})), y)$.

21. {Check the vertex condition} For the item $(\mathbf{y}, (b_o(\mathbf{y})), y)$, if $\min(b_o(\mathbf{y}))$ satisfies vertex condition (see Remark 3), then enter $(\mathbf{y}, (b_o(\mathbf{y})), y)$ in the solution list $\mathcal{L}^{sol}$ and return to step 5.

22. If $(w(\mathbf{y}) < \epsilon_x)$ & $(\max B_o(\mathbf{y}) - \min B_o(\mathbf{y})) < \epsilon_p$ then remove the item from the list $\mathcal{L}$, and enter it into the solution list $\mathcal{L}^{sol}$.

23. Go to step 5.

24. {Compute the global minimum} Set the global minimum $\widetilde{y}$ to the minimum of the third entries over all the items in $\mathcal{L}^{sol}$.

25. {Compute the global minimizers} Find all those items in $\mathcal{L}^{sol}$ for which the third entries are equal to $\widetilde{y}$. The first entries of these items contain the global minimizer(s) $\mathbf{x}^{(i)}$.

26. Return the lower bound $\widetilde{y}$ and upper bound $\widetilde{p}$ on the global minimum f^*, along with the set U containing all the global minimizer(s) $\mathbf{x}^{(i)}$.

4   Numerical Tests

In this work, we performed different numerical tests with our proposed algorithm IBBBU. For all computations, we used a desktop PC with Pentium IV 2.40 GHz

Table 1. Test problems with their characteristics and the global minimum obtained.

Example	l	l_i	f^*
Camel back	2	1	−1
Booth	2	1	0
Reaction diffusion	3	2	−10.32
Caprasse's	4	2	−2.87
Adaptive LV	4	2	−0.1
AH Wright	5	4	−30
Magnetism in physics (6)	6	3	−0.25
Butcher	6	1	−1.78
Magnetism in physics (7)	7	6	−0.25
Heart dipole problem	8	5	−5.50

l = Total number of decision variables.
l_i = Total number of integer decision variables.

processor with 2 GB RAM and all our presented algorithms are implemented in MATLAB [1]. We specify an accuracy $\epsilon = 10^{-6}$ for computing the global minimum and global minimizer(s) and allow a maximum number of 500 subdivisions in the proposed algorithm IBBBU.

We consider a set of 10 test problems. These test problems are taken from [12, 15]. These problems are appropriately modified to MINLPs assuming some of the decision variables are integer in nature. We report all these test problems in Appendix for the sake of completeness. For these test problems, we first compare the performance of the proposed algorithm IBBBU without accelerating devices, only cut-off test (that is, the Bernstein algorithm for unconstrained optimization reported in [9]), only monotonicity test, only concavity test, combinations of the cut-off, the monotonicity, and the concavity tests. We next compare the performance of the proposed algorithm IBBBU with the use of Bernstein hull consistency (BHC) and Bernstein box consistency (BBC) techniques (see Sect. 3.3) in combination with the three accelerating devices. Specifically, we shall apply the BHC and BBC techniques to the Eqs. (13) and (14) to delete nonoptimal points from the box **x** consequently reducing the overall width of the box **x**.

Table 1 reports the test problems, their dimensions (l), the total number of integer variables (l_i), and the global minimum obtained (f^*). Table 2 gives the performance comparison of the proposed algorithm IBBBU without accelerating devices, and the different combinations of the accelerating devices. We found that without accelerating devices the proposed algorithm IBBBU took maximum number of subdivisions for almost all the problems, except Adaptive LV, Butcher, and Magnetism in physics (7). Moreover, for the one test problem (AH Wright) the proposed algorithm IBBBU failed due to the out of memory error in MATLAB. In the sequel, we also observed the less improvement in the number of boxes processed with the use of concavity test. Finally, we found the combination

Table 2. Comparison of the number of boxes processed and computational time taken (in seconds) to find the global minimum with different combinations of the accelerating devices in the proposed algorithm IBBBU.

Example	Statistics	No accelerating devices	Only cut-off	Only monotonicity	Only concavity	Cut-off + Monotonicity	Cut-off + Monotonicity + Concavity
Camel back	Boxes	1004	386	682	1004	362	362
	Time	1.44	0.31	0.8	1.59	0.30	0.37
Booth	Boxes	1004	236	342	1004	152	152
	Time	1.59	0.18	0.28	1.73	0.11	0.14
Reaction diffusion	Boxes	1004	18	2	362	2	2
	Time	21.2	0.04	0.05	21.57	0.04	0.04
Caprasse's	Boxes	1004	634	1004	1004	384	384
	Time	3.64	1.44	9.79	3.81	2.09	2.12
Adaptive LV	Boxes	64	56	62	64	54	54
	Time	0.07	0.02	0.08	0.06	0.08	0.09
AH Wright	Boxes	*	42	2	20	2	2
	Time		0.13	0.11	0.11	0.2	0.12
Magnetism in physics (6)	Boxes	1004	610	1004	1004	610	610
	Time	9.92	1.93	8.18	10.22	2.21	3.20
Butcher	Boxes	10	10	6	10	6	6
	Time	0.02	0.02	0.04	0.02	0.04	0.05
Magnetism in physics (7)	Boxes	262	262	262	262	262	262
	Time	0.85	0.88	0.80	1.21	0.82	1.23
Heart dipole problem	Boxes	>5000	36	>5000	>5000	36	36
	Time		0.12			0.84	0.93

* The algorithm returned "out of memory error".
* Entry '1004' indicates the algorithm took maximum number of subdivisions to reach the solution.

Table 3. Comparison of the number of boxes processed and computational time taken (in seconds) to find the global minimum with use of the cut-off test, the Bernstein hull and Bernstein box consistencies (to the Eq. (14)) in the proposed algorithm IBBBU.

Example	Statistics	Cut-off test	Cut-off + BCF	Cut-off + HCF	Cut-off + (BCF + HCF)
Camel back	Boxes	386	254	380	232
	Time	0.31	0.20	0.32	0.18
Booth	Boxes	236	148	166	110
	Time	0.18	0.21	0.15	0.15
Reaction diffusion	Boxes	18	8	6	2
	Time	0.04	0.03	0.03	0.01
Caprasse's	Boxes	634	206	632	204
	Time	1.44	1.34	1.45	1.36
Adaptive LV	Boxes	56	26	26	26
	Time	0.05	0.05	0.04	0.01
AH Wright	Boxes	42	16	16	8
	Time	0.13	0.15	0.16	0.12
Magnetism in physics (6)	Boxes	610	504	504	390
	Time	1.93	9.02	22.62	1.10
Butcher	Boxes	10	10	6	6
	Time	0.02	0.03	0.01	0.02
Magnetism in physics (7)	Boxes	262	262	262	262
	Time	0.88	1.40	1.50	2.89
Heart dipole problem	Boxes	36	4	6	2
	Time	0.12	0.11	0.13	0.01

of the cut-off, and the monotonicity tests to be the most efficient amongst all others (nearly 50 % reduction in the average number of boxes processed and the computational time taken to found the global minimum).

We now present numerical findings which specifically reports the benefits obtained (in terms of the number of boxes processed) with pruning tools and trade-off associated in terms of the computational time to reach the solution. Tables 3, 4, and 5 give the performance comparison of the proposed algorithm IBBBU with the BHC and BBC techniques applied to the Eq. (14) (reported as flags HCF and BCF, respectively) with its combination with the three accelerating devices. Overall, we found the combination of all three accelerating devices with the BCF to be more efficient than HCF, resulting on an average 50 % reduction in the number of boxes processed and the computational time. Similarly,

Table 4. Comparison of the number of boxes processed and computational time taken (in seconds) to find the global minimum with use of the monotonicity test, the Bernstein hull and Bernstein box consistencies (to the Eq. (14)) in the proposed algorithm IBBBU.

Example	Statistics	Monotonicity test	Monotonicity + BCF	Monotonicity + HCF	Monotonicity + (BCF + HCF)
Camel back	Boxes	682	114	182	112
	Time	0.80	0.12	0.23	1.22
Booth	Boxes	342	96	122	80
	Time	0.28	0.20	0.19	0.31
Reaction diffusion	Boxes	2	2	2	2
	Time	0.05	0.06	0.08	0.12
Caprasse's	Boxes	1004	192	344	178
	Time	9.79	0.5	0.98	1.92
Adaptive LV	Boxes	62	44	28	26
	Time	0.08	0.4	0.50	1.42
AH Wright	Boxes	2	2	2	8
	Time	0.11	0.12	0.13	0.12
Magnetism in physics (6)	Boxes	1004	500	610	140
	Time	8.18	1.90	1.20	2.78
Butcher	Boxes	6	34	6	6
	Time	0.04	0.35	0.05	0.32
Magnetism in physics (7)	Boxes	262	262	262	262
	Time	0.80	4.47	2.50	5.91
Heart dipole problem	Boxes	>5000	4	4	2
	Time		0.45	0.60	0.67

we observed that the combination of BCF and HCF with all three accelerating devices to be more efficient for three problems (Camel back, Booth, and Magnetism in physics (6)). We found on an average more than 50 % reduction in number of processed boxes, but with an average 5–10 % increase in the computational time. Further, we note that the Bernstein box consistency algorithm to be more efficient when the domain is small (see, for instance, results for Caprasse's in Tables 3, 4, and 5). This is evident from the fact that it involves Newton operator for pruning which has good convergence properties near the solution point [8]. Similarly, we note that the Bernstein hull consistency algorithm to be more efficient when the domain is large (see, for instance, results for Adaptive LV in Tables 3, 4, and 5; result for Camel back in Table 6).

Table 6 reports for the 10 test problems the total number of boxes processed, and the computational time taken in seconds to find the global minimum by the

Table 5. Comparison of the number of boxes processed and computational time taken (in seconds) to find the global minimum with use of the concavity test, the Bernstein hull and Bernstein box consistencies (to the Eq. (14)) in the proposed algorithm IBBBU.

Example	Statistics	Concavity test	Concavity + BCF	Concavity + HCF	Concavity + (BCF + HCF)
Camel back	Boxes	1004	254	380	232
	Time	1.59	0.74	2.39	2.12
Booth	Boxes	1004	148	166	110
	Time	1.73	0.41	0.59	1.02
Reaction diffusion	Boxes	362	8	6	2
	Time	21.57	0.11	0.14	0.11
Caprasse's	Boxes	1004	180	564	134
	Time	3.81	0.90	4.81	4.54
Adaptive LV	Boxes	64	44	28	26
	Time	0.06	0.47	0.81	1.42
AH Wright	Boxes	20	16	16	8
	Time	0.11	0.19	0.21	0.25
Magnetism in physics (6)	Boxes	1004	504	504	390
	Time	10.22	3.63	5.38	7.92
Butcher	Boxes	10	10	6	6
	Time	0.02	0.03	0.16	0.25
Magnetism in physics (7)	Boxes	262	262	262	262
	Time	1.21	14.91	15.31	21.32
Heart dipole problem	Boxes	>5000	4	6	2
	Time		0.22	0.45	0.62

proposed algorithm IBBBU. Specifically, we compare the proposed algorithm IBBBU based on the three different flags (A, B, and C) explained as below[4]:

- A: Application of the Bernstein hull consistency to the Eq. (13). We also apply the cut-off and the monotonicity tests.
- B: Application of the Bernstein box consistency to the Eq. (13). We also apply the cut-off and the monotonicity tests.
- C: Application of both Bernstein hull and Bernstein box consistencies to the Eq. (13) along with the cut-off and the monotonicity tests.

We found the performance of the flags A and B almost similar in terms of number boxes processed with a very little variation in the computational time, except for the two test problems (Camel back and Magnetism in physics (6)). On the other

---

[4] We note that concacity test is found to give small improvement in the number of boxes processed. Hence, we skipped its application in this numerical experimentation.

Table 6. Comparison of the number of boxes processed and computational time taken (in seconds) to find the global minimum with use of the cut-off test, the monotonicity test, and the Bernstein hull and Bernstein box consistencies (to the Eq. (19)) in the proposed algorithm IBBBU.

Example	Statistics	A	B	C
Camel back	Boxes	18	170	15
	Time	0.28	0.32	0.21
Booth	Boxes	24	24	20
	Time	0.22	0.24	0.43
Reaction diffusion	Boxes	2	2	2
	Time	0.39	0.03	0.03
Caprasse's	Boxes	346	346	342
	Time	2.10	3.10	4.51
Adaptive LV	Boxes	50	50	50
	Time	0.71	0.72	0.85
AH Wright	Boxes	2	2	2
	Time	0.09	0.08	0.08
Magnetism in physics (6)	Boxes	290	382	289
	Time	0.58	0.48	0.67
Butcher	Boxes	6	6	6
	Time	0.01	0.03	0.04
Magnetism in physics (7)	Boxes	40	44	39
	Time	2.32	2.31	2.29
Heart dipole problem	Boxes	25	25	23
	Time	8.91	9.20	9.57

hand, we found very little improvement with the use of flag C, except for the two test problems (Camel back and Caprasse's). Overall, we found this combination of the pruning tool to be the most efficient one in terms of the number of boxes processed. However, a significant increase (more than 50 %) in the number of the boxes processed and the computational time was observed, specifically for a 8-dimensional heart dipole problem.

5   Conclusions

We presented a Bernstein algorithm for finding the global minimum of the unconstrained MINLP problems. The proposed algorithm was composed of several new tools, such as box partitioning procedure for integer variables, the Bernstein box (BBC) and the Bernstein hull consistency (BHC) techniques to prune the solution search space. Different numerical experiments were performed with the proposed algorithm on a collection of 10 test problems with dimensions ranging from 2 to 8 with the number of integer variables varying from 1 to 6. The findings revealed the proposed algorithm to be more efficient than the classical Bernstein algorithm reported in the literature. Specifically, it was noted that the proposed algorithm composed of different accelerating devices resulted on an average 50 % reduction of in the number of boxes processed and the computational time. Similarly, numerical investigations with the BBC and BHC resulted on an average 40–50 % reduction in the number of boxes processed and the computational time. In sequel, it was noted that the application of a BBC and BHC for an 8-dimensional heart dipole problem computationally resulted in a significant increase (more than 50 %) in the number of the boxes processed and the computational time. Finding the efficient ways to handle such problems is left as a future research direction.

Appendix

We list below the test problems studied in this work for conducting different numerical experiments. We denote the test function as $f(x_k)$, and the initial bounds as $\mathbf{x}_k$, $k = 1, 2, \ldots, l$.

1. Camel back: The six hump camel back function

$$\min \ f(x) = 4x_1^2 - 2.1x_1^4 + (1/3)x_1^6 + x_1x_2 - 4x_2^2 + 4x_2^4$$
$$x_1 \in \mathbb{Z}, \ x_2 \in \mathbb{R}$$

where $\mathbf{x}_k = [-3, 3], k = 1, 2$.

2. Booth: The function defined by Booth

$$\min \ f(x) = 74 - 38x_1 + 5x_1^2 - 34x_2 + 8x_1x_2 + 5x_2^2$$
$$x_1 \in \mathbb{Z}, \ x_2 \in \mathbb{R}$$

where $\mathbf{x}_k = [-5, 5], k = 1, 2$.

3. Reaction diffusion: A three dimensional reaction diffusion problem

$$\min \ f(x) = -x_1 + 2x_2 - x_3 - 0.835634534x_2(1 - x_2)$$
$$x_1, x_2 \in \mathbb{Z}, \ x_3 \in \mathbb{R}$$

where $\mathbf{x}_k = [-5, 5], k = 1, 2, 3$.

4. Caprasse's: The system defined by Caprasse

$$\min \ f(x) = -x_1x_3^3 + 4x_2x_3^2x_4 + 4x_1x_3x_4^2 + 2x_2x_4^3 + 4x_1x_3$$
$$+ 4x_3^2 - 10x_2x_4 - 10x_4^2 + 2$$
$$x_1, x_3 \in \mathbb{Z}, \ x_2, x_4 \in \mathbb{R}$$

where $\mathbf{x}_k = [-1, 1], k = 1, 3$ and $\mathbf{x}_k = [-0.5, 0.5], k = 2, 4$.

5. Adaptive LV: A neural network modeled by an adaptive Lotka-Volterra system

$$\min \ f(x) = x_1 x_2^2 + x_1 x_3^2 + x_1 x_4^2 - 1.1 x_1 + 1$$
$$x_1, x_2 \in \mathbb{Z}, \quad x_3, x_4 \in \mathbb{R}$$

where $\mathbf{x}_1 = [0, 1], \mathbf{x}_2 = [-20, 20], \mathbf{x}_k = [-2, 2], k = 3, 4$.

6. AH Wright: The system defined by Wright

$$\min \ f(x) = x_1 + x_2 + x_3 + x_4 - x_5 + x_5^2 - 10$$
$$x_1, x_2 \in \mathbb{R}, \quad x_3, x_4, x_5 \in \mathbb{Z}$$

where $\mathbf{x}_k = [-5, 5], k = 1, \ldots, 5$.

7. Magnetism in Physics (6): A six variable magnetism in physics problem

$$\min \ f(x) = 2x_1^2 + 2x_2^2 + 2x_3^2 + 2x_4^2 + 2x_5^2 + x_6^2 - x_6$$
$$x_1, x_2, x_3 \in \mathbb{Z}, \quad x_4, x_5, x_6 \in \mathbb{R}$$

where $\mathbf{x}_k = [-1, 1], k = 1, \ldots, 6$.

8. Butcher: A function defined by Butcher

$$\min \ f(x) = x_6 x_2^2 + x_5 x_3^2 - x_1 x_4^2 + x_4^3 + x_4^2 - (1/3)x_1 + (4/3)x_4$$
$$x_1 \in \mathbb{Z}, \quad x_k \in \mathbb{R}, k = 2, \ldots, 6$$

where $\mathbf{x}_k = [-1, 1], k = 1, 2, 3, \mathbf{x}_4 = [-0.1, 0.2], \mathbf{x}_5 = [-0.3, 1.1], \mathbf{x}_6 = [-1.1, -0.3]$.

9. Magnetism in physics (7): Seven variable magnetism in physics problem

$$\min \ f(x) = x_1^2 + 2x_2^2 + 2x_3^2 + 2x_4^2 + 2x_5^2 + 2x_6^2 + 2x_7^2 - x_1$$
$$x_1 \in \mathbb{R}, \quad x_k \in \mathbb{Z}, k = 2, \ldots, 7$$

where $\mathbf{x}_k = [-1, 1], k = 1, \ldots, 7$.

10. Heart dipole: A heart dipole problem

$$\min \ f(x) = -x_1 x_6^3 + 3x_1 x_6 x_7^2 - x_3 x_7^3 + 3x_3 x_7 x_6^2 - x_2 x_5^3$$
$$+ 3x_2 x_5 x_8^2 - x_4 x_8^3 + 3x_4 x_8 x_5^2 - 0.9563453$$
$$x_k \in \mathbb{Z}, k = 1, \ldots, 5, \quad x_k \in \mathbb{R}, k = 5, 7, 8$$

where $\mathbf{x}_k = [-1, 1], k = 1, 2, 3, \mathbf{x}_4 = [-1, 0], \mathbf{x}_5 = [0, 1], \mathbf{x}_6 = [-0.1, 0.2]$
$\mathbf{x}_7 = [-0.3, 1.1], \mathbf{x}_8 = [-1.1, -0.3]$.

References

1. The Mathworks Inc., MATLAB version 7.1 (R14), Natick, MA (2005)
2. Bonami, P., Biegler, L.T., Conn, A., Cornuejols, G., Grossmann, I.E., Laird, C., Lee, J., Lodi, A., Margot, F., Sawaya, N., Wächter, A.: An algorithmic framework for convex mixed integer nonlinear programs. Discrete Optim. 5(2), 186–204 (2008)
3. Buchheim, C., D'Ambrosio, C.: Box-constrained mixed-integer polynomial optimization using separable underestimators. In: Lee, J., Vygen, J. (eds.) IPCO 2014. LNCS, vol. 8494, pp. 198–209. Springer, Heidelberg (2014)
4. D'Ambrosio, C., Lodi, A.: Mixed integer nonlinear programming tools: an updated practical overview. Ann. Oper. Res. 204(1), 301–320 (2013)
5. Floudas, C.A.: Nonlinear and Mixed-integer Optimization: Fundamentals and Applications. Oxford University Press, New York (1995)
6. GAMS Development Corp.: GAMS-The solver manuals, Washington, DC (2009)
7. Garloff, J.: The Bernstein algorithm. Interval Comput. 2, 154–168 (1993)
8. Hansen, E., Walster, G.W.: Global Optimization Using Interval Analysis. CRC Press, New York (2004)
9. Nataraj, P.S.V., Arounassalame, M.: A new subdivision algorithm for the Bernstein polynomial approach to global optimization. Int. J. Autom. Comput. 4(4), 342–352 (2007)
10. Nataraj, P.S.V., Arounassalame, M.: Constrained global optimization of multivariate polynomials using Bernstein branch and prune algorithm. J. Glob. Optim. 49(2), 185–212 (2011)
11. Patil, B.V., Nataraj, P.S.V., Bhartiya, S.: Global optimization of mixed-integer nonlinear (polynomial) programming problems: the Bernstein polynomial approach. Computing 94(2–4), 325–343 (2012)
12. Ratz, D., Csendes, T.: On the selection of subdivision directions in interval branch-and-bound methods for global optimization. J. Glob. Optim. 7(2), 183–207 (1995)
13. Stahl, V.: Interval methods for bounding the range of polynomials and solving systems of nonlinear equations. Ph.D. thesis, Johannes Kepler University, Linz (1995)
14. Tawarmalani, M., Sahinidis, N.V.: Convexification and Global Optimization in Continuous and Mixed-Integer Nonlinear Programming: Theory, Algorithms, Software, and Applications. Nonconvex Optimization and its Applications. Kluwer Academic Publishers, Dordrecht (2002)
15. Verschelde, J.: PHC pack, the database of polynomial systems. Technical report, Mathematics Department, University of Illinois, Chicago, USA (2001)

Dynamical Systems

Interval Regularization Approach to the Firordt Method of the Spectrophotometric Analysis of the Non-separated Mixtures

Valentin Golodov[✉]

Faculty of Computational Mathematics and Informatics, South Ural State University
(National Research University), Lenin prospekt 76, 454080 Chelyabinsk, Russia
avaksa@gmail.com, golodovva@susu.ac.ru
http://computer.susu.ac.ru

Abstract. This paper describes the author's experiences with application the so-called interval regularization approach to one of the chemical analysis methods, specifically the Firordt method – the spectrophotometric analysis of non-separated mixtures. In our approach, the uncertainty is described using intervals. The solution can be found for well-determined and overdetermined systems (when the number of the measurements exceeds number of the mixture components), also interval statement considers measurements errors. Exact rational computations are important part of the technique of solving interval task.

Keywords: System of linear equations · Interval uncertainty · Interval regularization · Firordt method · Exact computations

1 Introduction

The Firordt method is one of the methods of the analysis of the non-separated mixtures [1]. According to the Firordt's method, we can determine the concentration c_j of the each of the m components by solving the following system of the equations:

$$b_i = \sum_{j=1}^{m} a_{ij} \cdot c_j \cdot l, \qquad (1)$$

where:

- b_i is the measured absorbancy of the analyzed mixture on the i-th analytical wave length (AWL),
- a_{ij} is an molar coefficient of the absorption (or extinction) of the j-th component on i-th AWL (measured in advance for each component),
- l is the thickness of the absorbing layer.

Number of the AWL(k) (number of the equations) usually is equal to the number of the components (m) in the mixture. Overdetermined systems with $k > m$ may be used for the enhanced accuracy.

© Springer International Publishing Switzerland 2016
M. Nehmeier et al. (Eds.): SCAN 2014, LNCS 9553, pp. 201–208, 2016.
DOI: 10.1007/978-3-319-31769-4_16

Spectrophotometric measurements are always performed with some measurement errors, so, we have some imprecise system of linear algebraic equations for analysis with equations of the form (1). Here and below in the paper we use the standard notation of interval analysis [2], particularly, all interval values are written using bold type, e.g. $\boldsymbol{A}, \boldsymbol{b}_i$ are interval matrix and interval value correspondingly. Non-interval values typed with mathematical italic type as usual.

$$\boldsymbol{b}_i = \sum_{j=1}^{m} \boldsymbol{a}_{ij} \cdot c_j \cdot l, \tag{2}$$

System 2 will become simpler if all measurements are performed using l equal to 1 centimeter, then system takes on the form

$$\boldsymbol{b}_i = \sum_{j=1}^{m} \boldsymbol{a}_{ij} \cdot c_j, \tag{3}$$

or, in matrix form, $\boldsymbol{A}x = \boldsymbol{b}$, where x – is the sought for vector of the components concentrations.

2   Interval Regularization Approach

2.1   Essence of the Method

We consider interval system of linear algebraic equations $\boldsymbol{A}x = \boldsymbol{b}$, with an interval matrix $\boldsymbol{A}$ and interval right-hand side vector $\boldsymbol{b}$, as a model of imprecise system of linear algebraic equations of the same form.

We use a new regularization procedure proposed in [5] that reduces the solution of the imprecise linear system to computing an point from the tolerable solution set $(\varXi_{tol}(\boldsymbol{A}, \boldsymbol{b}) = \{x \in \mathbb{R}^m \mid (\forall A \in \boldsymbol{A})(\exists b \in \boldsymbol{b})(Ax = b)\})$ of the interval linear system with a widened right-hand side.

Tolerable solution set is the least sensitive, among all the solution sets [3], to the change in the interval matrix of the system $\boldsymbol{A}x = \boldsymbol{b}$, it may be demonstrated by following representation of the $\varXi_{tol}(\boldsymbol{A}, \boldsymbol{b})$.

$$\varXi_{tol}(\boldsymbol{A}, \boldsymbol{b}) = \bigcap_{A \in \boldsymbol{A}} \left\{ x \in \mathbb{R}^m \mid (\exists b \in \boldsymbol{b}) (Ax = b) \right\}, \tag{4}$$

In the above formula, $\left\{ x \in \mathbb{R}^m \mid (\exists b \in \boldsymbol{b})(Ax = b) \right\}$ is the solution set to the interval system $Ax = \boldsymbol{b}$ with the interval uncertainty concentrated only in the right-hand side vector. We exploit this idea that may be called *interval regularization* for the system of equations of the Firordt method (3).

Straightforward replacement of the $Ax = b$ to the $\boldsymbol{A}x = \boldsymbol{b}$ often leads to empty $\varXi_{tol}(\boldsymbol{A}, \boldsymbol{b})$. Right-hand part of the interval system may be extended using some non-negative parameter $z \in R, z >= 0$.

So we have to find point from the tolerable solution set of the system $\varXi_{tol}(\boldsymbol{A}, \boldsymbol{b}(z))$ with the widened right-hand part. The form of the extension may vary from one task to another by using expertise. For the present proved that extension of the form $b(z) = [b - zp, b + zq]$, where $p, q \in \mathbb{R}^m, p > 0, q > 0$ enables to find point from the tolerable solution set $\Xi_{tol}(A, b(z))$ for any given A, b.

Minimization of the extension of the right-hand part leads to the smallest possible set $\Xi_{tol}(A, b(z))$. *Usually*, when $z \neq 0$, it contains only one point, although in the general case, of course, $\Xi_{tol}(A, b(z))$ may contains more than one point.

Minimum z^* of the parameter z, ensuring non-empty $\Xi_{tol}(A, b(z^*))$, jointly with the corresponding extension of the right-hard part of the system $b(z^*) = [b - z^*p, b + z^*q]$ also usable as the measure of degeneracy or the measure of instability of the initial task. Also big value of the z^* shows that form of the extension (vectors p, q) is unsatisfactory to the solving problem.

Well known regularization methods has the known failures. One well known method of solving system with non-square or degenerated matrix is normal pseudo-solution that is the solution of the system $A^T A x = A^T b$. However, e.g. the simple 2×2 system:

$$\begin{cases} (1 + \varepsilon)x + y = 1, & \varepsilon \geq 0 \\ x + y = 1, \end{cases} \tag{5}$$

have traditional solution $(x, y) = (0, 1)^T$ for any $\varepsilon \neq 0$ and normal pseudo-solution $(x, y) = (1/2, 1/2)^T$ for $\varepsilon = 0$, consequently we have no convergence of the traditional solution to the normal pseudo-solution when $\varepsilon \to 0$. For the non-degenerated systems pseudo-solution is equal to its traditional solution.

Other well known method is the Tikhonov regularization procedure, with regard to the linear system of equations it leads to solving of the system $(A^T A - \delta E)x = A^T b$, where $(A^T$ – is transposed matrix of the system, E – is unity matrix, δ – is the parameter of the regularization, selection of the δ is the theme of a lot of papers. However, e.g., for the system with Hilbert matrix the procedure doesn't leads to success [4].

The interval regularization not constrains type of the initial imprecise system of linear equations type or possible the system's degeneracy. The initial system of linear equations could be underdetermined, well-determined or overdetermined. Next section deals with essential aspects of the computing technique what allows to operate even with strongly ill-conditioned systems using exact computations [6].

For the consistency, solution in the traditional sense of the point system of linear equations $Ax = b$ is, obviously, agree with solution in the interval regularization sense of the interval linear system $Ax = b$, where $A = [A, A]$, $b = [b, b]$.

2.2 Computing Technique

Computing technique is based on the theorem [5]:

Theorem 1. *There exists a solution* x^{+^*} *and* $x^{-^*} \in \mathbb{R}^m$, $z^* \in \mathbb{R}$ *to the linear programming problem*

$$\min_{x^+,\, x^-,\, z} \quad z, \tag{6}$$

$$\sum_{j=1}^{m} (\underline{a}_{ij} x_j^+ - \overline{a}_{ij} x_j^-) \geq \underline{b}_i - z p_i, \qquad i = 1, 2, \ldots, k, \tag{7}$$

$$\sum_{j=1}^{m} (\overline{a}_{ij} x_j^+ - \underline{a}_{ij} x_j^-) \leq \overline{b}_i + z q_i, \qquad i = 1, 2, \ldots, k, \tag{8}$$

$$x_j^+, x_j^-, z \geq 0, \qquad j = 1, 2, \ldots, k. \tag{9}$$

In addition, the vector $x^* = x^{+^*} - x^{-^*}$ *belongs to* $\Xi_{tol}(A, b(z^*))$. *Vectors* $p, q \in \mathbb{R}^{+k}$ *are used to manipulate with form of the right-hand part extension.*

We use simplex method to solve linear programming task (6)–(9), as one of the simplest and practically fast, but typical implementations of the simplex method could fail because of the linear programming task's strong degeneracy.

In our approach, we use exact computations [6] and procedure described in [7] to prevent simplex method cycling, such synergy allows to solve ill-conditioned problems sensitive to the data precision.

Simplex method requires only basic arithmetic (plus, minus, multiply and divide) during calculation, so, for the linear programming task with rational coefficients all calculation are being performed in rational number field. Interval regularization approach is usable even in the cases where well known regularization methods are inapplicable, e.g. for the tasks with the disturbed Gilbert and Vandermonde matrix or tasks with the disturbed matrix of the Godunov task [4]. Some computing experiments with Gilbert matrix was introduced previously in [5].

Computational complexity of the interval regularization approach is defined by complexity of the simplex method solving the corresponding linear programming task (6)–(9). The discussion about the simplex method complexity is given in [7], in most cases method requires linear (depended on task size) number of the iterations. Theoretically proved that solving of any linear programming task with rational coefficients has polynomial complexity [7].

2.3   Applying the Approach to the Firordt Method

The Firordt method requires to measure the extinction a_{ij} for each individual component i and each analytical wave length λ_j. Values a_{ij} are calculated using proportion $A = a \cdot c \cdot l$ with $c = 1$ mole per liter and $l = 1$ centimeter. Because of measurements errors during solution preparation and error of the spectrophotometer during extinction measuring each value a_{ij} is actually the interval value a_{ij}. Minimum relative error of the data in the computing experiment below is 1 %, more realistically relative error 5 % of all measurements.

We apply our approach to the system $Ax = b$ where a_{ij} and b_i are data of the absorbancy measurements for the individual components and mixture correspondingly. Vector x is unknown concentrations of the components.

For the case when $\Xi_{tol}(A, b(z))$ is empty useful extension form is $b(z) = [b_i - z, b_i + z]_{i=1...n}$, but when absorption levels differ a lot for taken wave lengths proportional extension $b(z) = [b_i - z|b_i|, b_i + z|b_i|]$, $i = 1, \ldots, k$ could be preferred.

Advantages of the interval analog of the Firordt method are:

- The approach results in robust solutions, we can use larger or smaller error estimates in the measurements data and the fluctuations will be minor.
- The approach is useful for the raw data when optimal set of the AWL is unknown.
- The approach has more correct model for the solution of the overdetermined systems, than, e.g., normal pseudo-solution.
- Value z^* optimal parameter of the right-hand part extension could be indicator of the goodness of the linear model of absorbancy of the mixture.
- Using of the exact rational calculations gives absolute repeatability of the calculation experiments.

3   Computing Experiment

Tables 1, 2, 3, 4 and Fig. 1 demonstrate example of the spectrophotometric data for the model solutions with the ions of the Cu, Ni and its mixture.

Let us consider collection of examples with the given data. First, since given measurements was done on the model solution, we know real concentrations of the mixture components $c_{Cu} = 0.5$ $c_{Ni} = 0.5$ mole per liter, so, relation errors for the all results $\delta = (\delta_{Cu}, \delta_{Ni})^\top$ below will be calculated using this values.

To apply the traditional Firordt method we should choose two analytical wave lengths (AWL) and solve corresponding system with $A \in \mathbb{R}^{2 \times 2}$, $b \in \mathbb{R}^{2 \times 1}$. Proper selection of data subset is non-trivial problem, data set contains measurements

Table 1. Data of the model solutions with the ions of the Cu, Ni and its mixture (1).

λ AWL (nm)	410	420	430	440	450	460	470	480	490	500
a_{Cu}	0.044	0.038	0.036	0.032	0.03	0.03	0.028	0.028	0.03	0.034
a_{Ni}	1.252	0.832	0.45	0.244	0.162	0.124	0.088	0.054	0.034	0.026
b_{NiCu}	0.662	0.428	0.244	0.138	0.096	0.078	0.06	0.044	0.034	0.033

Table 2. Data of the model solutions with the ions of the Cu, Ni and its mixture (2).

λ AWL (nm)	510	520	530	540	550	560	570	580	590	600
a_{Cu}	0.04	0.048	0.064	0.084	0.114	0.15	0.198	0.266	0.352	0.45
a_{Ni}	0.026	0.032	0.04	0.046	0.052	0.062	0.082	0.11	0.148	0.198
b_{NiCu}	0.037	0.044	0.055	0.069	0.087	0.109	0.142	0.187	0.25	0.319

Table 3. Data of the model solutions with the ions of the Cu, Ni and its mixture (3).

λ AWL (nm)	610	620	630	640	650	660	670	680	690	700
a_{Cu}	0.59	0.754	0.942	1.21	1.46	1.81	2.158	2.564	2.95	3.388
a_{Ni}	0.254	0.33	0.398	0.484	0.546	0.572	0.562	0.566	0.588	0.616
b_{NiCu}	0.409	0.536	0.681	0.829	1.001	1.191	1.352	1.545	1.766	2.016

Table 4. Data of the model solutions with the ions of the Cu, Ni and its mixture (4).

λ AWL (nm)	710	720	730	740	750	760	770	780	790	800
a_{Cu}	3.792	4.21	4.546	4.546	5.024	5.15	5.25	5.262	5.228	5.206
a_{Ni}	0.64	0.654	0.646	0.61	0.572	0.512	0.45	0.39	0.338	0.288
b_{NiCu}	2.204	2.407	2.552	2.722	2.849	2.846	2.879	3.038	3.008	3.007

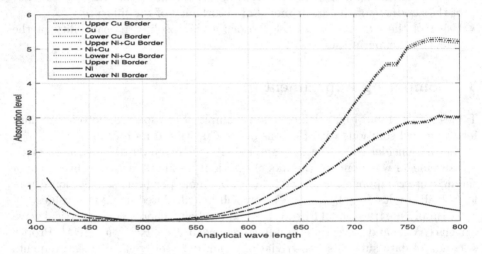

Fig. 1. Individual absorption levels of the mixture components (Cu, Ni) and mixture absorption level ($Ni+Cu$). Also allowable measurement error borders are given (*Upper *** Border, Lower *** Border*).

on 40 different AWL and different pairs of AWL could give *essentially* distinct results.

Papers devoted to selection of the best subset of the measurements, gives some tips [1]: e.g., data corresponding to the peaks of the absorbancy of the individual components or data corresponding to the measurements where absorbancy levels for individual components are most differ. More strictly matrix of the system should have least condition number between other possible matrixes [1].

Pair of equations which yields the matrix $A \in \mathbb{R}^{2\times 2}$ with the least possible condition number among other 2×2 matrixes, is the pair of equations corresponding to $\lambda_1 = 410$ (nm) and $\lambda_2 = 660$ (nm), condition number $\mu_{440,660} = 1.53$. For this system traditional solution of the system is: $c_{Cu} = 0.496427$ $c_{Ni} =$

0.511308 (relative errors are $\delta_{Cu} = 0.71\%$ and $\delta_{Ni} = 2.26\%$ correspondingly). Though it is good result best result is: $c_{Cu} = 0.5$ $c_{Ni} = 0.5$ (relative errors are $\delta_{Cu} = 0.0\%$ and $\delta_{Ni} = 0.0\%$ correspondingly) for pairs corresponding to the $\lambda_1 = 440$ (nm) and $\lambda_2 = 450$ (nm), condition number $\mu_{440,450} = 77.17$ or $\lambda_1 = 440$ (nm) and $\lambda_2 = 590$ (nm), condition number $\mu_{440,590} = 1.89$, and some other ones.

Interval regularization approach gives equal results for the specified pairs of AWL and consideration of the (1 %) and (5 %) measurement error does not affect to the result values, result are the same because of the small matrixes are not so strongly ill-conditioned.

So, traditional method is sensitive to selection of the wave lengths, the matrix with the least condition number does not guarantee best result and matrix with relatively great condition number may give very good results.

A lot of pairs of AWL gives results with relative error more than 100 % or gives negative values of concentrations of components. Interval regularization can not fully smooth over effect of *bad* data without additional *good* data but provide additional information in coefficient of the right-hand part extension z^*. E.g., for $\lambda_1 = 760$ (nm), $\lambda_2 = 780$ (see the Table 4), $c_{Cu} = 0.649$ $c_{Ni} = -0.977$ (relative errors are $\delta_{Cu} = 29.95\%$ and $\delta_{Ni} = 295.451\%$ correspondingly). Using interval regularization approach we can see that solution is very unstable, see Table 5: Data for the Cu component are relatively consistent with the model but for Ni component data are inconsistent. Here interval regularization gives some additional information, that may be used to further calculation experiments.

Table 5. Unstable system of equation. Dependence between supposed measurement error Δ and the solution of the interval task.

$\Delta(\%)$	(0 %)	(1 %)	(5 %)	(10 %)	(20 %)
c_{Cu}	0.649	0.648	0.641	0.641	0.563	0.562
c_{Ni}	-0.977	-0.95	-0.891	-0.891	0.000	0.000
z^*	0.0	0.0098	0.0716	0.0456	0.063	0.062

Further improvement of the result accuracy may be produced by using full set of the experimental data in calculation, so we have overdetermined system $Ax = b$, $A \in \mathbb{R}^{m \times 2}$. It may be solved, for example, using pseudo-solution, i.e. the solution of the system $A^\top Ax = A^\top b$ and we can use all available data corresponding to 40 different AWL. So, we have overdetermined system $Ax = b$, $A \in \mathbb{R}^{40 \times 2}$ solution is $c_{Cu} = 0.5234$ $c_{Ni} = 0.4390$ (relative error is 4.69 % and 12.21 % correspondingly).

Our approach gives robust solution $c_{Cu}^i = 0.5156$ $c_{Ni}^i = 0.5277$ (relative error is 3.12 % and 5.54 % correspondingly) as a point from the tolerable solution set $\Xi_{tol}(A, b(z))$ of the corresponding interval system $Ax = b(z)$ with $b(z) = [b - z, b + z]$ $z^* = 0.0716$. Parameter $z^* = 0.0716$ and corresponding $b(z^*) = [b - z^*, b + z^*]$ displays that data correlates accurately with the Firordt method model, the same result is for data with 1 % and 5 % measurement errors.

Results are consolidated in Table 6.

Table 6. Consolidated results. Value c^i corresponds to measurement error $\Delta = 1\%$.

$c^{*\mathsf{T}} = (0.5, 0.5)$	$\lambda_1 = 410,\ \lambda_2 = 660$	$\lambda_1 = 440,\ \lambda_2 = 450$, etc.	$A \in \mathbb{R}^{40 \times 2}$
$c^{\mathsf{T}} = (c_{Cu}, c_{Ni})$	$(0.496, 0.511)$	$(0.500, 0.500)$	—
$\delta^{\mathsf{T}} = (\delta_{Cu}, \delta_{Ni})$	$(0.71\,\%, 2.26\,\%)$	$(0.00, 0.00)$	—
$c^{ps\,\mathsf{T}} = (c_{Cu}^i, c_{Ni}^i)$	$(0.496, 0.511)$	$(0.500, 0.500)$	$(0.5234, 0.4390)$
$\delta^{ps\,\mathsf{T}} = (\delta_{Cu}^{ps}, \delta_{Ni}^{ps})$	$(0.71\,\%, 2.26\,\%)$	$(0.00\,\%, 0.00\,\%)$	$(4.69\,\%, 12.21\,\%)$
$c^{i\,\mathsf{T}} = (c_{Cu}^i, c_{Ni}^i)$	$(0.496, 0.511)$	$(0.500, 0.500)$	$(0.515, 0.527)$
$\delta^{i\,\mathsf{T}} = (\delta_{Cu}^i, \delta_{Ni}^i)$	$(0.71\,\%, 2.26\,\%)$	$(0.00\,\%, 0.00\,\%)$	$(3.12\,\%, 5.54\,\%)$
z	0.0	0.0	0.0716

4   Conclusions

With regards to the system of equations of the Firordt method (especially overdetermined) so-called *interval regularization* technique provides the enhanced
robustness and accuracy. Interval analog of the Firordt method gives robust
result and provides additional information about consistency between data
and the Firordt method's linear model in minimal right-hand part extension
coefficient z^*.

Proposed computing technique essentially uses exact rational computations,
it allows to solve sensitive and ill-conditioned problems [6] and provides full
repeatability of the computing experiment. Also, interval regularization approach may be useful for other linear and linearizable non-linear mathematical
models.

References

1. Vlasova, I.V., Vershinin, V.I.: Determination of binary mixture components by
   the Firordt method with errors below the specified limit. J. Anal. Chem. **64**(6),
   553–558 (2009)
2. Kearfott, R.B., Nakao, M.T., Neumaier, A., Rump, S.M., Shary, S.P., van Hentenryck, P.: Standardized notation in interval analysis. Comput. Technol. **15**(1), 7–13
   (2010)
3. Shary, S.P.: A new technique in systems analysis under interval uncertainty and
   ambiguity. Reliable Comput. **8**(5), 321–418 (2002)
4. Dikoussar, V.V.: Some numerical methods for solving linear algebraic equations.
   Soros Educ. J. **9**, 111–120 (1998). (in Russian)
5. Panyukov, A.V., Golodov, V.A.: Computing best possible pseudo-solutions to interval linear systems of equations. Reliable Comput. **19**(2), 215–228 (2013)
6. Golodov, V.A., Panyukov, A.V.: Library of classes "Exact Computation 2.0". State.
   Reg. 201361818, 14 March 2013: Official Bulletin of Russian Agency for Patents
   and Trademarks (in Russian). In: Series "Programs for Computers, Databases,
   Topology of VLSI", No. 2, Federal Service for Intellectual Property (2013)
7. Shrejver, A.: Theory of Linear and Integer Programming. Wiley, New York (1986)

Computing Capture Tubes

Luc Jaulin[1](✉), Daniel Lopez[4], Vincent Le Doze[2],
Stéphane Le Menec[2], Jordan Ninin[1], Gilles Chabert[3],
Mohamed Saad Ibnseddik[1], and Alexandru Stancu[4]

[1] Lab-STICC, ENSTA-Bretagne, Brest, France
lucjaulin@gmail.com
[2] MBDA-Airbus Group, Paris, France
[3] Ecole des Mines de Nantes, Nantes, France
[4] School of Electrical and Electronic Engineering,
Control Systems Research Group, University of Manchester,
Manchester M139PL, UK

Abstract. Many mobile robots such as wheeled robots, boats, or plane are described by nonholonomic differential equations. As a consequence, they have to satisfy some differential constraints such as having a radius of curvature for their trajectory lower than a known value. For this type of robots, it is difficult to prove some properties such as the avoidance of collisions with some moving obstacles. This is even more difficult when the initial condition is not known exactly or when some uncertainties occur. This paper proposes a method to compute an enclosure (a *tube*) for the trajectory of the robot in situations where a guaranteed interval integration cannot provide any acceptable enclosures. All properties that are satisfied by the tube (such as the non-collision) will also be satisfied by the actual trajectory of the robot.

Keywords: Capture tube · Contractors · Interval arithmetic · Robotics · Stability

1 Introduction

A dynamic system can generally be described a state equation of the form:

$$\mathcal{S}_{\mathbf{f}} : \dot{\mathbf{x}}(t) = \mathbf{f}(\mathbf{x}(t), t). \tag{1}$$

In the situation where the system is uncertain, the state equation becomes a time dependent differential inclusion:

$$\mathcal{S}_{\mathbf{F}} : \dot{\mathbf{x}}(t) \in \mathbf{F}(\mathbf{x}(t), t). \tag{2}$$

Validation of the stability properties of such systems is an important and difficult problem [15]. Most of the time, this problem can be transformed into proving the inconsistency of a *constraint network*. For invariant systems (*i.e.*, **f** or **F** do not depend on t), it has been shown [10] that the *V-stability* approach combined with

© Springer International Publishing Switzerland 2016
M. Nehmeier et al. (Eds.): SCAN 2014, LNCS 9553, pp. 209–224, 2016.
DOI: 10.1007/978-3-319-31769-4_17

interval analysis [16] can solve the problem efficiently. Here, we extend this work to systems where **f** depends on time. Moreover, we will show how to compute a *capture tube*, *i.e.*, a set-valued function which associate to each t a subset of $\mathbb{R}^n$ and such that a feasible trajectory cannot escape. For this, we will need to combine guaranteed integration and Lyapunov theory, such as in [19] or [13], in order to compute this capture tube.

The paper is organized as follows. Section 2 defines the notion of capture tube, which is a specific set of trajectories that encloses the unknown trajectory for the robot. Section 3 explains how tubes can be represented inside the computer and how we can calculate a tube for a trajectory which satisfies a differential inclusion. Section 4 provides a new algorithm that is able to calculate an interval of tubes which encloses the smallest capture tube which contains one candidate tube. An illustrative test-case is presented in Sect. 5 and a conclusion of the paper is given in Sect. 6.

2   Capture Tube

A *tube* $\mathbb{G}$ (see e.g., [1]) is a function which associates to each $t \in \mathbb{R}$ a subset of $\mathbb{R}^n$. Tubes are used for several applications in nonlinear control such as model predictive control [12] or state estimation [2].

Notations. Depending on the context, a tube $\mathbb{G}$ will be seen as a set-valued function $t \mapsto \mathcal{P}(\mathbb{R}^n)$, or also as a subset of $\mathbb{R} \times \mathcal{P}(\mathbb{R}^n)$, where $\mathcal{P}(\mathbb{R}^n)$ is the set of subsets of $\mathbb{R}^n$. It will often be written as $\mathbb{G}(\cdot)$ or also $\mathbb{G}(t)$ to recall that it is a function of t. For instance, when we write $\mathbf{x}(t) \in \mathbb{G}(t)$, we mean $\forall t, \mathbf{x}(t) \in \mathbb{G}(t)$ and when we write $(t_a, \mathbf{a}) \in \mathbb{G}(t)$, we mean $\mathbf{a} \in \mathbb{G}(t_a)$. ■

Consider an autonomous system described by a state equation $\mathcal{S}_{\mathbf{f}} : \dot{\mathbf{x}} = \mathbf{f}(\mathbf{x}, t)$ or a differential inclusion $\mathcal{S}_{\mathbf{F}} : \dot{\mathbf{x}} \in \mathbf{F}(\mathbf{x}, t)$. A tube $\mathbb{G}(t)$ is said to be a *capture tube* [5] (or also called *positive invariant tube*) for $\mathcal{S}_{\mathbf{f}}$ or $\mathcal{S}_{\mathbf{F}}$ if we have the following implication:

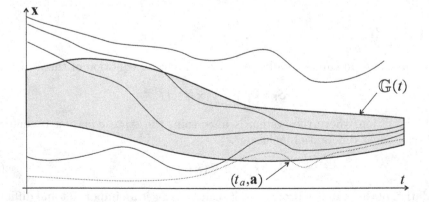

Fig. 1. A tube (painted gray) and possible trajectories for different initial conditions. If a trajectory such as the one represented by the dotted curve exists then the tube is not a capture tube

$$\mathbf{x}(t_a) \in \mathbb{G}(t_a), \tau > 0 \Rightarrow \mathbf{x}(t_a + \tau) \in \mathbb{G}(t_a + \tau). \tag{3}$$

Figure 1 gives some feasible trajectories and a tube $\mathbb{G}(t)$ (in gray). In this figure, all the trajectories are consistent with the assumption that $\mathbb{G}(t)$ is a capture tube, except the trajectory represented by the dotted curve at the bottom, which was able to escape from the tube for $t = t_a$. Consider the tube

$$\mathbb{G}(\cdot) : t \mapsto \{\mathbf{x} \mid \mathbf{g}(\mathbf{x}, t) \le \mathbf{0}\}, \tag{4}$$

where $\mathbf{g} : \mathbb{R}^n \times \mathbb{R} \to \mathbb{R}^m$ is assumed to be differentiable with respect to both $\mathbf{x}$ and t. The following theorem shows that the problem of proving that $\mathbb{G}(t)$ is a capture tube can be cast into proving that a set of inequalities has no solution.

Theorem 1a. If the system of constraints (called the *cross-out* conditions)

$$\begin{cases} (i) & \underbrace{\dfrac{\partial g_i}{\partial \mathbf{x}}(\mathbf{x}, t) \cdot \mathbf{f}(\mathbf{x}, t) + \dfrac{\partial g_i}{\partial t}(\mathbf{x}, t)}_{\dot{g}_i(\mathbf{x}, t)} \ge 0, \\ (ii) & g_i(\mathbf{x}, t) = 0, \\ (iii) & \mathbf{g}(\mathbf{x}, t) \le \mathbf{0}, \end{cases} \tag{5}$$

is inconsistent (*i.e.*, for all $\mathbf{x}$, all $t \ge 0$, and all $i \in \{1, \dots, m\}$, the inequalities are not satisfied), then $\mathbb{G}(\cdot) : t \mapsto \{\mathbf{x} \mid \mathbf{g}(\mathbf{x}, t) \le \mathbf{0}\}$ is a capture tube for the system $\dot{\mathbf{x}} = \mathbf{f}(\mathbf{x}, t)$.

Sketch of proof (see [21, 23] for more details). If $\mathbb{G}(t)$ is not a capture tube, it means that there exists one trajectory, which leaves $\mathbb{G}(t)$, *i.e.*, which crosses the ith boundary $g_i(\mathbf{x}, t) = 0$ from inside to outside. This means that there exists a time-space pair $(\mathbf{a}, t_a)$ on the boundary of $\mathbb{G}(t)$ (*i.e.*, such that *(ii)* and *(iii)* are satisfied) and such that $\dot{g}_i(\mathbf{x}, t) \ge 0$ (otherwise the trajectory cannot leave the tube). ∎

Example 1. Consider again Fig. 1 where we assume that the gray tube corresponds to $\mathbb{G}(\cdot) : t \mapsto \{x \mid g_1(x, t) \le 0\}$. The dotted trajectory leaves the tube at a time-space point (t_a, a), such that $g_1(a, t_a) = 0$ and $\dot{g}_1(a, t_a) > 0$. If such a trajectory is feasible, then $\mathbb{G}(\cdot)$ cannot be a capture tube.

Example 2. We now illustrate the difficulty to get a capture tube on the simple pendulum described by the state equations

$$\begin{cases} \dot{x}_1 = x_2 \\ \dot{x}_2 = -\sin x_1 - 0.15 \cdot x_2 \end{cases} \tag{6}$$

where x_1 is the position of the pendulum and x_2 its rotational speed (see Fig. 2). To find a positive invariant set (*i.e.*, a capture tube) for such a mechanical system the classical method is to take sublevel sets of the energy of the system. Indeed, since the energy of the system

$$E(\mathbf{x}) = \frac{1}{2}\dot{x}_1^2 - \cos x_1 + 1 = \frac{1}{2}x_2^2 - \cos x_1 + 1 \tag{7}$$

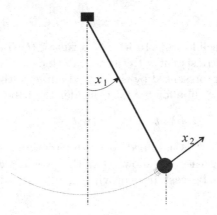

Fig. 2. Simple pendulum

is supposed to decrease with time, we may think that it may be a good candidate for the function g. Let us propose for $g(\mathbf{x}, t)$, which defines our candidate for the capture tube (or positive invariant tube):

$$g(\mathbf{x}, t) = E(\mathbf{x}) - 1 = \frac{1}{2}x_2^2 - \cos x_1, \qquad (8)$$

which is here time independent. The cross-out conditions of Theorem 1a are

$$\begin{cases} \text{(i)} \; (\sin x_1 \; x_2) \begin{pmatrix} x_2 \\ -\sin x_1 - 0.15 \cdot x_2 \end{pmatrix} = -0.15 \cdot x_2^2 \geq 0, \\ \text{(ii)} \qquad\qquad \frac{1}{2}x_2^2 - \cos x_1 = 0. \end{cases} \qquad (9)$$

Note that, since $g(\mathbf{x})$ is scalar, we have $i = 1$ and the condition (iii) is a consequence of (ii). This system has two solutions: $\mathbf{x} = (\pm\frac{\pi}{2}, 0)$. Therefore, Theorem 1a cannot conclude that our tube is positive invariant. Note that, even for this simple two dimensional example which is time-invariant and for which we have a good intuition of a function (the energy) which decreases (almost always), getting a capture tube is difficult. We will see in Sect. 3 how a capture tube can be computed automatically.

Theorem 1b. If the system of constraints (*cross-out* conditions)

$$\begin{cases} \text{(i1)} \; \frac{\partial g_i}{\partial \mathbf{x}}(\mathbf{x}, t) \cdot \mathbf{a} + \frac{\partial g_i}{\partial t}(\mathbf{x}, t) \; \geq 0, \\ \text{(i2)} \qquad\qquad \mathbf{a} \in \mathbf{F}(\mathbf{x}, t), \\ \text{(ii)} \qquad\qquad g_i(\mathbf{x}, t) = 0, \\ \text{(iii)} \qquad\qquad \mathbf{g}(\mathbf{x}, t) \leq 0, \end{cases} \qquad (10)$$

is inconsistent for all $\mathbf{x}$, all $\mathbf{a}$, all $t \geq 0$, and all $i \in \{1, \ldots, m\}$ then $\mathbb{G}(\cdot) : t \mapsto \{\mathbf{x} \mid \mathbf{g}(\mathbf{x}, t) \leq 0\}$ is a capture tube for the differential inclusion $\dot{\mathbf{x}} \in \mathbf{F}(\mathbf{x}, t)$.

Proof. The proof is a direct consequence of Theorem 1a. See also [23].

Consequence. From Theorems 1a and 1b, we conclude that checking that "a tube defined by inequalities is a capture tube" amounts to checking that a set of constraints (here (5) or (10)) is inconsistent. This type of results was already known since several decades [9,23]. Now, proving such an inconsistency can easily be performed [21] using ontractor-based methods [7].

We now have a procedure to prove that a tube is a capture tube. In practice, such a capture tube is difficult to obtain, especially for nonholonomic robots. Even if we have a good intuition of the system and if we are very confident on a potential tube, a contractor-based algorithm often finds a counterexample. In the following section, we will give a new method able to compute automatically capture tubes.

3   Computing with Tubes

3.1   Representation of Tubes

Recall that a tube is a function which associates to any $t \in \mathbb{R}$ a subset of $\mathbb{R}^n$. In the case where these subsets are intervals or boxes, a tube can be represented in the computer by stepwise functions (see [2,4]) as illustrated in Fig. 3.

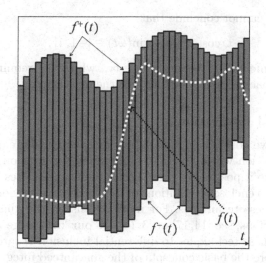

Fig. 3. In numerical computations, a tube $[f](t)$ can be approximated by a lower and an upper stepwise functions $f^-(t)$ and $f^+(t)$. The tube $[f](t)$ encloses an uncertain trajectory $f(t)$

Another possible representation of a tube (see [16]) is an interval expression, which depends on t. For instance,

$$[f](t) = [1,2] \cdot t + \sin([1,3] \cdot t) \tag{11}$$

corresponds to such a tube. Interval polynomials [16] also enter within this class. An example of a third degree polynomial tube is given by

$$[f](t) = [a_0] + [a_1] t + [a_2] t^2 + [a_3] t^3,$$ (12)

where the $[a_i]$ are known intervals. The advantage of interval polynomial is that all operations on scalar polynomials (such as integral, composition, *etc.*) can easily be extended to this class. For instance

$$\int_0^t [f](\tau) \, d\tau = [a_0] t + [a_1] \frac{t^2}{2} + [a_2] \frac{t^3}{3} + [a_3] \frac{t^4}{4}.$$ (13)

It has been proved [16] for the integration, for the composition, and other operations (such as $+, -, /, \cdot$) that the fundamental inclusion property is satisfied. More precisely, for the integration, this inclusion property is

$$f(\cdot) \in [f](\cdot) \;\Rightarrow\; \forall t, \int_0^t f(\tau) \, d\tau \in \int_0^t [f](\tau) \, d\tau.$$ (14)

Remark. For the derivative, this extension cannot be done. For a counterexample, consider the relation

$$\sin(\omega t) \cdot t \in [-1, 1] \cdot t.$$ (15)

It is clear that we cannot conclude that

$$\omega \cos(\omega t) \cdot t + \sin(\omega t) \in [-1, 1].$$ (16)

Thus, the fundamental inclusion property, which is required by all set-membership approaches, is not satisfied for the derivative.

3.2 Guaranteed Integration

For the problem we consider in this paper, *i.e.*, computing capture tubes, the guaranteed integration will be needed. Guaranteed integration is a set of techniques, which make it possible to compute a tube that encloses the solution of a state equation or to enclose all solutions of a differential inclusion. We here recall the principle of these techniques. For more details on the guaranteed integration of state equations, see [14,17] or [3,18]. To our knowledge in the literature, the extension of these techniques to differential inclusion is rarely done. This is why we present here the basic concepts of the guaranteed integration in order to show how they can be extended to the uncertain case, *i.e.*, to differential inclusions. More details and more efficient algorithms for the interval integration of differential inclusions can be found in [11,22]

Brouwer Theorem. Any continuous function f mapping a compact convex set $\mathbb{X}$ into itself has a fixed point, *i.e.*,

$$\exists x \in \mathbb{X} \mid f(x) = x.$$ (17)

Note that a direct corollary of this theorem is that these fixed points also belong to the set $f(\mathbb{X})$.

Example 3. Take $f(x) = \sin(x) \cdot \cos(x)$ and $\mathbb{X} = [-2, 2]$. Since

$$f([-2, 2]) \subset \sin([-2, 2]) \cdot \cos([-2, 2]) = [-1, 1] \cdot [-1, 1] = [-1, 1] \subset \mathbb{X}. \quad (18)$$

From the Brouwer theorem, we have

$$\exists x \in [-2, 2] \mid \sin(x) \cdot \cos(x) = x. \quad (19)$$

The Brouwer theorem is the corner stone that will make it possible to compute a tube containing the solution of a state equation. For its extension to differential inclusions, the uncertain case will be treated using a parametric version of the Brouwer theorem.

Parametric Brouwer Theorem. If $f : \mathbb{X} \times \mathbb{U} \to \mathbb{X}$, where $\mathbb{X}$ is a convex compact set and f is continuous with respect to $x \in \mathbb{X}$, then

$$\forall u \in \mathbb{U}, \ \exists x \in \mathbb{X} \mid f(x, u) = x. \quad (20)$$

Example 4. Take $f(x) = \sin(x + u) \cdot \cos(2x - u)$ and $\mathbb{X} = [-2, 2]$ and $u \in \mathbb{R}$. Since

$$f([-2, 2], \mathbb{R}) \subset [-1, 1] \subset \mathbb{X}, \quad (21)$$

we have

$$\forall u \in \mathbb{R}, \exists x \in [-2, 2] \mid \sin(x + u) \cdot \cos(2x - u) = x. \quad (22)$$

Guaranteed Integration of State Equations. Consider the system $\dot{\mathbf{x}} = \mathbf{f}(\mathbf{x})$, where $\mathbf{f}$ is Lipschitz continuous. The initial condition $\mathbf{x}_0^*$ is known. We want to have an interval enclosure for the trajectory $\mathbf{x}^*(\cdot)$[1]. Define the Picard-Lindelöf operator as

$$\mathcal{T} : \mathbf{x}(\cdot) \to \left(t \mapsto \mathbf{x}_0^* + \int_0^t \mathbf{f}(\mathbf{x}(\tau)) \, d\tau \right). \quad (23)$$

Since $\mathbf{f}$ is Lipschitz continuous, $\mathcal{T}$ has a unique fixed point which corresponds to the solution $\mathbf{x}^*(\cdot)$ of the state equation. Take an interval tube $[\mathbf{x}](\cdot)$. By interval tube, we mean that for all t, $[\mathbf{x}](t)$ is a box of $\mathbb{R}^n$ and not any subset of $\mathbb{R}^n$, as it is allowed for general tubes of $\mathbb{R}^n$. From the Brouwer theorem and since $\mathcal{T}$ has a unique fixed point, we have

$$\mathcal{T}([\mathbf{x}](\cdot)) \subset [\mathbf{x}](\cdot) \Rightarrow \mathbf{x}^*(\cdot) \in [\mathbf{x}](\cdot). \quad (24)$$

Figure 4 provides a representation of the tubes $[\mathbf{x}](\cdot)$ and $\mathcal{T}([\mathbf{x}](\cdot))$. Note that, due to the specific form of $\mathcal{T}$, around the initial instant $t = 0$, the tube $\mathcal{T}([\mathbf{x}](\cdot))$ is thin. Note also that we do not have $\mathcal{T}([\mathbf{x}](\cdot)) \subset [\mathbf{x}](\cdot)$ (*i.e.*, $\mathcal{T}([\mathbf{x}](t))$ is included in $[\mathbf{x}](t)$ only for $t \leq t_1$) and the trajectory may leave the tubes. If we restrict application of $\mathcal{T}$ over the interval $[0, t_1]$, we get the inclusion. Therefore,

$$\forall t \in [0, t_1], \mathbf{x}^*(t) \in \mathcal{T}([\mathbf{x}](t)), \quad (25)$$

---

[1] A trajectory $\mathbf{x}$, which is a function from $\mathbb{R}$ to $\mathbb{R}^n$, can be denoted equivalently $\mathbf{x}(t)$ or $\mathbf{x}(\cdot)$. When no ambiguity may exist, *i.e.*, when t is already used in the same paragraph, we shall often prefer $\mathbf{x}(t)$, for simplicity.

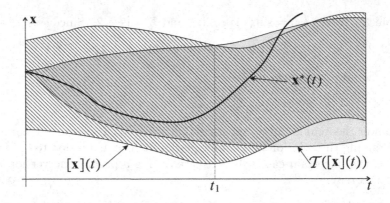

Fig. 4. Illustration of the Picard-Lindelöf operator to the tube $[\mathbf{x}](t)$

where
$$t_1 = \max\left\{t \in \mathbb{R}^+ \mid \forall \tau \in [0,t], \mathcal{T}([\mathbf{x}](\tau)) \subset [\mathbf{x}](\tau)\right\}. \tag{26}$$

Of course, the operator can be called several times, *i.e.*,
$$\forall i \geq 0, \forall t \in [0, t_1], \mathbf{x}^*(t) \in \mathcal{T}^i([\mathbf{x}](t)). \tag{27}$$

Case with Uncertainties. Assume now, that $\mathbf{x}_0$ is uncertain and that the system now depends on an uncertain input vector $\mathbf{u}(\cdot)$ More precisely, the system is described by
$$\dot{\mathbf{x}} = \mathbf{f}(\mathbf{x}, \mathbf{u}), \tag{28}$$

where $\mathbf{x}_0 \in [\mathbf{x}_0]$ and $\mathbf{u}(\cdot) \in [\mathbf{u}](\cdot)$. By setting $\mathbf{F}(\mathbf{x}, t) = \{\mathbf{f}(\mathbf{x}, \mathbf{u}) \mid \mathbf{u}(t) \in [\mathbf{u}](t)\}$, we obtain that a differential inclusion can be described with this formalism. We assume that $\mathbf{f}$ is Lipschitz continuous with respect to $\mathbf{x}$. The Picard operator
$$\mathcal{T}_{\mathbf{x}_0, \mathbf{u}} : \mathbf{x}(\cdot) \to \mathbf{x}_0 + \int_0^t \mathbf{f}(\mathbf{x}(\tau), \mathbf{u}(\tau)) \, d\tau, \tag{29}$$

has uncertainty now. For all $\mathbf{x}_0$, and all $\mathbf{u}(\cdot)$, the operator $\mathcal{T}_{\mathbf{x}_0, \mathbf{u}}$ has a unique fixed point $\mathbf{x}^*(t)$. Consider a tube $\mathbb{X}(\cdot)$. If
$$\mathcal{T}_{\mathbf{x}_0, \mathbf{u}}(\mathbb{X}(\cdot)) \subset \mathbb{X}(\cdot) \tag{30}$$

then, from the Brouwer theorem, $\mathbb{X}(\cdot)$ contains at least one fixed point, *i.e.*, $\mathbf{x}^*(\cdot) \in \mathbb{X}(\cdot)$.

Methodology. For a guaranteed integration, we first have to find a potential tube for which we think that it contains the unique solution of the state equation or contain all solutions of the differential inclusion. This candidate could be obtained using an Euler integration method from $[\mathbf{x}_0]$ followed by an inflation. Then we compute a tube $\mathcal{T}^+([\mathbf{x}](t))$ which encloses the tube
$$\mathcal{T}([\mathbf{x}](t)) = [\mathbf{x}(0)] + \int_0^t \mathbf{f}([\mathbf{x}](\tau), \tau) \, d\tau, \tag{31}$$

or the tube

$$\mathcal{T}\left(\left[\mathbf{x}\right]\left(t\right)\right) = \left[\mathbf{x}\left(0\right)\right] + \int_0^t \mathbf{F}\left(\left[\mathbf{x}\right]\left(\tau\right),\tau\right) d\tau, \tag{32}$$

in the case we have to deal with a differential inclusion. As illustrated in Fig. 4, we compute

$$t_1 = \max_{t \geq 0} \left\{ t \mid \forall \tau \in \left[0,t\right], \mathcal{T}^+\left(\left[\mathbf{x}\right]\left(\tau\right)\right) \subset \left[\mathbf{x}\right]\left(\tau\right)\right\}. \tag{33}$$

Within the interval $[0, t_1]$, from the Brouwer theorem, we conclude that the tube $\mathcal{T}^+\left(\left[\mathbf{x}\right]\left(\cdot\right)\right)$ encloses the solution.

High Order Taylor Method. For a more efficient integration [20], we can replace the Picard-Lindelöf fixed point equation:

$$\mathbf{x}\left(t\right) = \mathbf{x}_0 + \int_0^t \dot{\mathbf{x}}\left(\tau\right) d\tau \tag{34}$$

by the higher order fixed points Taylor equation with the integral remainder

$$\mathbf{x}\left(t\right) = \mathbf{x}_0 + \sum_{i=1}^k \frac{1}{i!}\left(\mathbf{x}^{(i)}\left(0\right)\right) t^i + \int_0^t \frac{\mathbf{x}^{(k+1)}\left(\tau\right)}{k!}\left(t - \tau\right)^k d\tau. \tag{35}$$

Note that for $k = 0$, we get the Picard-Lindelöf equation. This high order method is particularly suited to situations where $[\mathbf{x}_0]$ is known (or small). Indeed, when $\mathbf{x}_0$, is known, the fixed point Taylor operator becomes

$$\mathcal{T}\left(\left[\mathbf{x}\right]\left(t\right)\right) = \mathbf{x}_0 + \sum_{i=1}^k \frac{1}{i!}\left(\mathbf{x}^{(i)}\left(0\right)\right) t^i + \int_0^t \frac{\left[\mathbf{x}\right]^{(k+1)}\left(\tau\right)}{k!}\left(t - \tau\right)^k d\tau. \tag{36}$$

All uncertainties, stored inside $[\mathbf{x}]^{(k+1)}$, are divided by $k!$. Now, in practice, the width of $[\mathbf{x}]^{(k+1)}\left(\tau\right)$ increases polynomially with k, whereas $k!$ increases exponentially. Thus, the accuracy increases with k. The tube $[\mathbf{x}]^{(k+1)}\left(t\right)$ for $\mathbf{x}^{(k+1)}\left(t\right)$ is computed from the tube $[\mathbf{x}]\left(t\right)$ using the expression of the state equation $\dot{\mathbf{x}} = \mathbf{f}\left(\mathbf{x}, \mathbf{u}\right)$.

Remark. Consider the particular case where $k = 2$ and the system $\dot{\mathbf{x}} = \mathbf{f}\left(\mathbf{x}, \mathbf{u}\right)$. We have:

$$\ddot{\mathbf{x}} = \frac{\partial \mathbf{f}}{\partial \mathbf{x}}\left(\mathbf{x}, \mathbf{u}\right) \cdot \mathbf{f}\left(\mathbf{x}, \mathbf{u}\right) + \frac{\partial \mathbf{f}}{\partial \mathbf{x}} \cdot \dot{\mathbf{u}} = \psi^2\left(\mathbf{x}, \mathbf{u}, \dot{\mathbf{u}}\right). \tag{37}$$

For a more general $k \geq 0$, we get:

$$\mathbf{x}^{(k+1)} = \psi^{k+1}\left(\mathbf{x}, \mathbf{u}, \dot{\mathbf{u}}, \dots, \mathbf{u}^{(k)}\right). \tag{38}$$

We have an analytical expression $\psi^{k+1}\left(\mathbf{x}, \mathbf{u}, \dot{\mathbf{u}}, \dots, \mathbf{u}^{(k)}\right)$, but this expression depends on $\dot{\mathbf{u}}, \dots, \mathbf{u}^{(k)}$. Now, a tube for $\dot{\mathbf{u}}, \dots, \mathbf{u}^{(k)}$ is not available in the case of differential inclusions. More precisely, $\dot{\mathbf{x}} \in \mathbf{F}(\mathbf{x}, t)$ can be cast into the form $\dot{\mathbf{x}} = \mathbf{f}(\mathbf{x}, \mathbf{u})$, $\mathbf{u} \in [\mathbf{u}]$ but nothing can be deduced on $\dot{\mathbf{u}}, \ddot{\mathbf{u}}$, *etc.* Thus, high order methods will have difficulties to deal with differential inclusions. To deal with uncertain dynamics using a k-order fixed point Taylor method, we need to be able to express the system in the form $\dot{\mathbf{x}} = \mathbf{f}(\mathbf{x}, \mathbf{u})$ with $\mathbf{u} \in [\mathbf{u}], \ldots, \mathbf{u}^{(k)} \in [\mathbf{u}^{(k)}]$.

4   Computing Capture Tubes

4.1   Basic Idea

If a candidate $\mathbb{G}(t)$ for a capture tube is available, we can prove that $\mathbb{G}(t)$ is a capture tube by checking the inconsistency of a set of onlinear equations (see the previous sections). This inconsistency can then easily be checked using interval analysis. Now, for many systems such as for nonholonomic systems, we rarely have a candidate for a capture tube and we need to find one. The main contribution of this paper is to provide a method that can help us to find such a capture tube. The idea is to start with a non-capture tube $\mathbb{G}(t)$ (the *candidate*) and to try to characterize the smallest capture tube which encloses $\mathbb{G}(t)$. To do this, we predict for all $(\mathbf{x}, t)$, which satisfy the cross-out conditions, a guaranteed envelope for the trajectory within finite time-horizon window $[t, t + t_2]$ (where $t_2 > 0$ is fixed). If all corresponding $\mathbf{x}(t + t_2)$ belong to $\mathbb{G}(t + t_2)$, then the union of all trajectories and the initial $\mathbb{G}(t)$ (in the (x, t) space) corresponds to the smallest capture tube enclosing $\mathbb{G}(t)$.

4.2   Lattice and Capture Tubes

First, let us remark that since the set of subsets of $\mathbb{R}^n$ is a lattice with respect to the inclusion $\subset$, the set of tubes $(\mathbb{T}, \subset)$ is also a lattice. When we introduced the basic idea of how we could compute a capture tube, we wrote that we wanted to compute the smallest tube, which encloses the candidate $\mathbb{G}(t)$. This notion of the smallest tube makes sense because of the following theorem.

Theorem 2. Consider a state space system $\mathcal{S}_\mathbf{f} : \dot{\mathbf{x}} = \mathbf{f}(\mathbf{x}, t)$ or a differential inclusion $\mathcal{S}_\mathbf{F} : \dot{\mathbf{x}} \in \mathbf{F}(\mathbf{x}, t)$. The set of capture tubes $(\mathbb{T}_c, \subset)$ for $\mathcal{S}_\mathbf{f}$ or $\mathcal{S}_\mathbf{F}$ is a sublattice of the set of tubes $(\mathbb{T}, \subset)$.

Proof. Consider two captures tubes $\mathbb{G}_1(t)$ and $\mathbb{G}_2(t)$. If the trajectory $\mathbf{x}(t)$ belongs to both $\mathbb{G}_1(t)$ and $\mathbb{G}_2(t)$, then $\mathbf{x}(t)$ will leave neither $\mathbb{G}_1(t)$ nor $\mathbb{G}_2(t)$. Thus, the intersection $\mathbb{G}_1(t) \cap \mathbb{G}_2(t)$ is a capture tube. The same reasoning can be done for the union of the two tubes. Since $\mathbb{G}_1(t) \cap \mathbb{G}_2(t)$ is the largest tube included in $\mathbb{G}_1(t)$ and $\mathbb{G}_2(t)$ and since $\mathbb{G}_1(t) \cup \mathbb{G}_2(t)$ is the smallest tube which contains $\mathbb{G}_1(t)$ and $\mathbb{G}_2(t)$, we conclude that $(\mathbb{T}_c, \subset)$ is a lattice. Since all capture tubes are also tubes, we get that $(\mathbb{T}_c, \subset)$ is a sublattice of $(\mathbb{T}, \subset)$. ∎

Consequences. Since $\mathbb{T}_c$ is a sublattice of $\mathbb{T}$, for any tube $\mathbb{G}(t) \in \mathbb{T}$, we can define the following operator:

$$\text{capt}(\mathbb{G}(t)) = \bigcap \{\overline{\mathbb{G}}(t) \in \mathbb{T}_c \mid \mathbb{G}(t) \subset \overline{\mathbb{G}}(t)\}. \tag{39}$$

This set corresponds to the smallest capture tube which encloses $\mathbb{G}(t)$.

Interval of Tubes. The set of tubes is a lattice with respect to the inclusion $\subset$. Thus, we can define *intervals of tubes*. This notion is important in this paper, because we need to compute a tube, in a guaranteed way. Now, this tube may probably not be representable in the computer. This new notion of interval of tubes will be needed in order to characterize the tube we want to calculate.

4.3   Computing Capture Tubes

Since the set of tubes $(\mathbb{T}, \subset)$ is a lattice, we can define intervals of tubes as follows.

Definition. An interval of tubes $[\mathbb{G}]$ is a subset of the set of tubes $\mathbb{T}$ which satisfies

$$[\mathbb{G}] = \{\mathbb{G} \in \mathbb{T} \mid \mathbb{G} \subset \vee [\mathbb{G}] \text{ and } \mathbb{G} \supset \wedge [\mathbb{G}]\}. \tag{40}$$

Here, $\mathbb{G}^+ = \vee [\mathbb{G}]$ denotes the smallest outer bound of $[\mathbb{G}]$ and $\mathbb{G}^- = \wedge [\mathbb{G}]$ denotes the largest inner bound of $[\mathbb{G}]$. The set of intervals of tubes will be denoted by $\mathbb{IT}$. Note that we could also define the notion of interval of capture tubes, but this notion is not interesting in our context since it is very difficult to get (exactly) even one capture tube.

Problem to be Solved. Given a tube $\mathbb{G}(\cdot) : t \mapsto \{\mathbf{x} \mid \mathbf{g}(\mathbf{x}, t) \leq 0\}$ in $\mathbb{T}$, compute an interval $[\mathbb{C}^-(t), \mathbb{C}^+(t)] \in \mathbb{IT}$ such that

$$\text{capt}(\mathbb{G}(t)) \in [\mathbb{C}^-(t), \mathbb{C}^+(t)]. \tag{41}$$

This is illustrated in Fig. 5. Of course, since $\mathbb{G}(t) \subset \text{capt}(\mathbb{G}(t))$, we can take $\mathbb{C}^-(t) = \mathbb{G}(t)$. Thus, the main difficulty is to get a tube $\mathbb{C}^+(t)$, which is not too large.

Flow. The flow associated with the system $\mathcal{S}_{\mathbf{f}} : \dot{\mathbf{x}} = \mathbf{f}(\mathbf{x}, t)$ is a function $\phi_{t_0, t_1} : \mathbb{R}^n \to \mathbb{R}^n$ such that

$$\dot{\mathbf{x}} = \mathbf{f}(\mathbf{x}, t) \Rightarrow \phi_{t_0, t_1}(\mathbf{x}(t_0)) = \mathbf{x}(t_1). \tag{42}$$

This means that if the trajectory $\mathbf{x}(t)$ is a solution of $\mathcal{S}_{\mathbf{f}}$, we are able to go from the state at instant t_0 to the state at instant t_1 using the flow.

The flow associated with the differential inclusion $\mathcal{S}_{\mathbf{F}} : \dot{\mathbf{x}} \in \mathbf{F}(\mathbf{x}, t)$ is a function $\phi_{t_0, t_1} : \mathbb{R}^n \to \mathcal{P}(\mathbb{R}^n)$,

$$\dot{\mathbf{x}} \in \mathbf{F}(\mathbf{x}, t) \Rightarrow \mathbf{x}(t_1) \in \phi_{t_0, t_1}(\mathbf{x}(t_0)). \tag{43}$$

ϕ_{t_0, t_1} should also be the smallest with respect to the inclusion which satisfies this property. Equivalently, $\phi_{t_0, t_1}(\mathbf{x}(t_0))$ corresponds to the set of all states that can be reached at instant $t_1 \geq t_0$ by a trajectory consistent with $\mathcal{S}_{\mathbf{F}}$ and initialized at $\mathbf{x}(t_0)$ for $t = t_0$.

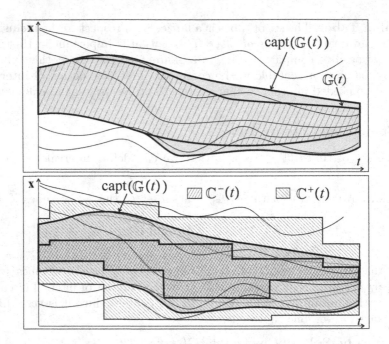

Fig. 5. The capture tube $\mathrm{capt}(\mathbb{G}(t))$, that we want to compute, will be enclosed by an interval of tubes $\left[\mathbb{C}^-(t), \mathbb{C}^+(t)\right]$

Theorem 3a. Consider the system $\mathcal{S}_\mathbf{f} : \dot{\mathbf{x}} = \mathbf{f}(\mathbf{x}, t)$. The tube

$$\mathbb{C}(\cdot) : t \rightarrow \{\mathbf{x} \mid \exists (\mathbf{x}_0, t_0),\ \mathbf{x}_0 \in \mathbb{G}(t_0),\ t \geq t_0,\ \mathbf{x} = \phi_{t_0,t}(\mathbf{x}_0)\}, \qquad (44)$$

where $\phi_{t_0,t}$ is the flow function of $\mathcal{S}_\mathbf{f}$, corresponds to $\mathrm{capt}(\mathbb{G}(t))$.

Proof of Theorem 3a. We will show that $\mathbb{C}(t)$, is the smallest capture tube which encloses $\mathbb{G}(t)$. For the proof, we will prove *(i)* that $\mathbb{C}(t)$ contains $\mathbb{G}(t)$, *(ii)* that $\mathbb{C}(t)$ is a capture tube and *(iii)* that $\mathbb{C}(t)$ is the smallest one.
(i) To prove that $\mathbb{G}(t) \subset \mathbb{C}(t)$, it suffices to take $t_0 = t$ and $\mathbf{x}_0 = \mathbf{x}$.
(ii) We now prove that $\mathbb{C}(t)$ is a capture tube. Take a pair $(\mathbf{x}^{t_a}, t_a)$ such that $\mathbf{x}^{t_a} \in \mathbb{C}(t_a)$. From (44), we have

$$\exists (\mathbf{x}_0, t_0),\ \mathbf{x}_0 \in \mathbb{G}(t_0),\ t_a \geq t_0,\ \mathbf{x}^{t_a} = \phi_{t_0,t_a}(\mathbf{x}_0). \qquad (45)$$

Take $\tau > 0$ and define the point $\mathbf{x}^{t_a+\tau} = \phi_{t_a,t_a+\tau}(\mathbf{x}^{t_a})$. From (45), we have

$$\exists (\mathbf{x}_0, t_0),\ \mathbf{x}_0 \in \mathbb{G}(t_0),\ t_a \geq t_0,\ \mathbf{x}^{t_a+\tau} = \phi_{t_0,t_a+\tau}(\mathbf{x}_0). \qquad (46)$$

Therefore, we have proved that

$$\mathbf{x}^{t_a} \in \mathbb{C}(t_a), \tau \geq 0 \Rightarrow \phi_{t_a,t_a+\tau}(\mathbf{x}^{t_a}) \in \mathbb{C}(t_a + \tau), \qquad (47)$$

i.e., $\mathbb{C}(t)$ is a capture tube.

(iii) We will now prove by contradiction that $\mathbb{C}(t)$ is the smallest capture tube that encloses $\mathbb{G}(t)$. Take a capture tube $\overline{\mathbb{G}}(t)$ such that $\overline{\mathbb{G}}(t) \supset \mathbb{G}(t)$ which is enclosed strictly in $\mathbb{C}(t)$. By strictly, we mean that $\exists\,(t_1, \mathbf{x}_1)$, $\mathbf{x}_1 \in \mathbb{C}(t_1)$ and $\mathbf{x}_1 \notin \overline{\mathbb{G}}(t_1)$. From (44), $\exists\,(\mathbf{x}_0, t_0)$, $\mathbf{x}_0 \in \mathbb{G}(t_0)$, $\mathbf{x}_1 = \phi_{t_0,t_1}(\mathbf{x}_0)$. The corresponding trajectory crosses the tube $\overline{\mathbb{G}}(t)$ from inside to outside which is inconsistent with the fact that $\overline{\mathbb{G}}(t)$ is a capture tube. ∎

Theorem 3b. Consider the system $\mathcal{S}_{\mathbf{F}} : \dot{\mathbf{x}} \in \mathbf{F}(\mathbf{x}, t)$. The tube

$$\mathbb{C}(t) : t \to \{\mathbf{x} \mid \exists\,(\mathbf{x}_0, t_0)\,, \mathbf{x}_0 \in \mathbb{G}(t_0), t \geq t_0,\ \mathbf{x} \in \phi_{t_0,t}(\mathbf{x}_0)\}, \qquad (48)$$

where $\phi_{t_0,t}$ is the set membership flow function of $\mathcal{S}_{\mathbf{F}}$, corresponds to $\mathrm{capt}(\mathbb{G}(t))$.

Proof. The proof is a direct consequence of Theorem 3a. ∎

Theorem 4a. Consider the system $\mathcal{S}_{\mathbf{f}} : \dot{\mathbf{x}} = \mathbf{f}(\mathbf{x}, t)$. We have

$$\mathrm{capt}(\mathbb{G}(t)) = \mathbb{G}(t) \cup \Delta\mathbb{G}(t), \qquad (49)$$

with

$$\Delta\mathbb{G}(t) = t \mapsto \{\mathbf{x} \mid \exists\,(\mathbf{x}_0, t_0)\ \text{satisfying (5)}, \qquad (50)$$
$$t \geq t_0,\ \mathbf{x} = \phi_{t_0,t}(\mathbf{x}_0)\ \text{and}\ \mathbf{x} \notin \mathbb{G}(t)\,\}.$$

Proof. To build $\mathrm{capt}(\mathbb{G}(t))$, it suffices to add to the tube $\mathbb{G}(t)$ all pairs $(\mathbf{x}_1, t_1)$ outside $\mathbb{G}(t)$ that can be reached from a pair $(\mathbf{x}_a, t_a)$ in $\mathbb{G}(t)$. The corresponding trajectory will cross the boundary of the tube $\mathbb{G}(t)$ at instant t_0 at the state $\mathbf{x}_0$, *i.e.*, $(\mathbf{x}_0, t_0)$ satisfies (5). This is illustrated in Fig. 6. ∎

Theorem 4b. Consider the differential inclusion $\mathcal{S}_{\mathbf{F}} : \dot{\mathbf{x}} = \mathbf{F}(\mathbf{x}, t)$. We have

$$\mathrm{capt}(\mathbb{G}(t)) = \mathbb{G}(t) \cup \Delta\mathbb{G}(t), \qquad (51)$$

with

$$\Delta\mathbb{G}(t) = t \mapsto \{\mathbf{x} \mid \exists\,(\mathbf{x}_0, t_0)\ \text{satisfying (10)}, \qquad (52)$$
$$t \geq t_0,\ \mathbf{x} \in \phi_{t_0,t}(\mathbf{x}_0)\ \text{and}\ \mathbf{x} \notin \mathbb{G}(t)\,\}.$$

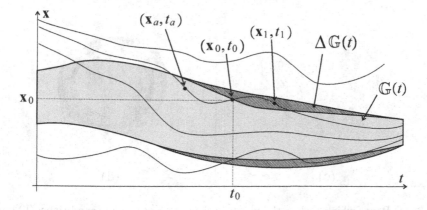

Fig. 6. $\Delta\mathbb{G}(t)$ contains all pairs $(\mathbf{x}_1, t_1)$ outside $\mathbb{G}(t)$ that can be reached from a pair $(\mathbf{x}_0, t_0)$ leaving $\mathbb{G}(t)$

Consequences. An interval $[\mathbb{C}^-(t), \mathbb{C}^+(t)]$ for $\mathrm{capt}(\mathbb{G}(t))$ will be composed by the tube $\mathbb{C}^-(t) = \mathbb{G}(t)$ and by adding to $\mathbb{C}^-(t)$ an enclosure of all trajectories generated from one pair $(\mathbf{x}_0, t_0)$ satisfying (5) or (10).

5   Test Case

Consider the pendulum presented in Sect. 2. Here, we do not consider the sublevel sets of the energy anymore, which only applies on a small class of systems. Instead, we consider, as an candidate tube, the one associated with the function

$$g(\mathbf{x}, t) = x_1^2 + x_2^2 - 1.$$

We have chosen here a time-invariant tube in order to be able to draw pictures. Indeed, both $\mathbb{G}(t)$ and $\Delta\mathbb{G}(t)$ do not depend on t and become subsets of $\mathbb{R}^2$. Our algorithm provides the results shown in Fig. 7. Subfigure (a) depicts a subpaving

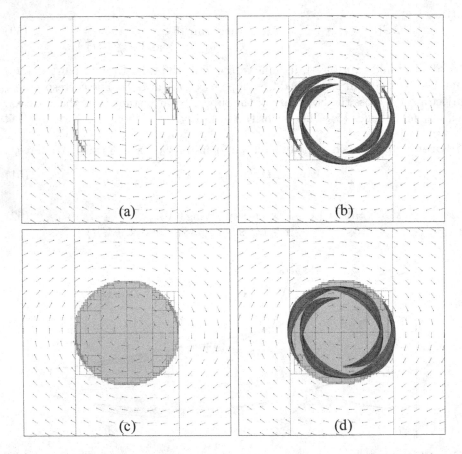

Fig. 7. (a) Boxes which enclose the points satisfying the cross-out conditions; (b) guaranteed integration $\Delta\mathbb{G}$ of these boxes; (c) inner approximation $\mathbb{C}^-$ of $\mathrm{Capt}(\mathbb{G})$; (d) outer approximation $\mathbb{C}^+$ of $\mathrm{Capt}(\mathbb{G})$

which encloses all points satisfying the cross-out conditions. The guaranteed integration $\Delta \mathbb{G}$ of all these boxes are shown on Subfigure (b). The integration has been performed using the DYNIBEX library [8]. Subfigure (c) represents a subpaving made with boxes shown to be inside $\mathbb{G}$. Since $\mathbb{G} \subset \text{capt}(\mathbb{G})$, this subpaving also corresponds to an inner approximation $\mathbb{C}^-$ of $\text{capt}(\mathbb{G})$. Subfigure (d) shows $\mathbb{C}^+$ which is the union of light gray boxes (back plane) and dark gray boxes (front plane). This union forms an outer approximation of $\text{capt}(\mathbb{G})$.

6 Conclusion

Proving that a controlled nonlinear system always stays inside a time moving bubble (or tube) amounts to proving a set of nonlinear inequalities. Now, in practice, even with a good intuition, finding such a significant capture tube is difficult. This paper proposes a new method for computing an approximation of the smallest tube, which encloses a candidate tube $\mathbb{G}(t)$. Even if $\mathbb{G}(t)$ is generally chosen as rather attractive, it is often possible to cross $\mathbb{G}(t)$ from inside to outside during the initialization of the system. Since this tube may not be representable in the computer, the method calculates an interval of tubes which encloses the capture tube we want to compute. The principle of the approach is to integrate (with a guaranteed interval integration) the state vectors that cross the candidate tube from inside to outside and to add all the corresponding trajectories to the candidate tube. Now, since the less we integrate, the more we are efficient, to deal with large scale systems, it should be necessary to limit the number of integration by giving more importance to the Lyapunov part of the resolution. This could be done, for instance, by computing barrier functions [6].

References

1. Aubin, J.P., Frankowska, H.: Set-Valued Analysis. Birkhäuser, Boston (1990)
2. Le Bars, F., Sliwka, J., Reynet, O., Jaulin, L.: State estimation with fleeting data. Automatica **48**(2), 381–387 (2012)
3. Berz, M., Makino, K.: Verified integration of ODEs and flows using differential algebraic methods on high-order Taylor models. Reliable Comput. **4**(3), 361–369 (1998)
4. Bethencourt, A., Jaulin, L.: Solving non-linear constraint satisfaction problems involving time-dependant functions. Math. Comput. Sci. **8**(3), 503–523 (2014)
5. Blanchini, F., Miani, S.: Set Theoretic Methods in Control. Birkhauser, Boston (2008)
6. Bouissou, O., Chapoutot, A., Djaballah, A., Kieffer, M.: Computation of parametric barrier functions for dynamical systems using interval analysis. In: IEEE CDC, Los Angeles, United States (2014)
7. Chabert, G., Jaulin, L.: Contractor programming. Artif. Intell. **173**, 1079–1100 (2009)
8. Chapoutot, A., Alexandre dit Sandretto, J., Mullier, O.: Dynibex. ENSTA (2015). http://perso.ensta-paristech.fr/~chapoutot/dynibex/
9. Fernandes, M.L., Zanolin, F.: Remarks on strongly flow-invariant sets. J. Math. Anal. Appl. **128**, 176–188 (1987)

10. Jaulin, L., Le Bars, F.: An interval approach for stability analysis: application to sailboat robotics. IEEE Trans. Robot. **27**(5), 282–287 (2012)
11. Kapela, T., Zgliczynski, P.: A lohner-type algorithm for control systems and ordinary differential inclusions. Discrete Continuous Dyn. Syst. **11**(2), 365–385 (2009)
12. Langson, W., Chryssochoos, I., Rakovic, S.V., Mayne, D.Q.: Robust model predictive control using tubes. Automatica **40**(1), 125–133 (2004)
13. Lhommeau, M., Jaulin, L., Hardouin, L.: Capture basin approximation using interval analysis. Int. J. Adap. Control Sig. Process. **25**(3), 264–272 (2011)
14. Lohner, R.: Enclosing the solutions of ordinary initial and boundary value problems. In: Kaucher, E., Kulisch, U., Ullrich, C.H. (eds.) Computer Arithmetic: Scientific Computation and Programming Languages, pp. 255–286. BG Teubner, Stuttgart (1987)
15. Le Menec, S.: Linear differential game with two pursuers and one evader. In: Breton, M., Szajowski, K. (eds.) Advances in Dynamic Games, vol. 11, pp. 209–226. Birkhauser, Boston (2011)
16. Moore, R.E.: Interval Analysis. Prentice-Hall, Englewood Cliffs (1966)
17. Nedialkov, N.S., Jackson, K.R., Corliss, G.F.: Validated solutions of initial value problems for ordinary differential equations. Appl. Math. Comput. **105**(1), 21–68 (1999)
18. Raissi, T., Ramdani, N., Candau, Y.: Set membership state and parameter estimation for systems described by nonlinear differential equations. Automatica **40**, 1771–1777 (2004)
19. Ratschan, S., She, Z.: Providing a basin of attraction to a target region of polynomial systems by computation of Lyapunov-like functions. SIAM J. Control Optim. **48**(7), 4377–4394 (2010)
20. Revol, N., Makino, K., Berz, M.: Taylor models and floating-point arithmetic: proof that arithmetic operations are validated in COSY. J. Logic Algebraic Program. **64**, 135–154 (2005)
21. Stancu, A., Jaulin, L., Bethencourt, A.: Stability analysis for time-dependent nonlinear systems: an interval approach. Internal report, University of Manchester (2015)
22. Wilczak, D., Zgliczynski, P.: Cr-Lohner algorithm. Schedae Informaticae **20**, 9–46 (2011)
23. Yorke, J.A.: Invariance for ordinary differential equations. Math. Syst. Theor. **1**(4), 353–372 (1967)

Some Remarks on the Rigorous Estimation of Inverse Linear Elliptic Operators

Takehiko Kinoshita[1,2]($\boxtimes$), Yoshitaka Watanabe[3], and Mitsuhiro T. Nakao[4]

[1] Center for the Promotion of Interdisciplinary Education and Research,
Kyoto University, Kyoto 606-8501, Japan
kinosita@kurims.kyoto-u.ac.ip
[2] Research Institute for Mathematical Sciences,
Kyoto University, Kyoto 606-8502, Japan
[3] Research Institute for Information Technology,
Kyushu University, Fukuoka 812-8581, Japan
[4] National Institute of Technology, Sasebo College, Nagasaki 857-1193, Japan

Abstract. This paper presents a new numerical method to obtain the rigorous upper bounds of inverse linear elliptic operators. The invertibility of a linearized operator and its norm estimates give important informations when analyzing the nonlinear elliptic partial differential equations (PDEs). The computational costs depend on the concerned elliptic problems as well as the approximation properties of used finite element subspaces, e.g., mesh size or so. We show the proposed new estimate is effective for an intermediate mesh size.

1 Introduction

The main aim of this paper is to provide an efficient estimates of a solution of the following linear elliptic partial differential equations (PDEs) with the Dirichlet boundary condition:

$$\begin{cases} -\Delta u + (b \cdot \nabla)u + cu = f & \text{in } \Omega, \\ u = 0 & \text{on } \partial\Omega, \end{cases} \tag{1a}$$
$$\tag{1b}$$

for an arbitrary $f \in L^2(\Omega)$. Here, $\Omega \subset \mathbb{R}^d$, $(d \in \{1,2,3\})$ is a bounded polygonal or polyhedral domains, $b \in L^\infty(\Omega)^d$, and $c \in L^\infty(\Omega)$. As well known, many physical problems have a linearized problem of the form (1a)-(1b), e.g., the stationary Burgers equations [7].

Now let $L^2(\Omega)$ be the set of all measurable functions from Ω to $\mathbb{C}$ with square integrable, which is a Hilbert space with associated inner product $(u,v)_{L^2(\Omega)} := \int_\Omega u(x)\overline{v(x)}\,dx$, where $\overline{}$ shows the complex conjugate. Let $H_0^1(\Omega) := \{u \in H^1(\Omega) | u = 0 \text{ on} \partial\Omega\}$ be the usual Sobolev space with respect to the inner product $(u,v)_{H_0^1(\Omega)} := (\nabla u, \nabla v)_{L^2(\Omega)^d}$. Let $L : H_0^1(\Omega) \times H_0^1(\Omega) \to \mathbb{C}$ be a bilinear form defined by

$$L(u,v) := (\nabla u, \nabla v)_{L^2(\Omega)^d} + ((b \cdot \nabla)u, v)_{L^2(\Omega)} + (cu, v)_{L^2(\Omega)}, \quad \forall u,v \in H_0^1(\Omega).$$

© Springer International Publishing Switzerland 2016
M. Nehmeier et al. (Eds.): SCAN 2014, LNCS 9553, pp. 225–235, 2016.
DOI: 10.1007/978-3-319-31769-4_18

We define the weak solution $u \in H_0^1(\Omega)$ of (1a)-(1b) by a solution of the following variational equation:

$$L(u,v) = (f,v)_{L^2(\Omega)}, \quad \forall v \in H_0^1(\Omega). \tag{2}$$

If we assume the coercivity of L, then, by the Lax-Milgram theorem, there exists a unique solution for (2). Moreover, it can be proved that this weak solution is a solution of (1a)-(1b) by a regularity argument (see e.g., [1]). This fact means that the linear elliptic operator $\mathscr{L} := -\triangle + b \cdot \nabla + c$ has the inverse operator.

On the other hand, Plum [10], Oishi [9], Nakao-Hashimoto-Watanabe [7] and Kinoshita-Watanabe-Nakao [5] proposed a computational technique to verify the existence of $\mathscr{L}^{-1}$ even though the coercivity of L is not assumed. In this paper, we also do not assume the coercivity to L at all. Moreover, we try to find the quantitative value of C_{L^2,H_0^1} satisfying

$$\left\| \mathscr{L}^{-1} \right\|_{\mathscr{L}(L^2(\Omega), H_0^1(\Omega))} \le C_{L^2,H_0^1}. \tag{3}$$

The constant C_{L^2,H_0^1} plays an essential role in the numerical verification of solutions for the boundary value problems for nonlinear elliptic PDEs [9,10] and it is desirable to compute C_{L^2,H_0^1} as small as possible. Particularly, the constant C_{L^2,H_0^1} proposed by Watanabe-Kinoshita-Nakao [13] is expected to converge to exact norm $\left\| \mathscr{L}^{-1} \right\|_{\mathscr{L}(L^2(\Omega), H_0^1(\Omega))}$ as the discretization parameter $h \to 0$ on the suitable assumptions. Therefore, in the asymptotic sense, the estimates of (3) by [13] would give better bounds than the results in [7]. Indeed, many numerical examples show this situation. However, in order to get successful calculation of C_{L^2,H_0^1} in [13], we often need smaller mesh size h than [7]. In other words, we could verify the existence of $\mathscr{L}^{-1}$ by the method in [7] with smaller computational costs than [13].

In this paper, we present a new method to compute the constant C_{L^2,H_0^1} in (3) based on the perturbation theory of linear operator with technique in [7]. The verification condition of the existence of $\mathscr{L}^{-1}$ by the proposed method is essentially same as in [7]. But as shown in the numerical results, the proposed C_{L^2,H_0^1} is often better.

The contents of this paper are as follows: In Sect. 2, we define the necessary notations and function spaces. In Sect. 3, we introduce previous results of the invertibility of $\mathscr{L}$ and its a posteriori estimates. In Sect. 4, we propose a new verification condition for the invertibility of $\mathscr{L}$ and its a posteriori estimates. In Sect. 5, we show several verification results for the proposed procedures.

2　Notations

Let $\mathscr{X}$ and $\mathscr{Y}$ be the Banach spaces. We represent the space of the bounded linear operators from $\mathscr{X}$ to $\mathscr{Y}$ by $\mathscr{L}(\mathscr{X}, \mathscr{Y})$. Especially, $\mathscr{L}(\mathscr{X})$ denotes $\mathscr{L}(\mathscr{X}, \mathscr{X})$. Let $\mathscr{L}_C(\mathscr{X}, \mathscr{Y}) \subset \mathscr{L}(\mathscr{X}, \mathscr{Y})$ be the space of the compact operators from $\mathscr{X}$ to $\mathscr{Y}$. Moreover, $\mathscr{L}_F(\mathscr{X}, \mathscr{Y}) \subset \mathscr{L}(\mathscr{X}, \mathscr{Y})$ denotes the set of the bounded Fredholm operators from $\mathscr{X}$ to $\mathscr{Y}$. For any linear operator $\mathscr{A} : \mathscr{X} \to \mathscr{Y}$, $D(\mathscr{A})$, $R(\mathscr{A})$, and $N(\mathscr{A})$ denote the domain, range, and kernel of $\mathscr{A}$, respectively. We define the norm of $D(\mathscr{A})$ by $\|u\|_{D(\mathscr{A})} := \|u\|_{\mathscr{X}} + \|\mathscr{A}u\|_{\mathscr{Y}}$, which is called graph norm. As well known, if $\mathscr{A}$ is a closed operator then $D(\mathscr{A})$ becomes a Banach space with respect to $\|\cdot\|_{D(\mathscr{A})}$.

Let $-\triangle : D(-\triangle) \subset L^2(\Omega) \to L^2(\Omega)$ be a Laplace operator, where the domain $D(-\triangle)$ is defined by

$$D(-\triangle) = \{u \in H_0^1(\Omega) \; ; \; -\triangle u \in L^2(\Omega)\}.$$

Then, $-\triangle$ is a closed operator from $L^2(\Omega)$ to $L^2(\Omega)$. We define the differential operator $B \in \mathscr{L}\big(H_0^1(\Omega), L^2(\Omega)\big)$ by $B := b \cdot \nabla + c$. The differential operators are treated as the closed operators in many cases. However, it is more convenient to treat the differential operators as the bounded operators in our verification method. Let $I_e : D(-\triangle) \hookrightarrow H_0^1(\Omega)$ be an embedding operator. Then, $I_e \in \mathscr{L}_C\big(D(-\triangle), H_0^1(\Omega)\big)$ is satisfied by the Rellich compactness theorem because Ω is in a class of the bounded domain with the Lipschitz continuous boundary. Moreover, $BI_e \in \mathscr{L}_C\big(D(-\triangle), L^2(\Omega)\big)$ is satisfied because composition operator of the bounded operator and compact operator is a compact operator. The bounded operator $\mathscr{L} \in \mathscr{L}\big(D(-\triangle), L^2(\Omega)\big)$ is represented by $\mathscr{L} := -\triangle + BI_e = -\triangle + b \cdot \nabla I_e + cI_e$. Especially, the domain of $\mathscr{L}$ is defined by $D(\mathscr{L}) = D(-\triangle)$. Then, $\mathscr{L} \in \mathscr{L}_F\big(D(-\triangle), L^2(\Omega)\big)$ and $\mathrm{ind}(\mathscr{L}) = 0$ by [5].

The norms of Banach space $L^\infty(\Omega)^d$ and $L^\infty(\Omega)$ are defined by

$$\|b\|_{L^\infty(\Omega)^d} := \operatorname*{ess\,sup}_{x \in \Omega} \sqrt{|b_1(x)|^2 + \cdots + |b_d(x)|^2}, \quad \|c\|_{L^\infty(\Omega)} := \operatorname*{ess\,sup}_{x \in \Omega} |c(x)|,$$

respectively. Let $C_{s,2}$ be a positive constant satisfying $\|u\|_{L^2(\Omega)} \le C_{s,2} \|u\|_{H_0^1(\Omega)}$ for all $u \in H_0^1(\Omega)$, which is called the Poincaré constant.

Let $S_h(\Omega)$ be an approximate finite dimensional subspace of $H_0^1(\Omega)$ dependent on the parameter h. For example, $S_h(\Omega)$ is considered to be a finite element subspace with the mesh size h or a set of polynomials less than a fixed degree. Let n be a degree of freedom for $S_h(\Omega)$ and $\{\phi_i\}_{i=1}^n$ be the basis functions of $S_h(\Omega)$. Namely, $S_h(\Omega) := \mathrm{span}_{1 \le i \le n}\{\phi_i\}$.

We denote the self-adjoint positive definite (SPD) matrices D_ϕ and L_ϕ in $\mathbb{C}^{n \times n}$ by

$$D_{\phi,i,j} := (\nabla \phi_j, \nabla \phi_i)_{L^2(\Omega)^d}, \quad L_{\phi,i,j} := (\phi_j, \phi_i)_{L^2(\Omega)}, \quad \forall i,j \in \{1, \ldots, n\}.$$

Since D_ϕ and L_ϕ are SPD, these have the Cholesky factorization. Let $D_\phi^{1/2}$ and $L_\phi^{1/2}$ be the Cholesky factors of D_ϕ and L_ϕ, respectively, i.e.,

$$D_\phi = D_\phi^{1/2} D_\phi^{H/2}, \quad \text{and} \quad L_\phi = L_\phi^{1/2} L_\phi^{H/2}$$

where $D_\phi^{H/2}$ shows the conjugate matrix of $D_\phi^{1/2}$. We define the H_0^1 projection $P_h^1 : H_0^1(\Omega) \to S_h(\Omega)$ by

$$\big(u - P_h^1 u, v_h\big)_{H_0^1(\Omega)} = 0, \quad \forall v_h \in S_h(\Omega). \tag{4}$$

Therefore, the problems of the solvability of the variational Eq. (4) and the nonsingularity of D_ϕ are equivalent. Because the matrix D_ϕ is positive definite, the projection P_h^1 is well defined. Now, we assume that the following error estimates of P_h^1 hold throughout this paper.

Assumption 1. *There exists a positive constant $C(h) > 0$ satisfying*

$$\left\|u - P_h^1 u\right\|_{H_0^1(\Omega)} \leq C(h) \left\|\triangle u\right\|_{L^2(\Omega)}, \quad \forall u \in D(-\triangle), \tag{5}$$

$$\left\|u - P_h^1 u\right\|_{L^2(\Omega)} \leq C(h) \left\|u - P_h^1 u\right\|_{H_0^1(\Omega)}, \quad \forall u \in H_0^1(\Omega). \tag{6}$$

Assumption 1 is the most basic error estimates in the Galerkin method. For example, in the case of the one dimensional bounded interval as Ω, if $S_h(\Omega)$ is a finite element space using piecewise linear polynomials, the value $C(h)$ is known by $C(h) = \frac{h}{\pi}$. Alternatively, in the case of piecewise quadratic polynomials, Assumption 1 is satisfied by $C(h) = \frac{h}{2\pi}$. Moreover, these approximations give the optimal constants (e.g., [6]). In case that N degree polynomials are used, Assumption 1 is satisfied by $C(h) = O(\frac{h}{N})$. However, in these cases, the optimal constants are unknown (e.g., [3]). In case of the two or three dimensional bounded rectangular or rectangular cuboid domain as Ω, if $S_h(\Omega)$ is a finite element space using the tensor product of one dimensional piecewise polynomial spaces, $C(h)$ is attained same constants in one dimensional case (e.g., [6]). In case of the two dimensional bounded polygonal domain as Ω, if $S_h(\Omega)$ is the P1 finite element space with triangular mesh, Assumption 1 is satisfied. The details of $C(h)$ are shown in e.g., [2].

Let G_ϕ be a matrix in $\mathbb{R}^{n \times n}$, where each elements are defined by

$$G_{\phi,i,j} := L(\phi_j, \phi_i) = (\nabla\phi_j, \nabla\phi_i)_{L^2} + ((b \cdot \nabla)\phi_j, \phi_i)_{L^2} + (c\phi_j, \phi_i)_{L^2}, \quad \forall i, j \in \{1, \ldots, n\}.$$

We assume that G_ϕ is nonsingular throughout this paper. Applying the proposed verification method, it is necessary to confirm the nonsingularity of G_ϕ by validated computations.

3 Previous Results

In this section, we introduce the results for the invertibility condition of the operator $\mathscr{L}$ and its a posteriori estimates. We define the following constants:

$$C_1 := \|b\|_{L^\infty(\Omega)^d} + C_{s,2}\|c\|_{L^\infty(\Omega)}, \qquad C_2 := \|b\|_{L^\infty(\Omega)^d} + C(h)\|c\|_{L^\infty(\Omega)},$$

$$M_\phi^{11}(h) := \left\|D_\phi^{H/2} G_\phi^{-1} D_\phi^{1/2}\right\|_2, \qquad M_\phi^{10}(h) := \left\|D_\phi^{H/2} G_\phi^{-1} L_\phi^{1/2}\right\|_2$$

where $\|\cdot\|_2$ is the matrix two-norm, i.e., the maximum singular value.

Theorem 1 ([7, **Theorem 2.1 & Corollary 1 & Theorem 2.3**] & [8]). *Let $\tilde{K}(h) > 0$ be defined by*

$$\tilde{K}(h) := \begin{cases} C(h)\left(C_{s,2}\|\operatorname{div} b\|_{L^\infty(\Omega)} + C_1\right) & \text{if } b \in W^{1,\infty}(\Omega)^d, \\ \\ C_{s,2}C_2 & \text{if } b \in L^\infty(\Omega)^d \backslash W^{1,\infty}(\Omega)^d. \end{cases}$$

And let $\tilde{\kappa}_\phi > 0$ be a constant satisfying

$$\tilde{\kappa}_\phi := C(h)\left(C_1 M_\phi^{11}(h)\tilde{K}(h) + C_2\right) < 1. \tag{7}$$

Then, there exists $\mathscr{L}^{-1} \in \mathscr{L}\left(L^2(\Omega), D(-\triangle)\right)$ and C_{L^2, H_0^1} in (3) can be taken as

$$C_{L^2, H_0^1} = \frac{C_{s,2}}{1 - \tilde{\kappa}_\phi} \left\| \begin{pmatrix} M_\phi^{11}(h)\left(1 - C_2 C(h)\right) & M_\phi^{11}(h)\tilde{K}(h) \\ M_\phi^{11}(h)C_1 C(h) & 1 \end{pmatrix} \right\|_2. \tag{8}$$

If b has sufficient regularity, from the fact that $\tilde{K}(h) = O\left(C(h)\right)$, the C_{L^2, H_0^1} defined (8) converges to:

$$C_{L^2, H_0^1} \to C_{s,2} \left\| \begin{pmatrix} M_\phi^{11}(0) & 0 \\ 0 & 1 \end{pmatrix} \right\|_2 = C_{s,2} \max\left\{M_\phi^{11}(0), 1\right\} \tag{9}$$

as $h \to 0$, where $M_\phi^{11}(0) := \lim_{h \to 0} M_\phi^{11}(h)$. This a posteriori estimates fails to converge to its exact operator norm. On the other hand, Watanabe-Kinoshita-Nakao proposed another a posteriori estimates in [5,13] as follows.

Theorem 2 ([13, **Theorem 4.2**] & [5, **Theorem 4.3**]). *Assume that $\hat{\kappa}_\phi > 0$ satisfy*

$$\hat{\kappa}_\phi := C(h)C_2\left(1 + M_\phi^{10}(h)C_1\right) < 1. \tag{10}$$

Then, there exists $\mathscr{L}^{-1} \in \mathscr{L}\left(L^2(\Omega), D(-\triangle)\right)$ and C_{L^2, H_0^1} in (3) can be taken as

$$C_{L^2, H_0^1} = \frac{\sqrt{M_\phi^{10}(h)^2 + C(h)^2\left(1 + M_\phi^{10}(h)C_1\right)^2}}{1 - \hat{\kappa}_\phi}. \tag{11}$$

The right hand side of (11) is expected to converge to the exact operator norm as $h \to 0$. Therefore, we expect that (11) would give better estimates than (8). In fact, we can prove $M_\phi^{10}(h) \leq C_{s,2} M_\phi^{11}(h)$ for arbitrary $h > 0$. However, in the actual verification process, we often meet the situation such that the criterion (10) is harder than (7) for a fixed h. Therefore, Theorem 1 should be effective for the problem that h cannot be taken so small. We now try to derive C_{L^2, H_0^1} smaller than (8) in Theorem 1 with the same criterion (7).

Note that, in order to obtain the values of $M_\phi^{11}(h)$ and $M_\phi^{10}(h)$, it is necessary to solve numerically some corresponding generalized matrix eigenvalue problems. If it succeeded in the verification of the finite upper bound of $M_\phi^{11}(h)$ or $M_\phi^{10}(h)$, it means that G_ϕ is nonsingular. Rump proposed an efficient method for solving this eigenvalue problem with result verification in [12].

4   Main Theorem

We describe a main theorem of this paper as Theorem 3 in this section. Before describing it, we need to get several lemmas as below.

Lemma 1. *Let $b \in L^\infty(\Omega)^d$ and $c \in L^\infty(\Omega)$. Then, we obtain the following estimates:*

$$\left\|P_h^1 u\right\|_{H_0^1(\Omega)} \leq M_\phi^{11}(h) \left\|P_h^1(-\triangle)^{-1}((b\cdot\nabla+c)(u-P_h^1 u)-\mathscr{L}u)\right\|_{H_0^1(\Omega)} \qquad (12)$$

for all $u \in D(-\triangle)$.

Proof. For an arbitrary $u \in D(-\triangle)$, let $u_\perp := u - P_h^1 u$ and $f := \mathscr{L}u = -\triangle u + b\cdot\nabla u + cu \in L^2(\Omega)$. Then, u satisfies (2). We take a test function v as $v = v_h \in S_h(\Omega) \subset H_0^1(\Omega)$ in (2), from the definition of H_0^1-projection, we have

$$(\nabla u, \nabla v_h)_{L^2(\Omega)^d} + (b\cdot\nabla u + cu, v_h)_{L^2(\Omega)} = (f, v_h)_{L^2(\Omega)}$$

$$L(P_h^1 u, v_h) = (-b\cdot\nabla u_\perp - cu_\perp + f, v_h)_{L^2(\Omega)}. \qquad (13)$$

We set $\psi := (-\triangle)^{-1}(-b\cdot\nabla u_\perp - cu_\perp + f) \in D(-\triangle)$. In (13), from the definition of H_0^1-projection, we obtain

$$\begin{aligned}
L(P_h^1 u, v_h) &= (-b\cdot\nabla u_\perp - cu_\perp + f, v_h)_{L^2(\Omega)}, \quad \forall v_h \in S_h(\Omega), \\
&= (-\triangle(-\triangle)^{-1}(-b\cdot\nabla u_\perp - cu_\perp + f), v_h)_{L^2(\Omega)} \\
&= (\nabla\psi, \nabla v_h)_{L^2(\Omega)^d} \\
&= (\nabla P_h^1\psi, \nabla v_h)_{L^2(\Omega)^d}. \qquad (14)
\end{aligned}$$

Since $P_h^1 u$ and $P_h^1\psi$ are elements of $S_h(\Omega)$, they are represented as linear combinations of the basis of $S_h(\Omega)$. Namely, there exist $\alpha = (\alpha_1, \ldots, \alpha_n)^T$, $\gamma = (\gamma_1, \ldots, \gamma_n)^T \in \mathbb{C}^n$ such that

$$P_h^1 u = \sum_{i=1}^n \alpha_i\phi_i, \quad P_h^1\psi = \sum_{i=1}^n \gamma_i\phi_i.$$

Then, (14) is rewritten using α and γ to have

$$G_\phi\alpha = D_\phi\gamma.$$

Therefore, we obtain

$$\begin{aligned}
\left\|P_h^1 u\right\|_{H_0^1(\Omega)}^2 &= \alpha^H D_\phi\alpha = \left(D_\phi^{H/2}\alpha\right)^H \left(D_\phi^{H/2}G_\phi^{-1}D_\phi^{1/2}\right)\left(D_\phi^{H/2}\gamma\right) \\
&\leq \left\|P_h^1 u\right\|_{H_0^1(\Omega)} \left\|D_\phi^{H/2}G_\phi^{-1}D_\phi^{1/2}\right\|_2 \left\|P_h^1\psi\right\|_{H_0^1(\Omega)},
\end{aligned}$$

which proves the lemma.

Let $L_{\mathrm{div}}^\infty(\Omega)^d := \{u \in L^\infty(\Omega)^d \; ; \; \mathrm{div}\, u \in L^\infty(\Omega)\}$. The right hand side of (12) can be estimated by the following lemma.

Lemma 2. *Let $b \in L_{\mathrm{div}}^\infty(\Omega)^d$ and $c \in L^\infty(\Omega)$. Then, we obtain the following estimates:*

$$\left\|P_h^1(-\triangle)^{-1}(b\cdot\nabla+c)(u-P_h^1 u)\right\|_{H_0^1(\Omega)} \leq K_1(h)\left\|u-P_h^1 u\right\|_{H_0^1(\Omega)} \qquad (15)$$

for all $u \in H_0^1(\Omega)$, where $K_1(h) := C(h)\left(C_{s,2}\|\mathrm{div}\, b\|_{L^\infty(\Omega)} + C_1\right)$.

Proof. For an arbitrary $u \in H_0^1(\Omega)$, let $u_\perp := u - P_h^1 u \in H_0^1(\Omega)$ and $\psi := (-\triangle)^{-1}$
$(b \cdot \nabla + c)u_\perp \in D(-\triangle)$. Then, we have

$$\left\| P_h^1 \psi \right\|_{H_0^1(\Omega)}^2 = (\nabla \psi, \nabla P_h^1 \psi)_{L^2(\Omega)^d}$$

$$= (-\triangle \psi, P_h^1 \psi)_{L^2(\Omega)}$$

$$= ((b \cdot \nabla)u_\perp, P_h^1 \psi)_{L^2(\Omega)} + (cu_\perp, P_h^1 \psi)_{L^2(\Omega)}$$

$$= -(u_\perp, \operatorname{div}(\bar{b} P_h^1 \psi))_{L^2(\Omega)} + (u_\perp, \bar{c} P_h^1 \psi)_{L^2(\Omega)}$$

$$\leq \left(\left\| \operatorname{div}(\bar{b} P_h^1 \psi) \right\|_{L^2(\Omega)} + \left\| \bar{c} P_h^1 \psi \right\|_{L^2(\Omega)} \right) \|u_\perp\|_{L^2(\Omega)}$$

$$\leq \left(\left\| P_h^1 \psi \operatorname{div} \bar{b} \right\|_{L^2(\Omega)} + \left\| (\bar{b} \cdot \nabla) P_h^1 \psi \right\|_{L^2(\Omega)} + \left\| \bar{c} P_h^1 \psi \right\|_{L^2(\Omega)} \right) \|u_\perp\|_{L^2(\Omega)}$$

$$\leq \left(\|\operatorname{div} b\|_{L^\infty} \left\| P_h^1 \psi \right\|_{L^2} + \|b\|_{L^\infty} \left\| \nabla P_h^1 \psi \right\|_{L^2} + \|c\|_{L^\infty} \left\| P_h^1 \psi \right\|_{L^2} \right) \|u_\perp\|_{L^2}$$

$$\leq (C_{s,2} \|\operatorname{div} b\|_{L^\infty} + \|b\|_{L^\infty} + C_{s,2} \|c\|_{L^\infty}) \left\| \nabla P_h^1 \psi \right\|_{L^2} \|u_\perp\|_{L^2}.$$

Applying (6), we obtain (15).

Even if the regularity of b is only $L^\infty(\Omega)^d$, there exists the following lemma by [4].

Lemma 3 ([4, **Theorem 3.3**]). *Let $b \in L^\infty(\Omega)^d$, $c \in L^\infty(\Omega)$ and let $W_h(\Omega)$ be a finite element space of $H(\operatorname{div}, \Omega) := \{\phi \in L^2(\Omega)^d; \operatorname{div}\phi \in L^2(\Omega)\}$. For an arbitrary $\psi_h \in S_h$, let $(w_h, v_h) \in W_h(\Omega) \times S_h(\Omega)$ be the solution of the following problem:*

$$\begin{cases} (w_h, w_h^*)_{L^2(\Omega)^d} + (\nabla v_h, w_h^*)_{L^2(\Omega)^d} = (b\psi_h, w_h^*)_{L^2(\Omega)^d} & \forall w_h^* \in W_h(\Omega), \\ (w_h, \nabla v_h^*)_{L^2(\Omega)^d} = 0 & \forall v_h^* \in S_h(\Omega), \end{cases}$$

And define $\sigma_0(h)$ and $\sigma_1(h)$ as follows

$$\sigma_0(h) := \sup_{S_h \ni \psi_h \neq 0} \frac{\|w_h + \nabla v_h - b\psi_h\|_{L^2(\Omega)^d}}{\|\nabla \psi_h\|_{L^2(\Omega)^d}}, \quad \sigma_1(h) := \sup_{S_h \ni \psi_h \neq 0} \frac{\|\operatorname{div} w_h\|_{L^2(\Omega)}}{\|\nabla \psi_h\|_{L^2(\Omega)^d}}.$$

Then, we have

$$\left\| P_h^1 (-\triangle)^{-1}(b \cdot \nabla + c)(u - P_h^1 u) \right\|_{H_0^1(\Omega)} \leq K_0(h) \left\| u - P_h^1 u \right\|_{H_0^1(\Omega)}$$

for all $u \in H_0^1(\Omega)$, where $K_0(h) := \sigma_0(h) + C(h)\sigma_1(h) + C(h)C_{s,2} \|c\|_{L^\infty(\Omega)}$.

Now, let $K(h)$ be a positive constant defined by:

$$K(h) := \begin{cases} K_1(h) & \text{if } b \in L_{\operatorname{div}}^\infty(\Omega)^d, \\ \min\{C_{s,2}C_2, K_0(h)\} & \text{if } b \in L^\infty(\Omega)^d \backslash L_{\operatorname{div}}^\infty(\Omega)^d. \end{cases} \tag{16}$$

From Lemmas 2 and 3, (12) is estimated by

$$\left\| P_h^1 u \right\|_{H_0^1(\Omega)} \leq M_\phi^{11}(h) \left\| P_h^1 (-\triangle)^{-1}((b \cdot \nabla + c)(u - P_h^1 u) - \mathscr{L}u) \right\|_{H_0^1(\Omega)}$$

$$\leq M_\phi^{11}(h)K(h) \left\| u - P_h^1 u \right\|_{H_0^1(\Omega)} + M_\phi^{11}(h) \left\| P_h^1 (-\triangle)^{-1} \mathscr{L}u \right\|_{H_0^1(\Omega)}$$

$$\leq M_\phi^{11}(h)K(h) \left\| u - P_h^1 u \right\|_{H_0^1(\Omega)} + M_\phi^{11}(h)C_{s,2} \|\mathscr{L}u\|_{L^2(\Omega)} \tag{17}$$

Lemma 4. *Let $b \in L^\infty(\Omega)^d$ and $c \in L^\infty(\Omega)$. Then, we obtain the following estimates:*

$$\left\| u - P_h^1 u \right\|_{H_0^1(\Omega)} \leq C(h) \left(C_1 \left\| P_h^1 u \right\|_{H_0^1(\Omega)} + C_2 \left\| u - P_h^1 u \right\|_{H_0^1(\Omega)} + \left\| \mathscr{L} u \right\|_{L^2(\Omega)} \right) \quad (18)$$

for all $u \in D(-\triangle)$.

Proof. For an arbitrary $u \in D(-\triangle)$, let $u_\perp := u - P_h^1 u$. From the Poincaré inequality and (6), we have

$$\|\triangle u\|_{L^2} = \|\mathscr{L} u - b \cdot \nabla u - c u\|_{L^2(\Omega)}$$

$$\leq \|\mathscr{L} u\|_{L^2(\Omega)} + \|b\|_{L^\infty(\Omega)^d} \|\nabla u\|_{L^2(\Omega)^d} + \|c\|_{L^\infty(\Omega)} \|u\|_{L^2(\Omega)}$$

$$\leq \|\mathscr{L} u\|_{L^2} + \|b\|_{L^\infty} \left(\left\| P_h^1 u \right\|_{H_0^1} + \|u_\perp\|_{H_0^1} \right) + \|c\|_{L^\infty} \left(\left\| P_h^1 u \right\|_{L^2} + \|u_\perp\|_{L^2} \right)$$

$$\leq \|\mathscr{L} u\|_{L^2} + \left(\|b\|_{L^\infty} + C_{s,2} \|c\|_{L^\infty} \right) \left\| P_h^1 u \right\|_{H_0^1} + \left(\|b\|_{L^\infty} + C(h) \|c\|_{L^\infty} \right) \|u_\perp\|_{H_0^1}.$$

Therefore, from (5), we obtain

$$\|\nabla u_\perp\|_{L^2(\Omega)^d} \leq C(h) \|\triangle u\|_{L^2(\Omega)}$$

$$\leq C(h) \left(C_1 \left\| P_h^1 u \right\|_{H_0^1(\Omega)} + C_2 \|u_\perp\|_{H_0^1(\Omega)} + \|\mathscr{L} u\|_{L^2(\Omega)} \right).$$

By the effective use of the above lemmas, we propose the following estimates based on the Fredholm theory.

Theorem 3. *Let $K(h) > 0$ be defined by (16). And let $\kappa_\phi > 0$ be a constant satisfying*

$$\kappa_\phi := C(h) \left(C_1 M_\phi^{11}(h) K(h) + C_2 \right) < 1. \quad (19)$$

Then, there exists $\mathscr{L}^{-1} \in \mathscr{L}\left(L^2(\Omega), D(-\triangle) \right)$ and C_{L^2, H_0^1} in (3) can be taken as

$$C_{L^2, H_0^1} = \frac{\sqrt{M_\phi^{11}(h)^2 \left(C_{s,2} + C(h) \left(K(h) - C_{s,2} C_2 \right) \right)^2 + C(h)^2 \left(1 + C_{s,2} M_\phi^{11}(h) C_1 \right)^2}}{1 - \kappa_\phi}. \quad (20)$$

Proof. For an arbitrary $u \in D(-\triangle)$, we set $u_\perp := u - P_h^1 u \in H_0^1(\Omega)$. From (17) and (18), we obtain

$$\begin{pmatrix} 1 & -K(h) M_\phi^{11}(h) \\ -C(h) C_1 & 1 - C(h) C_2 \end{pmatrix} \begin{pmatrix} \left\| P_h^1 u \right\|_{H_0^1(\Omega)} \\ \|u_\perp\|_{H_0^1(\Omega)} \end{pmatrix} \leq \begin{pmatrix} C_{s,2} M_\phi^{11}(h) \\ C(h) \end{pmatrix} \|\mathscr{L} u\|_{L^2(\Omega)}$$

where the inequality is meant componentwise. From the assumption (19),

$$\det \begin{pmatrix} 1 & -K(h) M_\phi^{11}(h) \\ -C(h) C_1 & 1 - C(h) C_2 \end{pmatrix} = 1 - \kappa_\phi > 0$$

is satisfied. Therefore, the solution of this simultaneous inequalities can be written as

$$\begin{pmatrix} \left\| P_h^1 u \right\|_{H_0^1(\Omega)} \\ \|u_\perp\|_{H_0^1(\Omega)} \end{pmatrix} \leq \frac{1}{1 - \kappa_\phi} \begin{pmatrix} 1 - C(h) C_2 & K(h) M_\phi^{11}(h) \\ C(h) C_1 & 1 \end{pmatrix} \begin{pmatrix} C_{s,2} M_\phi^{11}(h) \\ C(h) \end{pmatrix} \|\mathscr{L} u\|_{L^2(\Omega)}$$

$$= \frac{1}{1 - \kappa_\phi} \begin{pmatrix} C_{s,2} M_\phi^{11}(h) + C(h) M_\phi^{11}(h) \left(K(h) - C_{s,2} C_2 \right) \\ C(h) \left(1 + C_{s,2} M_\phi^{11}(h) C_1 \right) \end{pmatrix} \|\mathscr{L} u\|_{L^2(\Omega)}.$$

Then, we have

$$\|u\|_{H_0^1(\Omega)}^2 = \|P_h^1 u\|_{H_0^1(\Omega)}^2 + \|u_\perp\|_{H_0^1(\Omega)}^2$$

$$\leq \left(\frac{C_{s,2} M_\phi^{11}(h) + C(h) M_\phi^{11}(h) \left(K(h) - C_{s,2} C_2\right)}{1 - \kappa_\phi} \right)^2 \|\mathscr{L}u\|_{L^2(\Omega)}^2$$

$$+ \left(\frac{C(h) \left(1 + C_{s,2} M_\phi^{11}(h) C_1\right)}{1 - \kappa_\phi} \right)^2 \|\mathscr{L}u\|_{L^2(\Omega)}^2.$$

Finally, the invertibility of $\mathscr{L}$ is followed by the same arguments in [5, Theorem 4.3].

Remark 1. If $b \in W^{1,\infty}(\Omega)^d$, the criterion (19) is equal to (7) because $K(h) = \tilde{K}(h)$. Therefore, the attainability of criteria (19) and (7) are essentially same. On the other hand, even if the convergence order $K(h) = O(1)$, namely, independent of smoothness of the function b, the constant C_{L^2,H_0^1} of (20) converges to $C_{s,2} M_\phi^{11}(0)$ as $h \to 0$. Comparing this result with (9), we can say that (20) is better than (8) in the asymptotic sense as $h \to 0$.

5 Numerical Results

In this section, we show some verified computation results of constants C_{L^2,H_0^1} by (8), (11), and (20). Let $\mathscr{L} = -\triangle + b \cdot \nabla + c : D(-\triangle) \to L^2(\Omega)$ be a non-self-adjoint operator with $b := R \begin{pmatrix} -x_2 + 1/2 \\ x_1 - 1/2 \end{pmatrix}$, $R \in \mathbb{R}$, and $c \in \mathbb{C}$ on $\Omega := (0,1) \times (0,1) \subset \mathbb{R}^2$. We adopted P1 finite element space with uniform triangular meshes as $S_h(\Omega)$. Then, discretization parameter $h > 0$ is the element side length. In this case, Assumption 1 holds with $C(h) = 0.493h([2])$ and $C_{s,2} = \frac{1}{\pi\sqrt{2}}$. Note that, of course our arguments above can also be applied for not only P1 element but also any finite element spaces. We use the interval arithmetic toolbox INTLAB [11] Version 7 with MATLAB 8.0.0.783 (R2012b) on Intel Core i7 3.4 GHz with Mac OSX 10.8.3.

Table 1. $R = 10$, $c = 15$

$1/h$	$M_\phi^{11}(h)$	$M_\phi^{10}(h)$	Theorem 1		Theorem 2		Theorem 3	
			$\tilde{\kappa}_\phi$	C_{L^2,H_0^1}	$\hat{\kappa}_\phi$	C_{L^2,H_0^1}	κ_ϕ	C_{L^2,H_0^1}
5	0.9732	0.1270	1.8758	——	1.9610	——	1.8758	——
8	0.9903	0.1276	0.9032	3.3368	1.1493	——	0.9032	**2.6387**
10	0.9939	0.1277	0.6488	0.8671	0.8987	1.6951	0.6488	**0.6589**
20	0.9986	0.1279	0.2497	0.3543	0.4284	**0.2453**	0.2497	0.2760
50	0.9999	0.1279	0.0818	0.2632	0.1663	**0.1559**	0.0818	0.2316
100	1.0001	0.1279	0.0379	0.2426	0.0823	**0.1400**	0.0379	0.2267

In Table 1, the short line segment means that the corresponding criteria (7), (10), or (19) were not satisfied, which also implies we failed to compute the rigorous upper bounds C_{L^2,H_0^1}. From these results, we can say that, for sufficiently small h, the estimates (11) should be finest. On the other hand, if h is not so small, then our proposed estimates (20) is better than others. Therefore, we conclude that three kinds of methods would have their own ranges of suitable applicability depending on each problem.

6   Conclusion

We presented an alternative approach to the numerical verification method for linear ellitipc problems based on Theorem 3. It is proved that our new method gives a better results from the viewpoint in computational costs. As the future subjects, we will show that the present method can also be applied to fourth order elliptic problems or more general linear elliptic operators.

Acknowledgments. The authors are very grateful to two anonymous reviewers. This work was supported by the Grant-in-Aid from the Ministry of Education, Culture, Sports, Science and Technology of Japan (No. 23740074, No. 24340018, and No. 24540151) and supported by Program for Leading Graduate Schools "Training Program of Leaders for Integrated Medical System for Fruitful Healthy-Longevity Society."

References

1. Grisvard, P.: Singularities in Boundary Value Problems. Springer, New York (1992)
2. Kikuchi, F., Liu, X.: Determination of the Babuska-Aziz constant for the linear triangular finite element. Jpn. J. Ind. Appl. Math. **23**(1), 75–82 (2006)
3. Kimura, S., Yamamoto, N.: On the L^2 a priori error estimates to the finite element solution of elliptic problems with singular adjoint operator. Bull. Inform. Cybern. **31**(2), 109–115 (1999)
4. Kinoshita, T., Hashimoto, K., Nakao, M.T.: The L^2 a priori error estimates for singular adjoint operator. Numer. Func. Anal. Optim. **30**(3–4), 289–305 (2009)
5. Kinoshita, T., Watanabe, Y., Nakao, M.T.: An improvement of the theorem of a posteriori estimates for inverse elliptic operators. NOLTA **5**(1), 47–52 (2014)
6. Nakao, M.T., Yamamoto, N., Kimura, S.: On the best constant in the error bound for the H_0^1-projection into piecewise polynomial spaces. J. Approx. Theory **93**, 491–500 (1998)
7. Nakao, M.T., Hashimoto, K., Watanabe, Y.: A numerical method to verify the invertibility of linear elliptic operators with applications to nonlinear problems. Computing **75**, 1–14 (2005)
8. Nakao, M.T., Watanabe, Y., Kinoshita, T., Kimura, T., Yamamoto, N.: Some considerations of the invertibility verifications for linear elliptic operators. Jpn. J. Ind. Appl. Math. **32**(1), 19–31 (2015)
9. Oishi, S.: Numerical verification of existence and inclusion of solutions for nonlinear operator equations. J. Comput. Appl. Math. **60**(1–2), 171–185 (1995)
10. Plum, M.: Computer-assisted proofs for semilinear elliptic boundary value problems. Jpn. J. Ind. Appl. Math. **26**(2–3), 419–442 (2009)
11. Rump, S.M.: INTLAB - INTerval LABoratory. In: Csendes, T. (ed.) Developments in Reliable Computing, pp. 77–104. Kluwer Academic Publishers, Dordrecht (1999). http://www.ti3.tu-harburg.de/rump/

12. Rump, S.M.: Verified bounds for singular values, in particular for the spectral norm of a matrix and its inverse. BIT Numer. Math. **51**(2), 367–384 (2011)
13. Watanabe, Y., Kinoshita, T., Nakao, M.T.: A posteriori estimates of inverse operators for boundary value problems in linear elliptic partial differential equations. Math. Comput. **82**, 1543–1557 (2013)

Verified Parameter Identification for Dynamic Systems with Non-Smooth Right-Hand Sides

Andreas Rauh[(✉)], Luise Senkel, and Harald Aschemann

Chair of Mechatronics, University of Rostock,
Justus-von-Liebig-Weg 6, D-18059 Rostock, Germany
{andreas.rauh,luise.senkel,harald.aschemann}@uni-rostock.de

Abstract. Modeling of systems in engineering involves two major stages. First, a system structure is derived that is based on the fundamental laws from physics that characterize the relevant processes. Second, specific parameter values are determined by minimizing the distance between the measured and simulated system outputs. In previous work, strategies for verified parameter identification using techniques from interval analysis were developed. These techniques are extended in this paper to a verified estimation for systems with non-smooth ordinary differential equations. Suitable experimental results for parameter estimation of a mechanical system with friction conclude this contribution to highlight the practical applicability of the developed identification procedure.

Keywords: Non-smooth ordinary differential equations · Verified parameter identification · Interval analysis · Mechanical systems · Friction

1 Introduction

Dynamic system models given by ordinary differential equations (ODEs) with non-smooth right-hand sides are widely used in engineering. They can, for example, be employed to describe transitions between static and sliding friction in mechanical systems and to represent variable degrees of freedom for dynamic applications in robotics with contacts between at least two (rigid) bodies.

The verified simulation of such systems has to detect those points of time at which either one of the discrete model states (in a representation of the ODEs by means of a state transition diagram) becomes active or at which one of the discrete model states is deactivated [1,7,8,14]. As long as mechanical systems are taken into consideration that are described by position and velocity as corresponding state variables, it is guaranteed that the trajectories (i.e., the solutions of the ODE) remain continuous if switchings between different submodels occur.

For practical applications, however, it is on the one hand necessary to derive verified simulation techniques and to compute state variables that can be reached within a given time horizon under consideration of a predefined control law. Such a control law is usually given by the actuator signal (e.g. force or torque) acting

© Springer International Publishing Switzerland 2016
M. Nehmeier et al. (Eds.): SCAN 2014, LNCS 9553, pp. 236–246, 2016.
DOI: 10.1007/978-3-319-31769-4_19

onto the (mechanical) system [10,13]. On the other hand, a system identification is necessary to determine parameter values that comply with both the non-smooth system model and the measured data. In engineering applications, these measurements are usually subject to uncertainty that is often in the same order of magnitude as the measured data themselves. For large uncertainty, it is in general not reliable to determine point values for the system parameters. In [2], for example, it has been shown that the naive application of least squares techniques for the minimization of the distance between point-valued measured and simulated system outputs (computed in pure floating point arithmetic) may lead to results that do not comply with a verified set-valued enclosure. Such set-valued enclosures represent those parameter ranges that are at the same time compatible with the system model and bounded measurement uncertainty.

For this reason, two options for the verified parameter identification are discussed in this paper with respect to their applicability to systems with state-dependent transitions between different piecewise smooth ODE representations.

The identification makes use of a verified simulation of ODEs with non-smooth right-hand sides. This routine employs a generalized Taylor series-based integration to determine guaranteed state enclosures that are reachable over some time span. As shown by the parameter estimation of a test rig for the longitudinal dynamics of a vehicle, the minimum series expansion order leads already to state enclosures that are suitable for a verified identification. For this test rig, parameters related to the mass moment of inertia as well as the static and sliding friction coefficients are estimated. Besides verified integration of non-smooth ODEs, the reliable identification exploits an interval subdivision procedure.

This paper is structured as follows. Section 2 gives an overview of the class of systems for which parameters are estimated in this paper. In Sect. 3, a brief review of the verified interval-based simulation routine for ODEs with non-smooth right-hand sides is given. Section 4 describes different options for the implementation of verified identification procedures. A summary of identification results for a laboratory test rig at the Chair of Mechatronics, University of Rostock, is given in Sect. 5. Conclusions and an outlook on future work can be found in Sect. 6.

2   Dynamic Systems with Non-Smooth Right-Hand Sides

In this paper, parameter identification strategies are considered for (open-loop) dynamical systems with l different continuous-time models $\mathcal{S} = \{S_1, S_2, \ldots, S_l\}$, which are each given by the state-space representations

$$\dot{\mathbf{x}}(t) = \mathbf{f}_{S_i}(\mathbf{x}(t), \mathbf{p}, \mathbf{u}(t), t) \quad \text{for} \quad i \in \{1, \ldots, l\} . \tag{1}$$

In (1), the vector $\mathbf{x} \in \mathbb{R}^n$ denotes the state vector and $\mathbf{p} \in \mathbb{R}^{n_{\mathrm{p}}}$ the vector of uncertain parameters that are identified by the subsequent procedure. Moreover, $\mathbf{u} \in \mathbb{R}^{n_u}$ is the vector of control variables. In the case of open-loop systems, this vector is assumed to be piecewise constant for a time interval $t \in [t_k \; ; \; t_{k+1})$.

Because closed-loop feedback control procedures are assumed to be implemented in a discrete-time manner throughout this paper, the input $\mathbf{u}(t)$ is also piecewise constant in this case. The input is then given as a state-dependent function that is evaluated at each point of time $t = t_k$ for the current state vector $\mathbf{x}(t_k)$. The explicit time dependency of (1) is then used to describe the influence of a predefined reference trajectory $\mathbf{x}_\mathrm{d}(t)$ on the dynamic system.

For a complete specification of the system behavior, conditions $T_i^j(\mathbf{x}, \mathbf{u})$ for the transition from the model state S_i to S_j, $i,j \in \{1, \ldots, l\}$, have to be given additionally. These conditions are specified in the following by means of a state transition diagram, where all discrete model states S_i are assumed to be mutually exclusive in a real-life experiment. This also holds for simulations in the case of exactly known parameters and system states as long as no time discretization errors influence the system dynamics. However, multiple states S_i can be active simultaneously during simulations in the uncertain case, including the effect of time discretization. Then, a definite distinction between two different models S_i and S_j is no longer possible due to the before-mentioned uncertainties.

Note that the case $i = j$ refers to the operating conditions for which the current state S_i remains active. However, a verified simulation does not only have to account for scenarios in which the transition between two different states S_i and S_j occurs exactly at a sampling point t_k. The simulation also has to detect transitions that take place between two subsequent points t_k and t_{k+1}.

As a representative benchmark application, the drive train test rig depicted in Fig. 1 is considered. It represents a simplified model for the longitudinal dynamics of a vehicle. After introducing the benchmark application, a suitable verified simulation technique is briefly reviewed. It is the basis for the subsequent parameter estimation, where measurements of $\mathbf{y}(t_k) = \mathbf{g}(\mathbf{x}(t_k), \mathbf{p})$ are assumed to be available at discrete points of time $t = t_k$.

To describe the system dynamics in a reliable way, three different operating conditions are distinguished. The models S_1 and S_3 represent sliding friction for the motion in *backward* and *forward direction*, respectively. Obviously, these models are valid for non-zero motor angular velocities $\dot{\varphi}_\mathrm{M} = \omega_\mathrm{M} = x_2(t) \neq 0$. Additionally, the breakaway point is included as an activation condition in S_1 and S_3. This is the operating condition in which the actuator torque overcomes the static friction $T_{\mathrm{F},\mathrm{s}}$. The corresponding state equations are given by

$$\dot{\mathbf{x}}(t) = \mathbf{f}_\mathcal{I}(\mathbf{x}(t), \mathbf{p}, u(t), t) = \begin{bmatrix} x_2(t) \\ \alpha \cdot x_2(t) + \beta \cdot (u(t) - T_\mathrm{F}(t)) \end{bmatrix} \quad \text{for} \quad \mathcal{I} \in \{S_1, S_3\},$$

$$(2)$$

with the friction term $T_\mathrm{F}(t) = T_{\mathrm{F},\mathrm{s}} \cdot \mathrm{sign}(x_2(t))$, the parameters $\mathbf{p} = \begin{bmatrix} \alpha & \beta & T_{\mathrm{F},\mathrm{s}} \end{bmatrix}^T$, the state vector $\mathbf{x}(t) = \begin{bmatrix} x_1(t) & x_2(t) \end{bmatrix}^T = \begin{bmatrix} \varphi_\mathrm{M}(t) & \omega_\mathrm{M}(t) \end{bmatrix}^T$, and the motor torque as the piecewise constant control signal $u(t) = T_\mathrm{M}(t)$, $n_\mathrm{u} = 1$. For the static friction case, the angular velocity $x_2(t)$ becomes zero and the additional condition $|u(t)| \leq T_{\mathrm{F},\mathrm{s}}$ holds with the state equations $\dot{\mathbf{x}}(t) = \mathbf{f}_{S_2}(\mathbf{x}(t), \mathbf{p}, u(t), t) = \begin{bmatrix} 0 & 0 \end{bmatrix}^T$.

In detail, the parameter α represents the ratio between velocity-proportional friction and the overall mass moment of inertia; β is the reciprocal of the mass

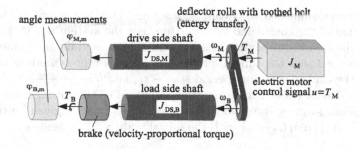

Fig. 1. Benchmark application: Test rig for the longitudinal dynamics of a vehicle.

moment of inertia; $T_{F,s}$ is the static friction coefficient, which may vary after each standstill of the test rig. A state transition diagram for the uncertain dynamic system is shown in Fig. 2. It contains the nominal system model if the parameter intervals $[\alpha]$, $[\beta]$, and $[T_{F,s}]$ as well as the control signal $u(t)$ are replaced by point values. Further generalizations of the modeling approach are described in [1].

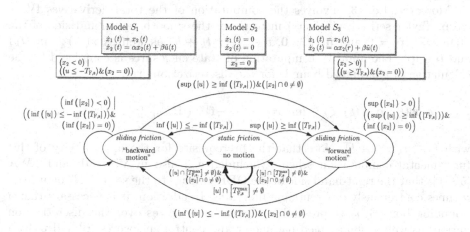

Fig. 2. State transition diagram of the benchmark application with the interval parameters $\alpha \in [\alpha] = [\underline{\alpha}\,;\,\overline{\alpha}]$, $\beta \in [\beta] = [\underline{\beta}\,;\,\overline{\beta}]$, $T_{F,s} \in [T_{F,s}] = [\underline{T}_{F,s}\,;\,\overline{T}_{F,s}]$, and the system input $\tilde{u}(t) := u(t) - T_F(t)$ with $[T_F^{\max}] := [-\overline{T}_{F,s}\,;\,\overline{T}_{F,s}]$.

3   Verified Simulation of ODEs with Non-Smooth Right-Hand Sides

The prerequisite for the following verified parameter identification scheme is the computation of guaranteed state enclosures for the uncertain system models given in the previous section. For that purpose, a generalization of a Taylor series-based enclosure technique is employed. Details about this simulation approach are published in [1,10,13]. Hence, only a short overview is given in this section.

The basic assumption of Taylor series-based simulation procedures is the discretization of the considered time horizon. For the system in Fig. 1, in which measured data and control variables are available at equidistant points of time t_k, it is assumed that the time discretization mesh is represented by an integer divisor of the control sampling period $t_{k+1} - t_k$, i.e., $h = \frac{t_{k+1} - t_k}{N}$, $N \in \mathbb{N}$.

Then, a Taylor series expansion of the solution of the initial value problem for a continuous-time system model with respect to time — given by the ODEs $\dot{\mathbf{x}}(t_k) = \mathbf{f}\left(\mathbf{x}(t_k), \mathbf{p}, \mathbf{u}(t_k), t_k\right)$ with the initial state $\mathbf{x}(0)$ — leads to

$$\mathbf{x}(t_k + h) = \mathbf{x}(t_k) + \sum_{i=1}^{\nu} \frac{h^i}{i!} \mathbf{f}^{(i-1)}\left(\mathbf{x}(t_k), \mathbf{p}, \mathbf{u}(t_k), t_k\right) + \mathbf{e}\left(\mathbf{x}(\xi), \mathbf{p}, \mathbf{u}(\xi), \xi\right) \quad (3)$$

with the before-mentioned integration step-size h. Since the uncertain parameters $\mathbf{p} \in [\mathbf{p}]$ are assumed to be constant and since changes of control signals $\mathbf{u}(t_k)$ only occur at the discrete points of time $t = t_k$, the expression (3) is evaluated recursively until the point of time $t = t_{k+1}$ is reached. Here, a new control signal $\mathbf{u}(t_{k+1})$ becomes active and new measured data are available.

Moreover, Eq. (3) involves the computation of the total derivatives $\mathbf{f}^{(i-1)}$ (resp. Taylor series coefficients) in terms of the smooth right-hand side of the ODE with $\dot{\mathbf{p}} = \mathbf{0}$ and $\dot{\mathbf{u}}(t) = \mathbf{0}$, $t \in (t_k \,;\, t_k + h)$ as well as $\mathbf{x}(t_k) \in [\mathbf{x}](t_k)$ and $\mathbf{p} \in [\mathbf{p}]$. The iterative computation of state enclosures is completed by the calculation of guaranteed bounds for the discretization error

$$\mathbf{e}\left(\mathbf{x}(\xi), \mathbf{p}, \mathbf{u}(\xi), \xi\right) \subseteq [\mathbf{e}_k] := \frac{h^{\nu+1}}{(\nu+1)!} \mathbf{f}^{(\nu)}\left([\mathbf{B}_{x,k}], [\mathbf{p}], \mathbf{u}([\tau_k]), [\tau_k]\right) \quad (4)$$

with $\xi \in [t_k \,;\, t_k + h]$. Note that the prerequisite for the applicability of this fundamental computation scheme (included e.g. also in VNODE and AWA [5,6]) is that the right-hand side of the ODE belongs to the set $\mathbf{f} \in C^\nu$ of at least ν times continuously differentiable functions. In addition, it is necessary that a bounding box $[\mathbf{B}_{x,k}]$ (representing all reachable states over the discretization period) as well as guaranteed parameter and control enclosures $[\mathbf{p}]$ and $\mathbf{u}([\tau_k])$ are available for the time interval $[\tau_k] := [t_k \,;\, t_k + h]$. To extend the use of (3), (4) to ODE systems with non-smooth right-hand sides, the following extensions are necessary for the iteration scheme as well as for the Picard iteration that is employed to determine the bounding box $[\mathbf{B}_{x,k}]$ (for details, cf. [1, Sec. 3.3]).

Step S1. Calculation of a bounding box $[\mathbf{B}_{x,k}] = [\mathbf{B}_{a,k}]$ for the time interval $[\tau_k]$, where $\mathbf{f}(\cdot) = \mathbf{f}_a(\cdot)$ is a continuously differentiable function describing the union of all system models from the set $\mathcal{S}$ which are *active* at $t = t_k$.

Step S2. Check, whether additional models from the set $\mathcal{S}$ are activated within the interval $[\tau_k]$: If additional models are activated, repeat **Step S1** after modifying the continuously differentiable enclosure $\mathbf{f}_a$ by consideration of all additionally activated models; otherwise, continue with **Step S3**.

Step S3. Interval evaluation of the series expansion for $\mathbf{f}(\cdot) = \mathbf{f}_a(\cdot)$ according to (3), (4). Note that generally $\nu > 1$ can be chosen. However, if measured data are available after a few discretization steps N (as in the case of the verified parameter identification that is considered in this paper), it is often sufficient to restrict the series expansion order to $\nu \equiv 1$.

Step S4. Deactivation of system models which can no longer be active at $t = t_k + h$ and continue with **Step S1** for the next time interval $[t_k + h\,;\,t_k + 2h]$.

4 Verified Parameter Identification

For verified parameter identification, two fundamentally different approaches exist. The first one is based on subdividing an initial parameter domain into subintervals, afterwards performing a verified integration of the ODEs for these subintervals, and subsequently checking the resulting enclosures for admissibility. A parameter box is treated as consistent with the measured data if the simulated state enclosures are subsets of intervals for the measured data $[y_{m,q}](t_k) = y_{m,q}(t_k) + [\Delta y_{m,q}](t_k)$ for each sampling time t_k and each sensor q, where $[\Delta y_{m,q}](t_k)$ is the measurement tolerance. All parameter intervals which lead to an enclosure that does not overlap with $[y_{m,q}](t_k)$ for at least one q and k are inconsistent. All remaining interval boxes can be divided further and investigated for consistency [9]. However, this procedure is disadvantageous if parameters are varying over time (depending on the states $\mathbf{x}$). This is the case for the application scenario considered in this paper. It is characterized by the fact that the static friction coefficient changes its value after each standstill of the drive and therefore has to be re-identified within the initial bounds for $[T_{F,s}]$.

In such cases, the second option for verified parameter identification is reasonable. It has the same structure as the well-known Kalman filter (or Luenberger observer) [4] for dynamic systems with stochastic disturbances, namely (i) a prediction phase in which the (nonlinear) system model is evaluated between two subsequent measurement points t_{k-1} and t_k and (ii) a correction step in which an intersection between the predicted state intervals and their bounds — resulting from the sensor information — is performed, see Fig. 3. In the correction step, parameter intervals which lead to an empty intersection of both before-mentioned estimates are guaranteed to be inconsistent and can, hence, be eliminated. All undecided interval boxes are evaluated by the following algorithm. For the sake of simplicity, it is assumed that the sensors q provide a direct measurement of selected state variables. If this was not the case, i.e., if $y_{m,q}(t_k)$ is a (generally nonlinear) function of (multiple) state variables, techniques for constraint propagation or verified Newton methods become necessary in the correction step [3]. Note that resetting parameter intervals to their initial domains is an easy task for this second type of identification procedure. In the following, the proposed parameter identification procedure is described in detail.

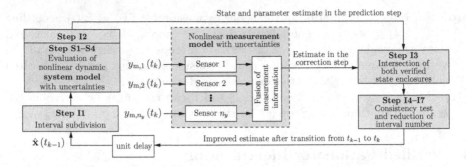

Fig. 3. Block diagram of the verified parameter identification procedure.

Verified Parameter Identification: Prediction-Correction-Framework

Step I1. Description of the state enclosure by a list of L interval boxes, where $\mathbf{z}$ is a vector containing time-varying state variables and constant parameters

$$\left[\mathbf{z}^{\langle l\rangle}\right](t_k) := \left[\left[\mathbf{x}^{\langle l\rangle}\right](t_k)^T \ \left[\mathbf{p}^{\langle l\rangle}\right](t_k)^T\right]^T, \quad l \in \{1, \ldots, L\}; \tag{5}$$

Perform M interval subdivisions, if at least one interval l is characterized by

$$\prod_{j=1}^{n+n_{\mathrm{p}}} \mathrm{diam}\left\{\left[z_j^{\langle l\rangle}\right](t_k)\right\} \neq 0, \tag{6}$$

leading to a new interval list of length $L + M - 1$. Here, the candidates to be subdivided are determined as the boxes with the largest pseudo-volume

$$l^* = \arg\max_{l=1,\ldots,L'} \prod_{j=1}^{n+n_{\mathrm{p}}} \mathrm{diam}\left\{\left[z_j^{\langle l\rangle}\right](t_k)\right\}, \quad L' \geq L. \tag{7}$$

Unnecessarily conservative interval bounds in the prediction step have to be avoided to reduce ambiguities between static and sliding friction in the computation of the state enclosures (multiple model states $\mathcal{S}$ may be active). Therefore, the following heuristic, application-dependent scheme for detecting the vector component of $\left[\mathbf{z}^{\langle l^*\rangle}\right]$ to be subdivided is used in the remainder of this paper:

(a) Split the static friction interval $\left[T_{\mathrm{F},\mathrm{s}}^{\langle l^*\rangle}\right]$ according to the following procedure if the condition $[u](t_k) \cap \mathrm{hull}\left\{-\left[T_{\mathrm{F},\mathrm{s}}^{\langle l^*\rangle}\right], \left[T_{\mathrm{F},\mathrm{s}}^{\langle l^*\rangle}\right]\right\} \neq \emptyset$ holds
   - Select the splitting point $\overline{u}(t_k) + \epsilon$, $\epsilon > 0$ for $[u](t_k) > 0$ with $\underline{T}_{\mathrm{F},\mathrm{s}}^{\langle l^*\rangle} < \underline{u}(t_k)$ and $\overline{T}_{\mathrm{F},\mathrm{s}}^{\langle l^*\rangle} > \overline{u}(t_k)$
   - Select the splitting point $\underline{u}(t_k) - \epsilon$, $\epsilon > 0$ for $[u](t_k) < 0$ with $-\overline{T}_{\mathrm{F},\mathrm{s}}^{\langle l^*\rangle} < \underline{u}(t_k)$ and $-\underline{T}_{\mathrm{F},\mathrm{s}}^{\langle l^*\rangle} > \overline{u}(t_k)$
   - Else: Splitting of $\left[T_{\mathrm{F},\mathrm{s}}^{\langle l^*\rangle}\right]$ at its midpoint

(b) Split the angular velocity interval[1] $\left[x_2^{\langle l^* \rangle}\right]$ if it is the major source for large interval diameters, i.e., if diam $\left\{\left[x_2^{\langle l^* \rangle}\right]\right\} \geq$ diam $\left\{\left[\beta^{\langle l^* \rangle}\right]\right\}$ holds

(c) Split the interval $\left[\beta^{\langle l^* \rangle}\right]$ (typically at its midpoint) if

$$\left(\left[\alpha^{\langle l^* \rangle}\right] \cdot \left[x_2^{\langle l^* \rangle}\right]\right) \cap \left(\left[\beta^{\langle l^* \rangle}\right] \cdot \left([u](t_k) - \left[T_{F,s}^{\langle l^* \rangle}\right]\right)\right) \neq \emptyset \qquad (8)$$

(d) Else: Split the interval $\left[\alpha^{\langle l^* \rangle}\right]$ (typically at its midpoint)

Step I2. Verified integration of the IVP[2] until the next measurement point t_{k+1} to compute the enclosures $\left[\mathbf{z}^{\langle l \rangle}\right](t_{k+1})$

Step I3. Intersection of all interval boxes with the measured data $z_1(t_{k+1}) \in [y_m](t_{k+1})$ (assuming that only a direct scalar measurement exists for the first state variable, i.e., the angle measurement $\varphi_{M,m}$ in Fig. 1)

$$\left[\tilde{z}_1^{\langle l \rangle}\right](t_{k+1}) := \left[z_1^{\langle l \rangle}\right](t_{k+1}) \cap [y_m](t_{k+1}) \qquad (9)$$

Step I4. Replace $\left[z_1^{\langle l \rangle}\right](t_{k+1})$ by $\left[\tilde{z}_1^{\langle l \rangle}\right](t_{k+1})$ for all $l \in \{1, \ldots, L+M-1\}$

Step I5. Delete all subintervals with $\left[\tilde{z}_1^{\langle l \rangle}\right](t_{k+1}) = \emptyset$ in (9) from the list

Step I6. Replace static friction intervals with the initial range $\left[T_{F,s}^{ini}\right]$ if standstill is detected for a minimum time span (detected by a binary signal from the velocity sensor $\omega_{M,m}$):

(a) For each list entry $l \in \{1, \ldots, L'\}$, define new static friction subintervals
$$\left[T_a^{\langle l \rangle}\right] := \left[\underline{T}_{F,s}^{ini} \, ; \, \underline{T}_{F,s}^{\langle l \rangle}\right], \, \left[T_b^{\langle l \rangle}\right] := \left[\underline{T}_{F,s}^{\langle l \rangle} \, ; \, \overline{T}_{F,s}^{\langle l \rangle}\right], \, \left[T_c^{\langle l \rangle}\right] := \left[\overline{T}_{F,s}^{\langle l \rangle} \, ; \, \overline{T}_{F,s}^{ini}\right]$$

(b) Create a list of up to $3L'$ subintervals[3], where $\left[T_{F,s}^{\langle l \rangle}\right]$ is replaced by each of the intervals $\left[T_a^{\langle l \rangle}\right], \left[T_b^{\langle l \rangle}\right], \left[T_c^{\langle l \rangle}\right]$ with non-zero diameter

(c) Subsequent merging of intervals avoids the combination of intervals with different active model states S_i, $i \in \{1, 2, 3\}$

Step I7. Reduce the number of subintervals[4] by a convex hull with sufficiently small overestimation [11]: New list length $L := L^*$ (for further extensions, see [12])

---

[1] Optional: Trisectioning of $\left[x_2^{\langle l^* \rangle}\right]$ around the value zero if static and sliding friction are possible simultaneously in the simulation of the uncertain system model.

[2] The presented integration procedure is implemented by using the toolbox INTLAB, where a parallelization of the evaluation can be achieved in a straightforward manner if the state equations are evaluated after a distribution onto multiple CPU cores. The PARALLEL COMPUTING TOOLBOX can be utilized for this purpose in MATLAB.

[3] The increase of the list length from L' to $3L'$ has the advantage that information about the parameter splitting before the reset is not lost. Usually, the static friction is similar after standstill, even if it does not remain identical. In this case, the splitting information speeds up the identification and elimination of inconsistent subdomains.

[4] Note that the interval replacement (**Step I6**) and the reduction of the interval number (**Step I7**) can be employed interchangeably.

5   Identification Results

In this section, the verified identification procedure is applied to determine enclosures for the parameters α, β, and $T_{\mathrm{F,s}}$ of the drive train test rig in Fig. 1. The initial domains are: $[\alpha] = -[1 \; ; \; 6] \frac{1}{\mathrm{s}}$, $[\beta] = [10 \; ; \; 400] \frac{\mathrm{rad}}{\mathrm{Nm \cdot s^2}}$, and $[T_{\mathrm{F,s}}] = [0.01 \; ; \; 0.30]$ Nm; measured data are available with a discretization period of 10 ms. In accordance with the available motor angle sensor, the tolerance bounds are $[\Delta y_{\mathrm{m}}] = [-0.1 \; ; \; 0.1]$ rad. Figure 4a gives an overview of the control signal, and the angle measurement for the identification time span of $t_{\mathrm{f}} = 80$ s.

Now, the identification procedure is run until the final point of time t_{f}, while the interval for the static friction coefficient is reset to its initial bound after each standstill. It can be seen that especially the interval for the parameter β is significantly reduced in this first run of the identification. Because the parameters α and β remain uncertain but constant, the identification is repeated five times, where the reinitialization values at $t = 0$ s correspond to the final parameter intervals $[\alpha]$, $[\beta]$ at the end of the previous run (Fig. 4b). Subdomains of α and β are classified as inconsistent with the measured data during the first four runs, while the results remain constant in run five. The corresponding results can, for example, be used for the specification of intervals in which control and online state estimation procedures have to be robust and asymptotically stable.

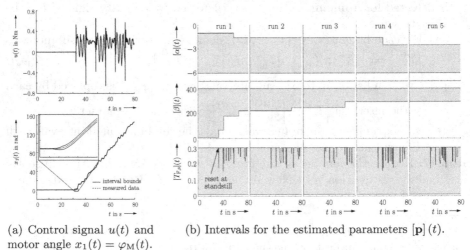

(a) Control signal $u(t)$ and motor angle $x_1(t) = \varphi_{\mathrm{M}}(t)$.

(b) Intervals for the estimated parameters $[\mathbf{p}](t)$.

Fig. 4. Identification results ($M = 50$, five repetitions of the procedure).

6   Conclusions and Outlook on Future Work

In this paper, verified integration of ODEs with non-smooth right-hand sides was combined in a novel way with an offline-applicable reliable parameter identification procedure. Real-life results were presented for a laboratory test rig at the Chair of Mechatronics, University of Rostock. Future work will deal with an extension of the identification procedure to systems with larger initial search domains for a-priori unknown parameters. Moreover, more complex friction and hysteresis models will be identified. For this purpose, it will be necessary to investigate how the interval subdivision routine — in combination with the parallelized integration of ODEs — scales to higher-dimensional sets of state equations. Finally, the computed intervals will be employed to initialize online-applicable estimation procedures which are based on sliding mode principles.

References

1. Auer, E., Kiel, S., Rauh, A.: A verified method for solving piecewise smooth initial value problems. Int. J. Appl. Math. Comput. Sci. AMCS **23**(4), 731–747 (2013)
2. Hofer, E.P., Rauh, A.: Applications of interval algorithms in engineering. In: CD-Proceedings of 12th GAMM-IMACS International Symposium on Scientific Computing, Computer Arithmetic, and Validated Numerics SCAN 2006. IEEE Computer Society, Duisburg, Germany (2007)
3. Jaulin, L., Kieffer, M., Didrit, O., Walter, É.: Applied Interval Analysis. Springer, London (2001)
4. Kalman, R.E.: A new approach to linear filtering and prediction problems. Trans. ASME - J. Basic Eng. **82**(Series D), 35–45 (1960)
5. Lohner, R.: Enclosing the solutions of ordinary initial and boundary value problems. In: Kaucher, E.W., Kulisch, U.W., Ullrich, C. (eds.) Computer Arithmetic: Scientific Computation and Programming Languages, pp. 255–286. Wiley-Teubner Series in Computer Science, Stuttgart (1987)
6. Nedialkov, N.S.: Interval tools for ODEs and DAEs. In: CD-Proceedings of 12th GAMM-IMACS International Symposium on Scientific Computing, Computer Arithmetic, and Validated Numerics SCAN 2006. IEEE Computer Society, Duisburg, Germany (2007)
7. Nedialkov, N.S., von Mohrenschildt, M.: Rigorous simulation of hybrid dynamic systems with symbolic and interval methods. In: Proceedings of the American Control Conference ACC, Anchorage, USA, pp. 140–147 (2002)
8. Ramdani, N., Nedialkov, N.S.: Computing reachable sets for uncertain nonlinear hybrid systems using interval constraint-propagation techniques. Nonlinear Anal. Hybrid Syst. **5**(2), 149–162 (2011)
9. Rauh, A., Dötschel, T., Auer, E., Aschemann, H.: Interval methods for control-oriented modeling of the thermal behavior of high-temperature fuel cell stacks. In: Proceedings of 16th IFAC Symposium on System Identification SysID 2012. Brussels, Belgium (2012)
10. Rauh, A., Kletting, M., Aschemann, H., Hofer, E.P.: Interval methods for simulation of dynamical systems with state-dependent switching characteristics. In: Proceedings of the IEEE International Conference on Control Applications CCA 2006, pp. 2243–2248. Munich, Germany (2006)
11. Rauh, A., Kletting, M., Aschemann, H., Hofer, E.P.: Reduction of overestimation in interval arithmetic simulation of biological wastewater treatment processes. J. Comput. Appl. Math. **199**(2), 207–212 (2007)
12. Rauh, A., Senkel, L., Aschemann, H.: Experimental comparison of interval-based parameter identification procedures for uncertain odes with non-smooth right-hand sides. In: CD-Proceedings of IEEE International Conference on Methods and Models in Automation and Robotics MMAR 2015. Miedzyzdroje, Poland (2015)

13. Rauh, A., Siebert, C., Aschemann, H.: Verified simulation and optimization of dynamic systems with friction and hysteresis. In: Proceedings of ENOC 2011. Rome, Italy (2011)
14. Rihm, R.: Enclosing solutions with switching points in ordinary differential equations. In: Computer Arithmetic and Enclosure Methods. Proceedings of SCAN 91. North-Holland, Amsterdam, pp. 419–425 (1992)

Exponential Enclosure Techniques for Initial Value Problems with Multiple Conjugate Complex Eigenvalues

Andreas Rauh[1]([✉]), Ramona Westphal[1],
Harald Aschemann[1], and Ekaterina Auer[2]

[1] Chair of Mechatronics, University of Rostock,
Justus-von-Liebig-Weg 6, 18059 Rostock, Germany
{andreas.rauh,harald.aschemann}@uni-rostock.de
[2] University of Applied Sciences Wismar,
Faculty of Engineering, 23952 Wismar, Germany
ekaterina.auer@hs-wismar.de

Abstract. The computation of guaranteed state enclosures has a large variety of applications in engineering if initial value problems for sets of ordinary differential equations are concerned. One possible scenario is the use of such state enclosures in the design and verification of linear and nonlinear feedback controllers as well as in predictive control procedures. In many of these applications, system models are characterized by a dominant linear part (commonly after a suitable coordinate transformation) and by a not fully negligible nonlinear part. To compute guaranteed state enclosures for such systems, general purpose approaches relying on a Taylor series expansion of the solution can be employed. However, they do not exploit knowledge about the specific system structure. The exponential state enclosure technique makes use of this structure, allowing users to compute tight enclosures that contract over time for asymptotically stable dynamics. This paper firstly gives an overview of exponential enclosure techniques, implemented in VALENCIA-IVP, and secondly focuses on extensions to dynamic systems with single and multiple conjugate complex eigenvalues.

Keywords: Ordinary differential equations · Initial value problems · Complex interval arithmetic · VALENCIA-IVP

1 Introduction

VALENCIA-IVP is a verified solver providing guaranteed enclosures for solutions to initial value problems (IVPs) for sets of ordinary differential equations (ODEs). In the basic version of this solver, the verified solution is computed as the sum of a non-verified approximate solution (computed, for example,

---

This work was performed while R. Westphal was with the Chair of Mechatronics, University of Rostock.

M. Nehmeier et al. (Eds.): SCAN 2014, LNCS 9553, pp. 247–256, 2016.
DOI: 10.1007/978-3-319-31769-4_20

by Euler's method) and additive guaranteed error bounds determined using a simple iteration scheme [1].

The disadvantage of this iteration scheme, however, is that the widths of the resulting state enclosures might get larger even for asymptotically stable ODEs [9]. This phenomenon is caused by the so-called wrapping effect which arises if non-axis-parallel state enclosures are described by axis-aligned interval boxes in a state-space of dimension $n > 1$. In general purpose solvers such as VNODE-LP [7] or VSPODE [5], the corresponding counter-measure against this type of overestimation is the preconditioning of state equations (e.g. using Lohner's QR decomposition approach) in combination with a high-order series expansion of the solution to the IVP over time. However, these approaches may become quite time consuming for large system orders n.

A possible approach to deal with the overestimation in such a way as to allow real-time implementations (e.g. for predictive control [10]), that is, without increasing the computational cost too much, is to transform the ODEs into a suitable canonical form. For the case of linear ODEs with real eigenvalues of multiplicity one, this is given by the Jordan canonical form. The transformation results in a decoupling of the vector-valued set of state equations. A solution of this transformed IVP can then be determined by an exponential enclosure technique which guarantees that asymptotically stable solutions are represented by contracting interval bounds if a suitable time discretization step size is chosen. For real eigenvalues, this property holds as long as the value zero is not included in any vector component of the solution interval.

As shown in [11, 12], the before-mentioned advantageous contraction property can be preserved for linear ODEs with conjugate complex eigenvalue pairs if a transformation into the complex Jordan canonical form is employed. Then, a complex-valued interval iteration scheme is used to determine state enclosures [12]. The corresponding solution procedure — originally derived for dynamic systems with eigenvalues of multiplicity one — is extended in this paper to more general situations with several multiple real and complex eigenvalues.

This paper is structured as follows. Sect. 2 gives an overview of the real-valued and complex-valued iteration schemes that are applicable inside the exponential state enclosure approach. Extensions to eigenvalues with multiplicity greater than one are discussed in Sect. 3. Representative simulation results for a technically motivated benchmark system from control engineering, typically containing bounded uncertainty in initial values and parameters, are presented in Sect. 4. Conclusions and an outlook on future work can be found in Sect. 5.

2   Basic Exponential State Enclosure Approach

Throughout this paper, it is assumed that dynamic system models are given by the set of ODEs

$$\dot{\mathbf{x}}(t) = \mathbf{f}(\mathbf{x}(t)), \quad \mathbf{x} \in \mathbb{R}^n, \quad \mathbf{f} : \mathbb{R}^n \mapsto \mathbb{R}^n, \tag{1}$$

with smooth right-hand sides $\mathbf{f}\left(\mathbf{x}\left(t\right)\right)$ and the uncertain initial conditions

$$\mathbf{x}\left(0\right) \in \left[\mathbf{x}_0\right] := \left[\mathbf{x}\right]\left(0\right) = \left[\underline{\mathbf{x}}\left(0\right); \overline{\mathbf{x}}\left(0\right)\right]. \tag{2}$$

In (1) and (2), external control (input) signals $\mathbf{u}(t) = \mathbf{u}(\mathbf{x}(t))$ are directly included in the corresponding expression for $\mathbf{f}\left(\mathbf{x}\left(t\right)\right)$. The same holds for time-invariant uncertain system parameters $p_j \in \left[\underline{p}_j; \overline{p}_j\right]$, $j = 1, \ldots, n_{\mathrm{p}}$, with the corresponding derivatives $\dot{p}_j = 0$. Together with the enclosures for the time-varying system states, they are contained in the component-wise defined interval vectors $[\mathbf{x}] = \left[[x_1] \ldots [x_n]\right]^T$ with the individual vector entries $[x_i] = [\underline{x}_i; \overline{x}_i]$, $\underline{x}_i \le x_i \le \overline{x}_i$, $i = 1, \ldots, n$.

In the basic implementation of VALENCIA-IVP, the state enclosure $[\mathbf{x}]\left(t\right)$ of the true solution $\mathbf{x}^*(t)$ to an IVP is defined by $\mathbf{x}^*(t) \in [\mathbf{x}]\left(t\right) := \tilde{\mathbf{x}}(t) + [\mathbf{R}]\left(t\right)$, where $\tilde{\mathbf{x}}(t)$ is an approximate solution computed in usual (non-verified) floating point arithmetic. Verified error bounds $[\mathbf{R}]\left(t\right)$ are then computed by an appropriate iteration scheme [9,12].

Note that without suitable counter-measures, the diameters of the solution enclosures may diverge even for asymptotically stable systems. This is mostly caused by the wrapping effect that can be compensated for systems with a dominant (locally) linear behavior by using the following exponential enclosure technique. As a fundamental ansatz for the representation of contracting state enclosures, the expression

$$\mathbf{x}^*(t) \in [\mathbf{x}_e]\left(t\right) := \exp\left([\boldsymbol{\Lambda}] \cdot t\right) \cdot [\mathbf{x}_e]\left(0\right) \tag{3}$$

is used, with $0 \notin [x_{e,i}]\left(0\right)$, $[\mathbf{x}_e]\left(0\right) = [\mathbf{x}_0]$, and the diagonal matrix $[\boldsymbol{\Lambda}] := \mathrm{diag}\left\{[\lambda_i]\right\}$, $i = 1, \ldots, n$, with the element-wise negative real entries λ_i.

After defining $\exp\left([\boldsymbol{\Lambda}] \cdot t\right) := \mathrm{diag}\left\{\exp\left([\lambda_1] \cdot t\right), \ldots, \exp\left([\lambda_n] \cdot t\right)\right\}$ as the corresponding interval matrix exponential, a Picard iteration scheme [2,6]

$$\mathbf{x}^*(t) \in [\mathbf{x}_e]^{(\kappa+1)}\left(t\right) := [\mathbf{x}_0] + \int_0^t \mathbf{f}\left([\mathbf{x}_e]^{(\kappa)}\left(s\right)\right) \mathrm{d}s \tag{4}$$

can be employed to determine intervals $[\lambda_i]$ so that all reachable states are contained in the time-dependent interval enclosure functions $[\mathbf{x}_e]\left(t\right)$. To derive the iteration scheme for $[\mathbf{x}_e]\left(t\right)$, the Picard iteration (4) is reformulated as the time-dependent expression

$$\mathbf{x}^*(t) \in \exp\left([\boldsymbol{\Lambda}]^{(\kappa+1)} \cdot t\right) \cdot [\mathbf{x}_e]\left(0\right) = [\mathbf{x}_e]^{(\kappa+1)}\left(t\right)$$

$$=: [\mathbf{x}_0] + \int_0^t \mathbf{f}\left(\exp\left([\boldsymbol{\Lambda}]^{(\kappa)} \cdot s\right) \cdot [\mathbf{x}_e]\left(0\right)\right) \mathrm{d}s. \tag{5}$$

Its differentiation with respect to time and the evaluation for $t \in [0; T]$ leads to

$$\dot{\mathbf{x}}^*\left([0; T]\right) \in \mathrm{diag}\left\{[\lambda_i]^{(\kappa+1)}\right\} \cdot \exp\left([\boldsymbol{\Lambda}]^{(\kappa+1)} \cdot [0; T]\right) \cdot [\mathbf{x}_e]\left(0\right)$$

$$\subseteq \mathbf{f}\left(\exp\left([\boldsymbol{\Lambda}]^{(\kappa)} \cdot [0; T]\right) \cdot [\mathbf{x}_e]\left(0\right)\right). \tag{6}$$

Suppose that the convergence condition

$$\exp\left([A]^{(\kappa+1)} \cdot t\right) \cdot [\mathbf{x}_e](0) \subseteq \exp\left([A]^{(\kappa)} \cdot t\right) \cdot [\mathbf{x}_e](0), \tag{7}$$

that is equivalent to $[\lambda_i]^{(\kappa+1)} \subseteq [\lambda_i]^{(\kappa)}$ and $[A]^{(\kappa+1)} \subseteq [A]^{(\kappa)}$, is fulfilled. Then, the final iteration formula is given by

$$[\lambda_i]^{(\kappa+1)} := \frac{f_i\left(\exp\left([A]^{(\kappa)} \cdot [0\,;\,T]\right) \cdot [\mathbf{x}_e](0)\right)}{\exp\left([\lambda_i]^{(\kappa)} \cdot [0\,;\,T]\right) \cdot [x_{e,i}](0)}, \quad i = 1, \ldots, n \tag{8}$$

with the guaranteed state enclosure at the point $t = T$

$$\mathbf{x}^*(T) \in [\mathbf{x}_e](T) := \exp\left([A] \cdot T\right) \cdot [\mathbf{x}_e](0). \tag{9}$$

A detailed derivation of this iteration approach is given in [12].

The above-mentioned iteration can be simplified for linear state equations

$$f_i(\mathbf{x}(t)) = \sum_{j=1}^{n} a_{ij} \cdot x_j(t) \qquad \text{according to} \tag{10}$$

$$[\lambda_i]^{(\kappa+1)} := \sum_{j=1, i \neq j}^{n} \left\{ a_{ij} \cdot \exp\left(\left([\lambda_j]^{(\kappa)} - [\lambda_i]^{(\kappa)}\right) \cdot [0\,;\,T]\right) \cdot \frac{[x_{e,j}](0)}{[x_{e,i}](0)} \right\}$$

$$+ a_{ii} \quad \text{with} \quad a_{ij} \in [a_{ij}]. \tag{11}$$

From this simplification, it becomes obvious that the computation of $[\lambda_i]$ is free of overestimation if the expressions in (10) are decoupled with $a_{ij} = 0$ for all $i \neq j$. In cases in which the linear parts of $f_i(\mathbf{x})$ represent the dominant features of the system dynamics, an (approximate) decoupling of the ODEs is only possible if pairwise different real eigenvalues are present. Then, the linear part of the system model is transformed into real-valued Jordan canonical form.

Already in the case of linear systems with conjugate complex eigenvalue pairs, there exist points of time at which the iteration (8) is no longer defined due to the fact that the value zero may be included in the true solution set and hence also in the denominator of (8). The latter problem can easily be solved by replacing the real-valued Jordan canonical form [3,4]

$$\boldsymbol{\Sigma} = \text{blkdiag}\{\ldots, \bar{\boldsymbol{\Sigma}}_i, \ldots\}, \quad \bar{\boldsymbol{\Sigma}}_i = \begin{bmatrix} \sigma_i & \omega_i \\ -\omega_i & \sigma_i \end{bmatrix} \tag{12}$$

by its complex-valued generalization [8] with the corresponding ODEs $\dot{\mathbf{z}}(t) = \boldsymbol{\Sigma} \cdot \mathbf{z}(t)$, the initial conditions $\mathbf{z}(0) \in \mathbb{C}^n$, $\mathbf{z}(0) \in [\mathbf{z}](0)$, and the diagonal matrices

$$\boldsymbol{\Sigma} = \text{blkdiag}\{\ldots, \boldsymbol{\Sigma}_i, \ldots\}, \quad \boldsymbol{\Sigma}_i = \begin{bmatrix} \sigma_i + \jmath\omega_i & 0 \\ 0 & \sigma_i - \jmath\omega_i \end{bmatrix}. \tag{13}$$

This transformation is possible if the linear state Eq. (10) have only eigen-values of multiplicity $\delta_i = 1$. For the exact solutions $z_i(t) = e^{(\sigma_i + \jmath\omega_i)\cdot t} \cdot z_i(0)$, $z_{i+1}(t) = e^{(\sigma_i - \jmath\omega_i)\cdot t} \cdot z_{i+1}(0)$ of the IVP, the iteration procedure (11) is always applicable for $0 \notin [z_i](0)$ due to

$$|z_i(t)|^2 = \left(e^{(\sigma_i + \jmath\omega_i)\cdot t} \cdot e^{(\sigma_i - \jmath\omega_i)\cdot t}\right) \cdot |z_i(0)|^2 = e^{2\sigma_i t} \cdot |z_i(0)|^2 \neq 0. \qquad (14)$$

As shown in [12], the corresponding enclosures show contracting behavior not only for purely linear asymptotically stable systems but also for nonlinear models if a linearization of the state equations approximates the dominant dynamic features sufficiently well for some finite time interval. However, according to [12], the complex-valued iteration (and also its real-valued counterpart) are applicable for *arbitrary* time spans only if the eigenvalue multiplicity is $\delta_i \equiv 1$. Therefore, novel extensions for $\delta_i > 1$ are derived in the following section.

3 Exponential State Enclosures for Multiple Eigenvalues

In this section, extensions are described for the exponential enclosure approach in the cases that linear state equations with multiple identical eigenvalues or nonlinear models with multiple eigenvalues of their linear parts are considered.

3.1 Linear State Equations

Even for linear system models, the dynamics can no longer be fully decoupled if eigenvalues have a multiplicity $\delta_i > 1$. In the case of real eigenvalues, the corresponding Jordan blocks are given by

$$\dot{\mathbf{z}}(t) = \mathbf{\Sigma} \cdot \mathbf{z}(t) \quad \text{with} \quad \mathbf{\Sigma} = \text{blkdiag}\{\lambda_1, \lambda_2 \ldots, \Sigma_i, \ldots \lambda_n\},$$

$$\mathbf{\Sigma}_i = \begin{bmatrix} \lambda_i & 1 & \ldots & 0 \\ 0 & \lambda_i & \ddots & \vdots \\ \vdots & \ddots & \ddots & 1 \\ 0 & \ldots & 0 & \lambda_i \end{bmatrix} \in \mathbb{R}^{\delta_i \times \delta_i} \quad \text{and} \quad \mathbf{z}(0) \in [\mathbf{z}](0), \qquad (15)$$

while the case of multiple complex eigenvalues leads to the canonical form

$$\mathbf{\Sigma} = \text{blkdiag}\{\ldots, \mathbf{\Sigma}_i^+, \mathbf{\Sigma}_i^-, \ldots\} \quad \text{with} \quad \lambda_i^+ = \sigma_i + \jmath\omega_i, \quad \lambda_i^- = \sigma_i - \jmath\omega_i ,$$

$$\mathbf{\Sigma}_i^+ = \begin{bmatrix} \lambda_i^+ & 1 & \ldots & 0 \\ 0 & \lambda_i^+ & \ddots & \vdots \\ \vdots & \ddots & \ddots & 1 \\ 0 & \ldots & 0 & \lambda_i^+ \end{bmatrix} \in \mathbb{C}^{\delta_i \times \delta_i} \quad \text{and} \quad \mathbf{\Sigma}_i^- = \begin{bmatrix} \lambda_i^- & 1 & \ldots & 0 \\ 0 & \lambda_i^- & \ddots & \vdots \\ \vdots & \ddots & \ddots & 1 \\ 0 & \ldots & 0 & \lambda_i^- \end{bmatrix} \in \mathbb{C}^{\delta_i \times \delta_i} \qquad (16)$$

for each eigenvalue pair $\lambda_i^\pm = \sigma_i \pm \jmath\omega_i$ with $\delta_i > 1$. In both the real and complex cases, all decoupled state equations can be solved independently from the Jordan block corresponding to the multiple eigenvalues. Overestimation is minimized if the enclosures for states of the Jordan blocks (15) and (16) are computed in a "bottom to top" manner, that is in the order $z_{i+\delta_i-1}, \ldots, z_{i+1}, z_i$.

Since the analytic representation of the solutions $z_{i+j}(t)$, $j = 0, \ldots, \delta_i - 1$, for the eigenvalue λ_i^+ can be stated explicitly as

$$z_{i+j}^*(t) = \left(\sum_{\zeta=j}^{\delta_i-1} \frac{t^{\zeta-j}}{(\zeta-j)!} \cdot z_{i+\zeta}(0) \right) \cdot e^{(\sigma_i + \jmath\omega_i) \cdot t} \tag{17}$$

in the case $\delta_i > 1$, the iteration scheme for computation of state enclosures is derived for a redefined enclosure that is given by

$$[z_{i+j}](t) = \left(\sum_{\zeta=j}^{\delta_i-1} \frac{t^{\zeta-j}}{(\zeta-j)!} [z_{i+\zeta}](0) \right) \cdot e^{[\lambda_{i+j}]t} \tag{18}$$

for all $j = 0, \ldots, \delta_i - 1$. The corresponding time derivative of (18) is

$$[\dot{z}_{i+j}](t) = [\lambda_{i+j}] \cdot \left(\sum_{\zeta=j}^{\delta_i-1} \frac{t^{\zeta-i}}{(\zeta-i)!} [z_{i+\zeta}](0) \right) \cdot e^{[\lambda_{i+j}]t}$$

$$+ \left(\sum_{\zeta=j+1}^{\delta_i-1} \frac{t^{\zeta-(j+1)}}{(\zeta-(j+1))!} [z_{i+\zeta}](0) \right) \cdot e^{[\lambda_{i+j}]t}. \tag{19}$$

Evaluating these enclosures for the interval initial conditions $z_\zeta(0) \in [z_\zeta](0)$ and for the solution parameter $\lambda_{i+j} \in [\lambda_{i+j}]$ with the discretization time span $t \in [0; T]$, a modified iteration scheme is obtained by following exactly the same arguments as in Sect. 2. The interval enclosures $[\lambda_i], \ldots, [\lambda_{i+\delta_i-1}]$ are given by

$$[\lambda_{i+j}]^{(\kappa+1)} := \frac{\lambda_i^* \cdot \left(\sum_{\zeta=j}^{\delta_i-1} \frac{t^{\zeta-i}}{(\zeta-i)!} z_{i+\zeta}(0) \right) \cdot e^{[\lambda_{i+j}]^{(\kappa)}t}}{\left(\sum_{\zeta=j}^{\delta_i-1} \frac{t^{\zeta-i}}{(\zeta-i)!} z_{i+\zeta}(0) \right) \cdot e^{[\lambda_{i+j}]^{(\kappa)}t}}$$

$$+ \frac{\left(\sum_{\zeta=j+1}^{\delta_i-1} \frac{t^{\zeta-(j+1)}}{(\zeta-(j+1))!} z_{i+\zeta}(0) \right) \cdot \left(e^{[\lambda_{i+j+1}]t} - e^{[\lambda_{i+j}]^{(\kappa)}t} \right)}{\left(\sum_{\zeta=j}^{\delta_i-1} \frac{t^{\zeta-i}}{(\zeta-i)!} z_{i+\zeta}(0) \right) \cdot e^{[\lambda_{i+j}]^{(\kappa)}t}} \tag{20}$$

for each subsystem model

$$\dot{z}_i(t) = \lambda_i^* \cdot z_i(t) + z_{i+1}(t)$$
$$\dot{z}_{i+1}(t) = \lambda_i^* \cdot z_{i+1}(t) + z_{i+2}(t)$$
$$\vdots \tag{21}$$
$$\dot{z}_{i+\delta_i-1}(t) = \lambda_i^* \cdot z_{i+\delta_i-1}(t).$$

Here, the one-sided decoupling in (21) can be exploited efficiently, since $[\lambda_{i+j}]$ depends on the result for $[\lambda_{i+j+1}]$ but not vice versa. Note that iteration (20) satisfies the inclusion property $\lambda_i^* \in [\lambda_{i+j}]$, $z_\zeta(0) \in [z_\zeta](0)$, $t \in [0; T]$, where λ_i^* is the true multiple eigenvalue.

If the system models (15), (16), or (21) are linear, the iteration formula (20) can be simplified symbolically as

$$
[\lambda_{i+j}]^{(\kappa+1)} := \lambda_i^* + \frac{\left(\sum_{\zeta=j+1}^{\delta_i-1} \frac{t^{\zeta-(j+1)}}{(\zeta-(j+1))!} z_{i+\zeta}(0) \right) \cdot \left(e^{\left([\lambda_{i+j+1}] - [\lambda_{i+j}]^{(\kappa)} \right) t} - 1 \right)}{\left(\sum_{\zeta=j}^{\delta_i-1} \frac{t^{\zeta-j}}{(\zeta-j)!} z_{i+\zeta}(0) \right)}.
$$

(22)

In this way, the overestimation due to multiple dependencies on common interval variables is reduced as much as possible. As before, (20) and (22) have to be evaluated for all $\lambda_{i+j} \in [\lambda_{i+j}]$, $z_\zeta(0) \in [z_\zeta](0)$, and $t \in [0; T]$.

3.2   Generalization to Nonlinear State Equations

The iteration procedure introduced in the previous subsection needs to be generalized in the practically important case of nonlinear terms on the right-hand sides of (21). In particular, system models given by $\dot{z}_i = \lambda_i^* \cdot z_i + z_{i+1} + g_i(\mathbf{z})$, $\dot{z}_{i+1} = \lambda_i^* \cdot z_{i+1} + z_{i+2} + g_{i+1}(\mathbf{z})$, ..., $\dot{z}_{i+\delta_i-1} = \lambda_i^* \cdot z_{i+\delta_i-1} + g_{i+\delta_i-1}(\mathbf{z})$ with $g_i(\mathbf{z}), \ldots, g_{i+\delta_i-1}(\mathbf{z}) : \mathbb{C}^n \mapsto \mathbb{C}$ are considered subsequently. Then, a vector-valued iteration has to be performed with the convergence condition

$$
\left[[\lambda_i]^{(\kappa+1)} [\lambda_{i+1}]^{(\kappa+1)} \ldots [\lambda_{i+\delta_i-1}]^{(\kappa+1)} \right]^T \overset{!}{\subset} \left[[\lambda_i]^{(\kappa)} [\lambda_{i+1}]^{(\kappa)} \ldots [\lambda_{i+\delta_i-1}]^{(\kappa)} \right]^T
$$

(23)

and the modified iteration scheme

$$
[\lambda_{i+j}]^{(\kappa+1)} := \lambda_i^* + \frac{\left(\sum_{\zeta=j+1}^{\delta_i-1} \frac{t^{\zeta-(j+1)}}{(\zeta-(j+1))!} z_{i+\zeta}(0) \right) \cdot \left(e^{\left([\lambda_{i+j+1}]^{(\kappa)} - [\lambda_{i+j}]^{(\kappa)} \right) t} - 1 \right)}{\left(\sum_{\zeta=j}^{\delta_i-1} \frac{t^{\zeta-j}}{(\zeta-j)!} z_{i+\zeta}(0) \right)}
$$

$$
+ \tilde{g}_{i+j} \left([\mathbf{z}]^{(\kappa)}(t) \right)
$$

(24)

with the nonlinear state-dependent enclosure term

$$
\tilde{g}_{i+j} \left([\mathbf{z}]^{(\kappa)}(t) \right) := \frac{g_{i+j} \left([\mathbf{z}]^{(\kappa)}(t) \right)}{\left(\sum_{\zeta=j}^{\delta_i-1} \frac{t^{\zeta-i}}{(\zeta-i)!} z_{i+\zeta}(0) \right) \cdot e^{[\lambda_{i+j}]^{(\kappa)} t}}, \quad z_\zeta(0) \in [z_\zeta](0), \; t \in [0; T].
$$

(25)

3.3   Simplified Enclosures

The procedure described in Sect. 3.2 can be simplified for systems with small time constants, especially if $-\sigma_i \gg \omega_i$ holds in (16). Then, it is sufficient to restrict the analytic expression for the enclosure and its time derivative to the terms

$$z_{i+j} = (z_{i+j}(0) + t \cdot z_{i+j+1}(0)) \cdot e^{\lambda_{i+j}t}$$
$$\dot{z}_{i+j} = \lambda_{i+j}(z_{i+j}(0) + t \cdot z_{i+j+1}(0)) \cdot e^{\lambda_{i+j}t} + z_{i+j+1}(0) \cdot e^{\lambda_{i+j}t}, \ j < \delta_i - 1.$$
(26)

In analogy to (20) and (22), this leads to the simplified iteration procedure

$$[\lambda_{i+j}]^{(\kappa+1)} := \lambda_i^* + \frac{e^{([\lambda_{i+j+1}] - [\lambda_{i+j}]^{(\kappa)})t} - 1}{\frac{z_{i+j}(0)}{z_{i+j+1}(0)} + t}, \quad z_\varsigma(0) \in [z_\varsigma](0), \quad t \in [0; T], \quad (27)$$

which typically yields wider interval bounds than the exact representation from the previous subsections. However, the simplification of the expressions leads to a reduction of the computational cost in the iteration. This simplified iteration is equally applicable to both the linear and nonlinear cases studied in this section.

4   Simulation Results

Linear dynamic systems with conjugate complex eigenvalues are a common model for a large variety of control systems. A real-life application scenario, namely oscillation damping for flexible high-bay rack feeder systems, was discussed in [11,12]. This system is characterized by the fact that — after feedback control design — only asymptotically stable eigenvalues of multiplicity $\delta_i = 1$ occur. However, many system models in drive trains with elasticities, e.g., series connections of several identical mass-spring-damper elements or series connections of electric oscillators have clusters of multiple identical eigenvalues. Transforming these state equations into Jordan canonical form (15) or (16) yields system models that are similar to the following illustrative application scenario.

Assume that — after transformation into Jordan canonical from (16) — the benchmark system ($\delta_i = 2$) is given by the initial conditions and system matrix

$$\mathbf{z}(0) \in \begin{bmatrix} \langle -5, 0.1 \rangle \\ \langle -4, 0.1 \rangle \\ \langle -5, 0.1 \rangle \\ \langle -4, 0.1 \rangle \end{bmatrix}, \quad \mathbf{\Sigma} \in \begin{bmatrix} \langle \lambda^+ \rangle & 1 & 0 & 0 \\ 0 & \langle \lambda^+ \rangle & 0 & 0 \\ 0 & 0 & \langle \lambda^- \rangle & 1 \\ 0 & 0 & 0 & \langle \lambda^- \rangle \end{bmatrix}, \quad (28)$$

with the uncertain eigenvalues $\langle \lambda^+ \rangle = \langle -2 + 3\jmath, 0.1 \rangle$ and $\langle \lambda^- \rangle = \langle -2 - 3\jmath, 0.1 \rangle$.

All uncertain initial conditions and eigenvalues are given in the complex-valued midpoint-radius notation that is, e.g., available in the MATLAB toolbox INTLAB [13]. Since the subsystems for eigenvalues with positive and negative imaginary parts are decoupled, the complete system model can be split up into two independent processes for simulation purposes. In the following, simulation results are only summarized for the state variables z_1 and z_2 since they show the same principle behavior as the remaining states z_3 and z_4.

It can be seen that despite the non-diagonal structure of the matrix Σ, leading to a one-sided coupling of the state equations, no relevant wrapping effect occurs and that the asymptotic stability of the dynamic system is preserved in the computed state enclosures despite the fact that the exact solutions of the IVP are no longer pure exponential functions as it has been shown in (17).

For practical applications, the state enclosures in Fig. 1 further have to be transformed back into the original real-valued coordinates $\mathbf{x}(t) \in [\mathbf{x}](t)$ by left-multiplying the enclosures $[\mathbf{z}](t)$ with the matrix of eigenvectors that has been used for the transformation into Jordan canonical form and — subsequently — taking the real part of the resulting interval boxes. However, this transformation preserves the presented contraction properties for sufficiently large $t > 0$.

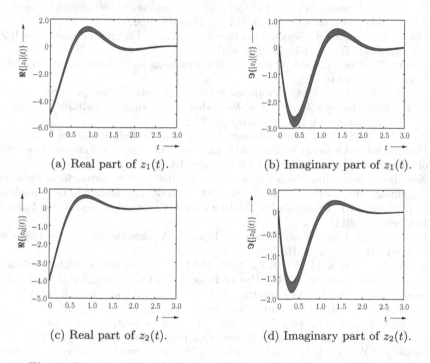

(a) Real part of $z_1(t)$.   (b) Imaginary part of $z_1(t)$.

(c) Real part of $z_2(t)$.   (d) Imaginary part of $z_2(t)$.

Fig. 1. Guaranteed state enclosures for the illustrative example (28).

5   Conclusions and Outlook on Future Work

In this paper, practically relevant extensions for exponential state enclosure techniques were presented for dynamic systems with both real and conjugate complex eigenvalues of a multiplicity larger than one. By the proposed extensions it is possible to compute guaranteed state enclosures for asymptotically stable systems which converge towards the steady-state operating points despite uncertainties and oscillations in the solution. The corresponding complex-valued interval iteration procedure will be included in future work in the design of sensitivity-based predictive control and path planning procedures [10]. Moreover, an analysis of the possible step sizes for which the suggested iteration scheme converges will be performed. It can be expected that the computational effort is significantly lower than for Taylor series-based approaches, since the computation of Taylor coefficients can be avoided and only a few evaluations of the functions, describing the ODEs, are necessary with a suitable initialization of the iteration scheme.

References

1. Auer, E., Rauh, A., Hofer, E.P., Luther, W.: Validated modeling of mechanical systems with SMARTMOBILE: improvement of performance by VALENCIA-IVP. In: Hertling, P., Hoffmann, C.M., Luther, W., Revol, N. (eds.) Real Number Algorithms. LNCS, vol. 5045, pp. 1–27. Springer, Heidelberg (2008)
2. Deville, Y., Janssen, M., van Hentenryck, P.: Consistency techniques for ordinary differential equations. Constraint 7(3–4), 289–315 (2002)
3. Hairer, E., Nørsett, S., Wanner, G.: Solving Ordinary Differential Equations I, 2nd edn. Springer, Berlin Heidelberg (2000)
4. Jordan, C.: Traité des substitutions et des équations algébriques. Gauthier-Villars, Paris (1870). in French
5. Lin, Y., Stadtherr, M.A.: Validated solution of initial value problems for odes with interval parameters. In: NSF Workshop Proceeding on Reliable Engineering Computing. Savannah GA, February 22–24 2006
6. Nedialkov, N.S.: Computing Rigorous Bounds on the Solution of an Initial Value Problem for an Ordinary Differential Equation. Ph.D. thesis, Graduate Department of Computer Science, University of Toronto (1999)
7. Nedialkov, N.S.: Implementing a rigorous ODE solver through literate programming. In: Rauh, A., Auer, E. (eds.) Modeling, Design, and Simulation of Systems with Uncertainties. Mathematical Engineering, pp. 3–19. Springer, Heidenberg (2011)
8. Petković, M., Petković, L.: Complex Interval Arithmetic and Its Applications. Wiley-VCH Verlag GmbH, Berlin (1998)
9. Rauh, A., Auer, E., Hofer, E.P.: ValEncIA-IVP: a comparison with other initial value problem solvers. In: CD-Proceedings of 12th GAMM-IMACS Intenational Symposium on Scientific Computing, Computer Arithmetic, and Validated Numerics SCAN 2006. IEEE Computer Society, Duisburg, Germany (2007)
10. Rauh, A., Kersten, J., Auer, E., Aschemann, H.: Sensitivity-based feedforward and feedback control for uncertain systems. Computing 2–4, 357–367 (2012)
11. Rauh, A., Westphal, R., Aschemann, H.: Verified simulation of control systems with interval parameters using an exponential state enclosure technique. In: CD-Proceedings of IEEE International Conference on Methods and Models in Automation and Robotics MMAR. Miedzyzdroje, Poland (2013)
12. Rauh, A., Westphal, R., Auer, E., Aschemann, H.: Exponential enclosure techniques for the computation of guaranteed state enclosures in ValEncIA-IVP. In: Proceedings of 15th GAMM-IMACS International Symposium on Scientific Computing, Computer Arithmetic, and Validated Numerics SCAN 2012, vol. 19(1), pp. 66–90. Novosibirsk, Russia, Special Issue of Reliable Computing (2013)
13. Rump, S.M.: IntLab - INTerval LABoratory. In: Csendes, T. (ed.) Developments in Reliable Computing, pp. 77–104. Kluver Academic Publishers, Dordrecht (1999)

PDE

Curve Veering for the Parameter-dependent Clamped Plate

Henning Behnke[✉]

Institut für Mathematik, TU Clausthal, Erzstraße 1,
38678 Clausthal-Zellerfeld, Germany
behnke@math.tu-clausthal.de

Abstract. The computation of vibrations of a thin rectangular clamped plate results in an eigenvalue problem with a partial differential equation of fourth order. If we change the geometry of the plate for fixed area, this results in a parameter-dependent eigenvalue problem. For certain parameters, the eigenvalue curves seem to cross. We give a numerically rigorous proof of curve veering, which is based on the Lehmann-Goerisch inclusion theorems and the Rayleigh-Ritz procedure.

Keywords: Partial differential equations · Paramenter-dependent eigenvalue problem · Upper and lower eigenvalue bounds · Interval arithmetic

1 Parameter-dependent Eigenvalue Problems

Parameter-dependent eigenvalue problems occur in many applications, for example in the computation of vibrations of turbine blades [2], in computing sloshing frequencies of a liquid in a container, in studying molecule geometries, in computing vibrations of free plates [4] or in computing the vibrations of a thin rectangular clamped plate.

The lattermost problem is described by a partial differential equation of fourth order:

$$\frac{\partial^4}{\partial x^4}\varphi + P\frac{\partial^4}{\partial x^2 \partial y^2}\varphi + Q\frac{\partial^4}{\partial y^4}\varphi = \lambda\varphi \text{ in } \Omega, \tag{1}$$

$$\varphi = 0 \text{ and } \frac{\partial\varphi}{\partial n} = 0 \text{ on } \partial\Omega,$$

$$\varphi(x,y) = \varphi(-x,y) = \varphi(x,-y) \text{ in } \Omega,$$

here $P, Q \in \mathbb{R}$, $P > 0$, $Q > 0$, and $\Omega = (-\frac{a}{2}, \frac{a}{2}) \times (-\frac{b}{2}, \frac{b}{2}) \subseteq \mathbb{R}^2$. The differential operator in (1) is self-adjoint and the eigenvalues are positive.

We consider the eigenvalues as functions of $s = a/b$ for $F = ab = 4$. Here a and b are the side lengths of Ω. An approximate computation is shown in Fig. 1. The locations left of #3 and right of #4 are remarkable. It is not clear whether

© Springer International Publishing Switzerland 2016
M. Nehmeier et al. (Eds.): SCAN 2014, LNCS 9553, pp. 259–268, 2016.
DOI: 10.1007/978-3-319-31769-4_21

the veering of the eigenvalue curves is an effect introduced by the discretization or whether the eigenvalues of (1) intersect. (There are further "possible crossings" #1 and #2 near $s = 1.6$ and $s = 1.3$ which are not marked in Fig. 1, since #3 and #4 seem to be more interesting.)

The situation is even more astonishing, if we consider eigenfunctions. Figure 2 shows, that the shape of eigenfunctions is preserved along the eventual "crossings". In this paper we give a numerically rigorous proof for curve veering at positions #3 and #4. This paper shows that the resulting algebraic eigenvalue problems which can be treated successfully, can be considerably larger than in [2].

There are several papers dealing with curve veering for plate problems, but so far there exists no numerically rigorous proof.

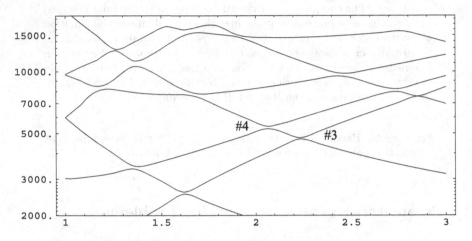

Fig. 1. Eigenvalues $\lambda_3, ..., \lambda_9$ as functions of $s = a/b$ with $F = ab = 4$

2   Inclusion Methods

In this section we briefly describe the Rayleigh-Ritz and the Lehmann-Goerisch methods for computing upper and lower eigenvalue bounds, respectively. For proofs see [9].

Let $(H, (.|.))$ be a Hilbert space with inner product $(.|.)$ and norm $\|.\|$ and V be a densely defined subspace of H. Denote by $[.|.]$ the inner product in V and let $(V, [.|.])$ be a Hilbert space ($|.|$ denotes the norm in V), the embedding $V \hookrightarrow H$ is assumed to be compact.

Then the eigenvalue problem reads as follows

$$\text{Determine } \lambda \in \mathbb{R} \text{ and } \varphi \in V, \varphi \neq 0 \text{ such that}$$
$$[\varphi|v] = \lambda(\varphi|v) \text{ for all } v \in V. \tag{2}$$

Problem (2) has a countable spectrum which consists of eigenvalues

$$0 < \lambda_1 \leq \lambda_2 \leq \cdots \quad , \quad \lim_{j \to \infty} \lambda_j = \infty.$$

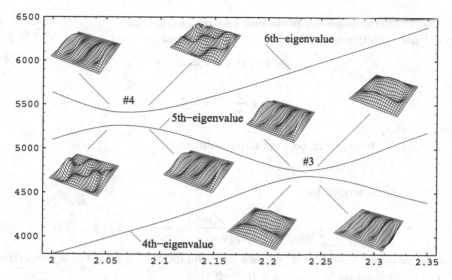

Fig. 2. Eigenvalues $\lambda_4, \ldots, \lambda_6$ and approximate eigenfunctions as functions of $s = a/b$ with $F = a\,b = 4$

The Rayleigh-Ritz procedure is a discretization of the Poincaré principle:

$$\lambda_j = \min_{\substack{E \subset V \\ \dim E = j}} \max_{\substack{u \in E \\ u \neq 0}} \frac{[u|u]}{(u|u)}, \quad j \in \mathbb{N}. \tag{3}$$

Now choose $u_1, \ldots, u_n \in V$, $n \in \mathbb{N}$, linearly independent, and define

$$V_n := \text{span}(u_1, \ldots, u_n).$$

A restriction of (3) to V_n instead of V results in

$$\Lambda_1^{[n]} \leq \Lambda_2^{[n]} \leq \cdots \leq \Lambda_n^{[n]},$$

and the upper bounds

$$\lambda_j \leq \Lambda_j^{[n]}, \quad j = 1, \ldots, n.$$

We call $\Lambda_j^{[n]}$ Rayleigh-Ritz bound for λ_j, the $\Lambda_j^{[n]}$ can be computed easily. Define

$$A_0 := \Big((u_i|u_k) \Big)_{i,k=1,\ldots,n},$$
$$A_1 := \Big([u_i|u_k] \Big)_{i,k=1,\ldots,n},$$

then the Rayleigh-Ritz bounds are the eigenvalues of

$$A_1 x = \Lambda^{[n]} A_0 x, \quad (\Lambda^{[n]}, x) \in \mathbb{R} \times \mathbb{R}^n.$$

The Rayleigh-Ritz bounds are monotonously decreasing in $n \in \mathbb{N}$.

The Lehmann-Goerisch procedure can be interpreted as discretization of a variational principle as well. Let $\rho \in \mathbb{R}$ be a spectral parameter, for a $N \in \mathbb{N}$, let

$$\lambda_N < \rho < \lambda_{N+1}. \tag{4}$$

Let

$$\lambda_{N+1-i} = \rho + \frac{1}{\sigma_i}, \quad i = 1, \ldots, N.$$

We assume $\sigma_i < 0$.

For $u \in V$, let $w_u \in H$ be the unique solution of

$$[u|v] = (w_u|v) \quad \text{for all } v \in V,$$

then σ_i is characterized by

$$\sigma_i = \inf_{\substack{E \subset V \\ \dim E = i}} \max_{\substack{u \in E \\ u \neq 0}} \frac{[u|u] - \rho(u|u)}{(w_u|w_u) - 2\rho[u|u] + \rho^2(u,u)}, \quad i = 1, \ldots, N. \tag{5}$$

A negative upper bound for σ_i yields a lower bound for λ_{N+1-i}. For a discretization of (5) determine $w_1, \ldots, w_n \in H$ such that

$$[u_i|v] = (w_i|v) \quad \text{for all } v \in V.$$

Let

$$A_2 := \left((w_i|w_k) \right)_{i,k=1,\ldots,n},$$

$$\left(A_1 - \rho A_0 \right) x = \tau \left(A_2 - 2\rho A_1 + \rho^2 A_0 \right) x, \quad (\tau, x) \in \mathbb{R} \times \mathbb{R}^n. \tag{6}$$

If the condition $\Lambda_N^{[n]} < \rho$ holds true for some $n \in \mathbb{N}$, (6) has exactly N negative eigenvalues

$$\tau_1 \leq \tau_2 \leq \ldots \leq \tau_N < 0 \leq \ldots \leq \tau_n.$$

We have ($\sigma_i \leq \tau_i$, $i = 1, \ldots, N$)

$$\Lambda_j^{\rho[n]} := \rho + \frac{1}{\tau_{N+1-j}} \leq \lambda_j, \quad j = 1, \ldots, N.$$

This discretization is the Lehmann-Goerisch procedure. We call $\Lambda_j^{\rho[n]}$ Lehmann-Goerisch bound for λ_j.

3   Application to the Plate Problem

In order to apply the inclusion theorems to (1), we define

$$H := \{ u \in L_2(\Omega) \mid u(x,y) = u(-x,y) = u(x,-y), \ (x,y) \in \Omega \},$$
$$V := \{ u \in H_0^2(\Omega) \mid u(x,y) = u(-x,y) = u(x,-y), \ (x,y) \in \Omega \},$$
$$(f|g) := \int_\Omega f \, g \, d\Omega \quad \text{for } f, g \in H,$$
$$[f|g] := \int_\Omega \left(\frac{\partial^2 f}{\partial x^2} \frac{\partial^2 g}{\partial x^2} + P \frac{\partial^2 f}{\partial x^2} \frac{\partial^2 g}{\partial y^2} + Q \frac{\partial^2 f}{\partial y^2} \frac{\partial^2 g}{\partial y^2} \right) d\Omega \quad \text{for } f, g \in V.$$

The eigenvalue problem reads as follows:
Determine $\lambda(s) \in \mathbb{R}$, $\phi \in V$ such that

$$[f|\phi] = \lambda(s)\,(f|\phi) \text{ for all } f \in V.$$

Let $u_i \in V \cap C^4(\Omega)$ and

$$M f := \frac{\partial^4}{\partial x^4} f + P \frac{\partial^4}{\partial x^2 \partial y^2} f + Q \frac{\partial^4}{\partial y^4} f,$$

now define

$$w_i \in H \quad \text{by } w_i := M\,u_i.$$

Then

$$[f|u_i] = (f|w_i) \text{ for all } f \in V.$$

Define

$$\tilde{f}_i(x) := \left(\frac{a^2}{4} - x^2\right)^{(i+1)}, \tilde{g}_j(y) := \left(\frac{b^2}{4} - y^2\right)^{(j+1)}, i, j \in \mathbb{N},$$

$$\tilde{u}_{i,j}(x,y) := \tilde{f}_i(x)\,\tilde{g}_j(y) \quad i, j \in \mathbb{N}.$$

The polynomials $\tilde{u}_{i,j}$ satisfy the boundary and symmetry conditions, i.e. $\tilde{u}_{i,j} \in V$.

To avoid the well known numerical problems with ill-conditioned matrices (see e.g. [5]), we construct orthogonal polynomials f_i and g_j from the $\tilde{f}_i$ and $\tilde{g}_j$ using the Gram-Schmidt process and a computer algebra system (for example *Mathematica* [8]). Then we define

$$u_{i,j}(x,y) := f_i(x)g_j(y) \text{ for } i, j \in \mathbb{N}.$$

In order to determine ρ (see (4)), we need rough lower eigenvalue bounds (the Rayleigh-Ritz procedure yields upper bounds). These can be obtained from the eigenvalues of the problem

$$\frac{\partial^4}{\partial x^4}\varphi + Q\frac{\partial^4}{\partial y^4}\varphi = \lambda\varphi \text{ in } \Omega,$$

$$\varphi = 0 \text{ and } \frac{\partial\varphi}{\partial n} = 0 \text{ on } \partial\Omega,$$

$$\varphi(x,y) = \varphi(-x,y) = \varphi(x,-y) \text{ in } \Omega,$$

which can be solved in closed form by separation of variables. If these rough bounds are too crude, a homotopy-method can be used ([3,7]). Using the described methods and interval arithmetic, verified bounds for eigenvalues of (1) for *fixed s* can be obtained, see Table 1. The results have been obtained using PROFIL/BIAS, see [6].

Table 1. Bounds for eigenvalues of the orthotropic plate, $\lambda_i \in [\inf[\Lambda_i^{LG}], \sup[\Lambda_i^{RR}]]$

λ_i	$n = 66$	$n = 136$	$n = 210$
	$P = 2,\quad Q = 2,\quad F = 4,\quad s = 1,\quad \rho = 30404.9542$		
1	$112.343^{80300370}_{29815764}\ E \pm 0$	$112.343^{8024882}_{48900881}\ E \pm 0$	$112.343^{80248479}_{51494059}\ E \pm 0$
2	$1.1217^{536545890}_{047907552}\ E + 3$	$1.1217^{536102349}_{451413903}\ E + 3$	$1.1217^{536090521}_{49228742}\ E + 4$
3	$2.0009^{984490746}_{895785651}\ E + 3$	$2.0009^{84420846}_{68829397}\ E + 3$	$2.0009^{84418602}_{79054004}\ E + 3$
4	$3.94^{5291688717}_{3268707890}\ E + 3$	$3.94^{5290445259}_{4883713625}\ E + 3$	$3.945^{29039462}_{02810079}\ E + 3$
5	$6.0^{10039214042}_{0911402918}\ E + 3$	$6.0^{1003872475}_{09841992629}\ E + 3$	$6.0^{1003870327}_{0999971107}\ E + 3$
6	$1.0^{613069606701}_{59334754544}\ E + 4$	$1.06^{13062508540}_{0821906680}\ E + 4$	$1.061^{306213463}_{214563549}\ E + 4$
7	$1.15^{4007769061}_{377723907}\ E + 4$	$1.15^{4007687367}_{3951000071}\ E + 4$	$1.15^{4007683119}_{399698021}\ E + 4$
8	$1.52^{44047455105}_{087332590}\ E + 4$	$1.52^{44038538533}_{348240199}\ E + 4$	$1.524^{4038028899}_{224606136}\ E + 4$
9	$2.00^{32337136971}_{176596685}\ E + 4$	$2.00^{3233489129}_{287182259}\ E + 4$	$2.003^{23347626}_{16212849}\ E + 4$
10	$2.^{5347032604487}_{47927946537}\ E + 4$	$2.5^{34699761186}_{199957246}\ E + 4$	$2.53^{4699534913}_{167375697}\ E + 4$
11	$2.^{7233110451279}_{6757640428}\ E + 4$	$2.7^{23308076912}_{074138273}\ E + 4$	$2.7^{23307897462}_{195766257}\ E + 4$

4 The Parameter-dependent Problem

4.1 Temple Quotients

In order to prove the curve veering for an interval $[\alpha, \beta]$, we need to calculate verified bounds for $\lambda_i(s)$ of the form

$$q_i(s) - \varepsilon \le \lambda_i(s) \le q_i(s) + \varepsilon \quad \text{for all } s \in [\alpha, \beta],$$

for ε small and q_i "simple" explicitly known functions. Since the application of the inclusion theorems to parameter-dependent eigenvalue problems results in parameter-dependent matrices A_0, A_1 and A_2, this can be achieved by treating parameter-dependent matrix eigenvalue problems [1].

For this aim consider the matrix eigenvalue problem

$$A\,x = \Lambda\,B\,x,$$

for $A, B \in \mathbb{R}^{n \times n}$, $A = A^T$, $B = B^T$, B positive definite. Let $\rho \in \mathbb{R}$, $u \in \mathbb{R}^n$, $u \neq 0$ and $v = B^{-1}A\,u$. Define

$$a_{0,A,B} := u^T B\,u,$$
$$a_{1,A,B} := u^T A\,u,$$
$$a_{2,A,B} := v^T B\,v.$$

For $a_{1,A,B} - \rho a_{0,A,B} \neq 0$, define the Temple Quotient

$$\tau_{A,B}(\rho) := \frac{a_{2,A,B} - \rho a_{1,A,B}}{a_{1,A,B} - \rho a_{0,A,B}}$$

and

$$\tau_{A,B}(\pm\infty) := \frac{a_{1,A,B}}{a_{0,A,B}}.$$

Then the interval

$$\begin{cases} (\rho, \tau_{A,B}(\rho)] & \text{if} \quad \rho < \tau_{A,B}(\rho) \\ [\tau_{A,B}(\rho), \rho) & \text{if} \ \tau_{A,B}(\rho) < \rho \end{cases}$$

contains at least one eigenvalue of the eigenvalue problem $A\,x = \Lambda\,B\,x$. (The case $\rho = \tau_{A,B}(\rho)$ can not happen, since $\rho = \tau_{A,B}(\rho)$ implies $v = \rho u$ and this results in $a_{1,A,B} - \rho a_{0,A,B} = 0$.)

These enclosures will be applied to parameter-dependent matrices, hence v and $a_{2,A,B}$ are not known exactly. Thus we mention, that the statement remains valid if $a_{2,A,B}$ is replaced by an upper bound [2]

$$a_{2,A,B} \leq \tilde{a}_{2,A,B}.$$

Let $c \in \mathbb{R}$, $0 < c \leq \Lambda_{\min(B)}$, $\tilde{v} \in \mathbb{R}^n$.
Then [1] or [3]

$$a_{2,A,B} = v^T B\,v$$

$$\leq \tilde{v}^T A\,u - \tilde{v}^T (B\,\tilde{v} - A\,u)$$
$$+\frac{1}{c}(B\,\tilde{v} - A\,u)^T (B\,\tilde{v} - A\,u)$$
$$=: \tilde{a}_{2,A,B}.$$

In the applications we choose

$$\tilde{v} \approx B^{-1}A\,u.$$

Based on Temple-Quotients, an easy procedure for computing upper eigenvalue bounds for

$$\Lambda_1 < \Lambda_2 < \ldots < \Lambda_n$$

can be constructed:

Procedure for upper bounds:

1. Calculate $0 < c \leq \Lambda_{\min}(B)$
2. Let $\rho := -\infty$ and $i := 1$.
3. Choose an appropriate $u \in \mathbb{R}^n$, let $v \approx B^{-1}A\,u$ and calculate $\tau_{A,B}(\rho)$ using $\tilde{a}_{2,A,B}$.
4. If $\tau_{A,B}(\rho) \leq \rho$ then break down.
5. Set the interval $\overline{\Lambda}_i$ to $(\rho, \tau_{A,B}(\rho)]$.
6. If $i < n$ let $\rho := \tau_{A,B}(\rho)$ and $i := i + 1$, goto 3.

If this procedure does *not* break down:

$$\Lambda_i \in \overline{\Lambda}_i \quad \text{for} \quad i = 1, \ldots, n.$$

In general the bound

$$\Lambda_i \leq \max(\overline{\Lambda}_i)$$

is very sharp, if

$$u = \tilde{x}_i,$$

where $\tilde{x}_i$ is an approximate eigenvector for Λ_i.

The computation of sharp lower eigenvalue bounds can be done similarily: Start the procedure with $\rho = +\infty$ and $i = n$.

4.2   Application to Parameter-dependent Matrices

If $A(s)$ and $B(s)$ are parameter-dependent matrices, then $a_{0,A,B}$, $a_{1,A,B}$ and $\tilde{a}_{2,A,B}$ depend on s. Thus, $(\tau_{A,B}(\rho))(s)$ is also a real function.

If in the i-th step bounds for Λ_i are to be calculated, we determine an approximation polynomial q_i for Λ_i in $[\alpha, \beta]$ and define

$$H_i(s) := A(s) - q_i(s) B(s).$$

The eigenvalues of

$$H_i(s) \, x(s) = \Lambda_{H_i}(s) \, B(s) \, x(s)$$

and

$$A(s) \, x(s) = \Lambda(s) \, B(s) \, x(s)$$

are closely related. Indeed, we have

$$\Lambda_{j,H_i}(s) = \Lambda_j(s) - q_i(s),$$

and

$$\Lambda_{i,H_i} \approx 0 \quad \text{in} \ [\alpha, \beta].$$

Now we compute $\tau_{H_i,B}(\rho - q_i)$ and determine bounds for the range of $\tau_{H_i,B}(\rho - q_i)$.

The elements of this range are close to zero if the approximation polynomial is good:

$$-\underline{\varepsilon}_i \ \leq \ \{ \, (\tau_{H_i,B}(\rho - q_i))(s) \mid s \in [\alpha, \beta] \, \} \ \leq \ \overline{\varepsilon}_i$$

This results in the bounds

$$q_i(s) - \underline{\varepsilon}_i \ \leq \ \Lambda_i(s) \ \leq \ q_i(s) + \overline{\varepsilon}_i \ \text{for all} \ s \in [\alpha, \beta].$$

5   Results

Now we can apply the theory to the clamped plate problem. For the upper bounds we compute upper bounds for the eigenvalues of (1) with the Rayleigh-Ritz procedure and upper bound for the eigenvalues of the resulting parameter-dependent matrix eigenvalue problem. For the lower bounds we apply the Lehmann-Goerisch method to (1) and compute lower bounds to the eigenvalues of the resulting parameter-dependent matrix eigenvalue problem. Since the

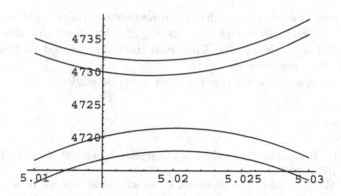

Fig. 3. Bounds for eigenvalues λ_4 and λ_5 near #3

Table 2. Bounds for eigenvalues λ_4 and λ_5 near #3

i	c_0	c_1	c_2	ε_i
	Parameter-interval $s^2 \in [5.01, 5.03]$, i.e. $s \in [2.23, 2.24]$;		location #3	
4	$-1.1728135\,E+6$	$4.6910569\,E+5$	$-4.6720581\,E+4$	1.7336
5	$1.1851818\,E+6$	$-4.7045191\,E+5$	$4.6872968\,E+4$	1.1327
		$n=128;\ \rho = 20000$		

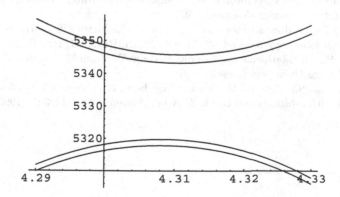

Fig. 4. Bounds for eigenvalues λ_5 and λ_6 near #4

Table 3. Bounds for eigenvalues λ_5 and λ_6 near #4

Parameter-interval $s^2 \in [4.29, 4.33]$, i.e. $s \in [2.07, 2.09]$; location #4			
i	c_0	c_1	c_2	ε_i
5	$-4.5180060\ E+5$	$2.1221805\ E+5$	$-2.4630601\ E+4$	0.95431
6	$4.6776638\ E+5$	$-2.1453928\ E+5$	$2.4883714\ E+4$	1.2748
n=400; $\rho = 12800$			

matrices depend on s^2, the resulting approximation polynomial is a polynomial in s^2. The coefficients and $\varepsilon_i = \max(\underline{\varepsilon}_i, \overline{\varepsilon}_i)$ are given in Tables 2 and 3. Figures 3 and 4 show the results. Thus it is rigorously proved that no crossing of the eigenvalue curves occurs.

The author thanks the referees for their valuable advice.

References

1. Behnke, H.: Bounds for Eigenvalues of Parameter-dependent Matrices. Computing **49**, 159–167 (1992)
2. Behnke, H.: A Numerically Rigorous Proof of Curve Veering in an Eigenvalue Problem for Differential Equations. Z. Anal. Anwendungen **15**, 181–200 (1996)
3. Behnke, H., Goerisch, F.: Inclusions for eigenvalues of selfadjoint problems. In: Herzberger, J. (ed.) Topics in validated computations, pp. 277–322. Elsevier, Amsterdam, Lausanne, New York, Oxford, Shannon, Tokyo (1994)
4. Behnke, H., Mertins, U.: Eigenwertschranken für das Problem der freischwingenden rechteckigen Platte und Untersuchungen zum Ausweichphänomen. ZAMM **75**, 342–363 (1995)
5. Fried, I.: Numerical Solution of Differential Equations. Academic Press [Harcourt Brace Jovanovich, Publishers], New York, London (1979). Computer Science and Applied Mathematics
6. Knüppel, O.: PROFIL/BIAS–a fast interval library. Computing **53**(3–4), 277–287 (1994). International Symposium on Scientific Computing, Computer Arithmetic and Validated Numerics (Vienna, 1993)
7. Plum, M.: Eigenvalue inclusions for second-order ordinary differential operators by a numerical homotopy method. Z. Angew. Math. Phys. **41**(2), 205–226 (1990)
8. Wolfram, S.: The Mathematica® book, 4th edn. Wolfram Media, Inc., Champaign, IL; Cambridge University Press (1999)
9. Zimmermann, S., Mertins, U.: Variational bounds to eigenvalues of self-adjoint problems with arbitrary spectrum. Z. Anal. Anwend. **14**, 327–345 (1995)

Numerical Verification for Elliptic Boundary Value Problem with Nonconforming $\mathcal{P}_1$ Finite Elements

Tomoki Uda$^{(\boxtimes)}$

Department of Mathematics, Kyoto University, Kitashirakawa Oiwake-cho, Sakyo-ku, Kyoto 606-8502, Japan
uda@math.kyoto-u.ac.jp

Abstract. We propose a numerical method with the nonconforming $\mathcal{P}_1$ FEM to verify the existence of solutions to an elliptic boundary value problem. Formulating the boundary value problem as a fixed-point problem on the sum space of the nonconforming $\mathcal{P}_1$ finite element space with the Sobolev space of 1st order with zero Dirichlet condition, we construct the numerical verification method based on the Schauder fixed-point theorem. We show a constructive inequality for a boundary integral that appears due to the discontinuity of a nonconforming $\mathcal{P}_1$ finite element function. Finally, we present a numerical example to show our proposed method works well.

Keywords: Nakao's method · Numerical verification · Elliptic boundary value problem · Nonconforming $\mathcal{P}_1$ finite element

1 Introduction

A finite element method (FEM), which is based on piecewise polynomial approximation, is widely used to solve partial differencial equations numerically. In a certain boundary value problem (BVP), when we use a lower-order conforming FEM, the numerical solution converges to the exact solution slowly, which is known as a *locking effect* [1]. To avoid the locking effect for such a BVP, it is known that a certain nonconforming FEM is helpful. For example, Lee et al. [5] proposed an optimal and robust nonconforming FEM for the planar linear elasticity problem.

In 1988, M. T. Nakao [7] developed a method based on a FEM to verify the existence of solutions to an elliptic BVP. Nakao's method is useful not only in verifying the existence of solutions mathematically but also in estimating an a-posteriori error for a numerical solution to the BVP. Nakao's theory implicitly assumed the finite element (FE) space to be conforming in order to deduce verification conditions [7]. Especially it relies on continuity of a conforming FE function, whereas a nonconforming FE function is discontinuous in general. Therefore, it is not obvious how to apply Nakao's method to a nonconforming FEM.

© Springer International Publishing Switzerland 2016
M. Nehmeier et al. (Eds.): SCAN 2014, LNCS 9553, pp. 269–279, 2016.
DOI: 10.1007/978-3-319-31769-4_22

Thus, in this paper, we generalize Nakao's method for the nonconforming $\mathcal{P}_1$ FEM, where $\mathcal{P}_1$ means piecewise linear approximation over a triangulation.

The structure of the paper is given as follows. In Sect. 2 we introduce the notations for several function spaces. Section 3 is the main part of the paper. In Sect. 3.1 we show some basic facts about the nonconforming $\mathcal{P}_1$ FEM. In Sect. 3.2 we formulate a BVP as a fixed-point problem in order to apply the Schauder fixed-point theorem. In Sects. 3.3, 3.4 and 3.5 we derive sufficient conditions for applying the Schauder fixed-point problem. In Sect. 3.6 we show a constructive inequality for a boundary integral arising due to the discontinuity of the nonconforming $\mathcal{P}_1$ FE function. A numerical test is presented in Sect. 4 and finally in Sect. 5 we summarize our proposed method.

2   Preliminaries

For a bounded open convex polygon $\Omega \subsetneq \mathbf{R}^2$, $L^2(\Omega)$ and $H^k(\Omega)$ $(k = 1, 2)$ denote the set of all $\mathbf{R}$-valued square-integrable functions and the Sobolev space of order k, respectively. Each function $v \in L^2(\Omega)$ belongs to $H^k(\Omega)$ if and only if the distributional derivative $\partial^\alpha v$ belongs to $L^2(\Omega)$ for all multi-indices $|\alpha| \leq k$. $H_0^1(\Omega)$ is the subspace of $H^1(\Omega)$ endowed with zero Dirichlet boundary conditions. Namely, $H_0^1(\Omega) := \{v \in H^1(\Omega) \mid v = 0 \text{ on } \partial\Omega\}$. $L^2(\Omega)$ is equipped with the standard inner product $(\cdot, \cdot)_{L^2}$ and the standard norm $\|\cdot\|_{L^2}$. Similarly, $H^k(\Omega)$ is equipped with the standard norm $\|\cdot\|_{H^k}$. Especially, the H^2 seminorm and the H^2 norm are defined as follows.

$$|u|_{H^2(\Omega)}^2 := \sum_{i,j=1}^{2} \left\|\partial_{x_i}\partial_{x_j} u\right\|_{L^2(\Omega)}^2 \qquad (u \in H^2(\Omega)),$$

$$\|u\|_{H^2(\Omega)}^2 := \|u\|_{L^2(\Omega)}^2 + \|\nabla u\|_{L^2(\Omega)}^2 + |u|_{H^2(\Omega)}^2 \qquad (u \in H^2(\Omega)).$$

In what follows, we omit the domain Ω in these notations unless otherwise stated.

We sometimes use the following Proposition 1 in estimating the norm $\|\triangle \cdot\|_{L^2}$ where $\triangle := \partial_{x_1}^2 + \partial_{x_2}^2$ denotes the Laplace operator $\triangle : H^2 \cap H_0^1 \to L^2$. See Theorem 4.3.1.4 of [4].

Proposition 1. $|u|_{H^2(\Omega)} = \|\triangle u\|_{L^2(\Omega)}$ holds for all $u \in H^2(\Omega) \cap H_0^1(\Omega)$.

3   Nakao's Method with Nonconforming $\mathcal{P}_1$ FEM

We consider the following (nonlinear) elliptic boundary value problem.

Problem 1 (Elliptic BVP). For given $f : H^1 \to L^2$, find $u \in H^2(\Omega)$ such that

$$\begin{cases} -\triangle u = f(u) & \text{in } \Omega, \\ u = 0 & \text{on } \partial\Omega. \end{cases} \qquad (1)$$

A nonlinear BVP is also considered in our formulation, but in Sect. 4 we give a numerical test only for the linear case.

Since the Laplace operator $\triangle$ endowed with zero Dirichlet boundary condition is invertible, we define the inverted operator $\triangle^{-1}$ from L^2 to $H^2 \cap H_0^1$. Thus, the BVP (1) is reforumulated as the fixed-point problem on H_0^1.

Problem 2 (Fixed-point Problem). For given $f : H_0^1 \to L^2$, find $u \in H_0^1$ such that

$$u = -\triangle^{-1} f(u) . \tag{2}$$

In order to apply the Schauder fixed-point theorem to the fixed-point problem (2), M. T. Nakao [7] assumed that the map f satisfies the following property.

Assumption 1. The map $f : H_0^1 \to L^2$ is continuous and bounded where H_0^1 is equipped with the inner product $(\nabla \cdot, \nabla \cdot)_{L^2}$. That is to say, for any bounded subset U in H_0^1, the image $f(U)$ is also bounded in L^2.

3.1 Nonconforming $\mathcal{P}_1$ Finite Elements

Let $(\mathcal{T}_h)_{h>0}$ be a regular family of triangulations of the domain Ω. Here, the subscript h denotes the mesh size of the triangulation $\mathcal{T}_h$. We denote by e_i $(i = 1, \ldots, N_h)$ a side of each triangle $T \in \mathcal{T}_h$. Here N_h denotes the number of all sides e_i. Let N_{h0} be the number of all internal sides e_i and we suppose that each e_i is numbered so that $e_i \not\subseteq \partial\Omega$ holds for $i = 1, \ldots, N_{h0}$ and $e_i \subseteq \partial\Omega$ for $i = N_{h0} + 1, \ldots, N_h$. For each side e_i $(i = 1, \ldots, N_h)$, let M_i be the midpoint of e_i. Define by X_{h0} the nonconforming $\mathcal{P}_1$ FE space over $\mathcal{T}_h$:

$$X_{h0} := \left\{ \varphi \in L^2 \,\middle|\, \begin{array}{l} \varphi \text{ is piecewise linear over the triangulation } \mathcal{T}_h, \\ \text{continuous on each midpoint } M_i \text{ for } i = 1, \ldots, N_h \\ \text{and } \varphi(M_i) = 0 \text{ for } N_{h0} < i \leq N_h \end{array} \right\} .$$

We call $\varphi \in X_{h0}$ a nonconforming $\mathcal{P}_1$ FE function. We denote by $\{\varphi_i\}_{i=1}^{N_{h0}} \subsetneq X_{h0}$ the basis functions of X_{h0} with $\varphi_j(M_i) = \delta_{ij}$ for $i, j = 1, \ldots, N_{h0}$, where δ_{ij} is the Kronecker delta.

Remark 1. X_{h0} is **not** a subspace of H_0^1 because $\varphi \in X_{h0}$ is **discontinuous** on some edges e_i in general. Each φ vanishes only on all the boundary midpoints $\{M_i \mid i > N_{h0}\}$, and thus φ does not satify zero Dirichlet boundary condition exactly.

We introduce an inner product and a norm for $u, v \in X_{h0} + H_0^1(\Omega)$ as follows.

$$(u, v)_h := \sum_{T \in \mathcal{T}_h} (\nabla u, \nabla v)_{L^2(T)} , \|u\|_h^2 := (u, u)_h .$$

Note that, if $u, v \in H_0^1(\Omega)$, this inner product $(u, v)_h$ coincides with $(\nabla u, \nabla v)_{L^2}$. It is well-known that $(X_{h0} + H_0^1, (\cdot, \cdot)_h)$ is a Hilbert space. For details, see Theorem 4.2.4 of [2].

We now define the nonconforming $\mathcal{P}_1$ interpolator $\Pi_h^{nc} : H_0^1 \to X_{h0}$ so that

$$\int_{e_i} u \, ds = \int_{e_i} (\Pi_h^{nc} u)|_T \, ds \tag{3}$$

holds for all functions $u \in H_0^1$, each side e_i ($i = 1, \ldots, N_{h0}$) and each associated triangle $T \in \mathcal{T}_h$ with $e_i \subsetneq \partial T$. Equation (3) uniquely determines the FE function $\Pi_h^{nc} u \in X_{h0}$ and Π_h^{nc} is thus well-defined. A constructive error estimate for the nonconforming $\mathcal{P}_1$ interpolator Π_h^{nc} has been given by Liu [6].

Proposition 2 (Liu '09). *Let $C_P(\mathcal{T}_h)$ be the maximum of the optimal Poincaré constant on $T \in \mathcal{T}_h$ that is,*

$$C_P(\mathcal{T}_h) := \max_{T \in \mathcal{T}_h} \sup \left\{ \frac{\|v - \overline{v}\|_{L^2}}{\|\nabla v\|_{L^2}} \,\middle|\, v \in H^1(T) \setminus \{0\} , \overline{v} := \frac{1}{|T|} \int_T v \, dx \, dy \right\} .$$

Then, for all $u \in H^2(\Omega) \cap H_0^1(\Omega)$, the following inequality holds.

$$\|u - \Pi_h^{nc} u\|_h \le C_P(\mathcal{T}_h) |u|_{H^2} .$$

Note that the optimal Poincaré constant on T is $O(\mathrm{diam}(T))$ where $\mathrm{diam}(T) := \sup_{x,y \in T} |x - y|$ is the diameter of T, from which it follows $C_P(\mathcal{T}_h) = O(h)$. We can see an exact value of $C_P(\mathcal{T}_h)$ in [6].

In fact, Π_h^{nc} coincides with the orthogonal projection P_h from $X_{h0} + H_0^1$ onto X_{h0}. Let $u \in X_{h0} + H_0^1$. Owing to Gauss-Green Theorem and (3), we have

$$(u - \Pi_h^{nc} u, \varphi_i)_h = \sum_T \left[-(u - \Pi_h^{nc} u, \triangle \varphi_i)_{L^2} + \int_{\partial T} (u - \Pi_h^{nc} u) \frac{\partial \varphi_i}{\partial \nu} \, ds \right] = 0 ,$$

where ν denotes the outward unit normal vector on ∂T. Since this holds for all basis functions φ_i, we obtain $P_h = \Pi_h^{nc}$. The orthogonal space of X_{h0} in the Hilbert space $(X_{h0} + H_0^1, (\cdot, \cdot)_h)$ is represented by $X_{h0}^\perp := (I - P_h)(H_0^1)$.

3.2 Extended Problem

In Nakao's theory, the conforming $\mathcal{P}_1$ FE space, which is a subspace of $H_0^1(\Omega)$, plays an important role in verifying the existence of solutions to (2). However, the theory is not applicable directly to the fixed-point problem with the nonconforming $\mathcal{P}_1$ FE space X_{h0}, since it is not a subspace of H_0^1. We thus consider the following extended fixed-point problem on the sum space $X_{h0} + H_0^1(\Omega)$ for fixed $h > 0$ under the following Assumption 2.

Assumption 2. There exists a continuous and bounded map $f_h : X_{h0} + H_0^1 \to L^2$ such that f_h is an extended map of f, that is, $f_h|_{H_0^1} = f$.

Problem 3 (Extended Fixed-point Problem). Find $u \in X_{h0} + H_0^1$ such that

$$u = -\triangle^{-1} f_h(u) \equiv \mathcal{F}(u) . \tag{4}$$

Remark 2. If f is of the form $f(u) = g(u, \nabla u)$ with a continuous map $g : \mathbf{R}^3 \to \mathbf{R}$, then we always have the natural extended map f_h satisfying Assumption 2. For this, we just replace ∇u by $\nabla_h u$ where ∇_h denotes the discrete differencial operator $\nabla_h : X_{h0} + H_0^1 \to L^2$ with the property $(\nabla_h u)|_T = \nabla(u|_T)$ for any $T \in \mathcal{T}_h$.

Remark 3. This extended fixed-point problem (4) is equivalent to (2), because, if there exists a solution $u \in X_{h0} + H_0^1$ to (4), u is an element of $\mathcal{D}(\triangle) = H^2 \cap H_0^1$ and thus (2) holds owing to Assumption 2.

Remark 4. The Schauder fixed-point theorem states that if U is a nonempty bounded closed convex subset of a Banach space X and $\mathcal{F} : U \to U$ is a continuous compact map, then $\mathcal{F}$ has a fixed point. In (4), $\mathcal{F} = -\triangle^{-1} f_h : X_{h0} + H_0^1 \to X_{h0} + H_0^1$ is a compact map, which is the assumption of the Schauder fixed-point theorem. Thus, in order to show the existence of a fixed-point to (4), it is sufficient to prove that there exists a certain bounded, convex and closed subset $U \subsetneq X_{h0} + H_0^1$ that includes its image $\mathcal{F}(U)$.

3.3   Candidate Set

In validated numerical computations, one needs to introduce a computable set in order to derive sufficient conditions that satisfy the assumption of the Schauder fixed-point thorem. Let $\mathbf{IR}$ denote the set of all bounded closed intervals in $\mathbf{R}$, that is, $\mathbf{IR} := \{[\underline{x}, \overline{x}] \mid \underline{x}, \overline{x} \in \mathbf{R}, \underline{x} \le \overline{x}\}$. We identify an interval vector $\boldsymbol{x} = (\boldsymbol{x}_i)_{i=1}^N \in \mathbf{IR}^N$ with a closed set $\prod_{i=1}^N \boldsymbol{x}_i \subsetneq \mathbf{R}^N$. In this paper, letters denoting an interval and an interval vector are boldface.

Definition 1 (Candidate Set). *For $u_h \in X_{h0}$, $\boldsymbol{v} \in \mathbf{IR}^{N_{h0}}$ and $\alpha > 0$, define a bounded, convex and closed subset $U = u_h + U_h + U_* \subsetneq X_{h0} + H_0^1$ such that*

$$U_h = \left\{ \sum_{i=1}^{N_{h0}} u_i \varphi_i \,\middle|\, (u_i) \in D^{-1}\boldsymbol{v} \right\}, U_* = \{u_* \in X_{h0}^\perp \| \|u_*\|_h \le \alpha\}, \tag{5}$$

where $D_{ij} := (\varphi_j, \varphi_i)_h$ is the $N_{h0} \times N_{h0}$ matrix. We call U a candidate set.

We usually take a numerical solution obtained by nonconforming $\mathcal{P}_1$ FEM as u_h. Then, the candidate set U is a closed neighborhood of the approximate solution in which we expect the existence of the exact solution. On the other hand, $\boldsymbol{v}$ and α are iteration parameters in our method. We practically choose them sufficiently small at the first iteration.

   Using the orthogonal decomposition of $X_{h0} + H_0^1$, we get the following two sufficient conditions for the existence of a fixed-point (recall Ramark 4).

$$(I - P_h)\mathcal{F}(U) \subseteq U_*, \tag{A}$$
$$P_h\mathcal{F}(U) \subseteq u_h + U_h. \tag{B}$$

Note that (A) and (B) are inclusion relations in the infinite dimensional space $X_{h0}^{\perp}$ and the finite dimensional space X_{h0}, respectively. Since it is not possible to verify these conditions directly in computers, we derive computable sufficient conditions instead of the conditions (A) and (B).

3.4   Sufficient Condition for Infinite-Dimensional Part

The following theorem gives a sufficient condition for (A), which allows us to verify (A) numerically rigorously in computers.

Theorem 1 (Sufficient Condition for (A)). *The condition* (A) *holds, if*

$$C_{\mathrm{P}}(\mathcal{T}_h) \sup_{u \in U} \|f_h(u)\|_{L^2} \le \alpha. \tag{6}$$

We can prove Theorem 1 in the same way as in [7].

Proof. Assume $C_{\mathrm{P}}(\mathcal{T}_h) \sup_{u \in U} \|f_h(u)\|_{L^2} \le \alpha$. Let u be an arbitrary element of U. Then, owing to Propositions 1 and 2 and the assumption, we have

$$\begin{aligned} \|(I - P_h)\mathcal{F}(u)\|_h &= \left\|(I - \Pi_h^{\mathrm{nc}})\triangle^{-1} f_h(u)\right\|_h \\ &\le C_{\mathrm{P}}(\mathcal{T}_h) \left|\triangle^{-1} f_h(u)\right|_{H^2} = C_{\mathrm{P}}(\mathcal{T}_h) \|f_h(u)\|_{L^2} \le \alpha. \end{aligned}$$

Consequently, $(I - P_h)\mathcal{F}(u) \in U_*$ holds.   □

Since $C_{\mathrm{P}}(\mathcal{T}_h) = O(h)$, we expect that the inequality (6) holds for a sufficiently small mesh size h.

3.5   Sufficient Condition for Finite-Dimensional Part

We define by K_i the support of φ_i. Using the integration by parts, we have, for each φ_i and all $v \in H^2(\Omega)$,

$$\int_{K_i} (\triangle v)\varphi_i \, \mathrm{d}x \, \mathrm{d}y = \int_{\partial K_i} \frac{\partial v}{\partial \nu} \varphi_i \, \mathrm{d}s - \sum_{T \in \mathcal{T}_h, T \subsetneq K_i} \int_T \nabla v \cdot \nabla \varphi_i|_T \, \mathrm{d}x \, \mathrm{d}y. \tag{7}$$

Note that, in general, the boundary integral in the right hand side of (7) does not vanish, since φ_i is discontinuous on ∂K_i. We thus define a continuous linear operator $b(\,\cdot\,; \varphi_i) : H^2 \cap H_0^1 \to \mathbf{R}$ by

$$b(v; \varphi_i) := \int_{\partial K_i} \frac{\partial v}{\partial \nu} \varphi_i \, \mathrm{d}s,$$

with which we can rewrite (7) as follows.

$$(\triangle v, \varphi_i)_{L^2} = b(v; \varphi_i) - (v, \varphi_i)_h. \tag{8}$$

Now the following theorem yields a computable sufficient condition for (B).

Theorem 2 (Sufficient Condition for (B)). *Take an interval vector* $d := (d_i) \in IR^{N_{h0}}$ *such that*

$$\{(f_h(u), \varphi_i)_{L^2} - (u_h, \varphi_i)_h - b(\triangle^{-1} f_h(u) ; \varphi_i) \mid u \in U\} \subseteq d_i, \qquad (9)$$

for each $i = 1, \ldots, N_{h0}$. *Then, the condition* (B) *holds, if*

$$d \subseteq v. \qquad (10)$$

We also prove Theorem 2 in the same way as is done in [7], but owing to the discontinuity of φ_i, the formula (9) is slightly different from the one in [7]. We remark that the sufficient condition (10) and the one in [7] are also different, because the finite-dimensional part (5) in our candidate set is modified from the original one.

Proof. Assume $d \subseteq v$. Let u be an arbitrary element of the candidate set U. Then, by the definition of P_h, we get $-P_h \triangle^{-1} f(u) = u_h + v_h$ for some $v_h \in X_{h0}$. It is sufficient to show $v_h \in U_h$ for any $u \in U$. We set by $v := (v_i)$ the coefficient vector of v_h, that is, $v_h = \sum_{i=1}^{N_{h0}} v_i \varphi_i$. Then, owing to (8) and (9), we have

$$(v_h, \varphi_i)_h = (-P_h \triangle^{-1} f(u) - u_h, \varphi_i)_h = (-\triangle^{-1} f(u), \varphi_i)_h - (u_h, \varphi_i)_h$$
$$= (f(u), \varphi_i)_{L^2} - b(\triangle^{-1} f(u) ; \varphi_i) - (u_h, \varphi_i)_h \in d_i,$$

which means $Dv \in d$ holds. Therefore, it follows from the assumption that $Dv \in v_h$. Consequently, owing to (5), we conclude $v_h \in U_h$.   □

3.6   Estimates for Boundary Integrals

In (9), we have the boundary integral term $b(w; \varphi_i)$ with $w := \triangle^{-1} f_h(u)$. Hence, we still have a difficulty in computing the left hand side of (9) on computers rigorously, since the condition (9) contains the term $b(\triangle^{-1} f_h(u) ; \varphi_i)$ with the inverse operator $\triangle^{-1}$. Instead, we get $O(h)$ estimates for $|b(w; \varphi_i)|$ so that we can replace the set $\{b(w; \varphi_i)\}$ with an interval of $O(h)$ width.

Lemma 1 (Dupont et al. '79). *Let K be a convex polygon in $\mathbf{R}^2$. Denote by $P_1(K)$ the set of all polynomial functions on K of at most 1st degree. Define the optimal Poincaré constant $C_P(K)$ as follows.*

$$C_P(K) := \sup \left\{ \frac{\|w\|_{L^2}}{\|\nabla w\|_{L^2}} \,\middle|\, w \in H^1(K) \setminus \{0\} , \int_K w \, dx \, dy = 0 \right\}.$$

Then, there exists a positive constant $C_{BH}(K)$ such that, for every $w \in H^2(K)$,

$$\inf_{p \in P_1(K)} \|w + p\|_{H^2(K)} \leq C_{BH}(K) |w|_{H^2(K)} . \qquad (11)$$

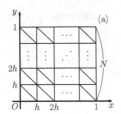
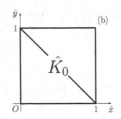
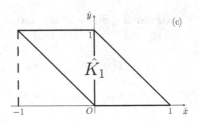

Fig. 1. (a) Triangulation $\mathcal{T}_h$, (b) $\hat{K}_0$ and (c) $\hat{K}_1$

Inequality (11) is well-known as Bramble-Hilbert Lemma, but the constructive proof for Lemma 1 has been given by Dupont et al. [3, '79].

For simplicity, we consider the case $\Omega := (0,1)^2$ and the triangulation $\mathcal{T}_h$ as shown in Fig. 1. (a). We divide the interval $(0,1)$ into N sections both in x-axis and in y-axis. We assume for $h = 1/N$. With this triangulation $\mathcal{T}_h$, the support K_i of φ_i is similar to either a rectangle $\hat{K}_0 := [0,1]^2$ or a parallelogram $\hat{K}_1 := \{0 \leq x+y \leq 1, 0 \leq y \leq 1\}$. See Figs. 1. (b) and (c). We introduce an affine map $\Phi_i : K_i \xrightarrow{1:1} \hat{K}_{k(i)}$ that induces the change of variable $(\hat{x}, \hat{y}) = \Phi_i(x,y)$. Here, we choose $k(i)$ as 0 or 1 so that K_i and $\hat{K}_{k(i)}$ are similar to each other. Let $\hat{w} := w \circ \Phi_i^{-1} \in H^2(\hat{K}_{k(i)})$ and $\hat{\varphi}_{k(i)} := \varphi_i \circ \Phi_i^{-1} \in H^1(\hat{K}_{k(i)})$. Since the basis function $\hat{\varphi}_k$ depends only on k and not on i, we omit i and denote $k = k(i)$ later.

By using the change of variable Φ_i, we can rewrite the boundary integral term $b(w; \varphi_i)$ as follows.

$$b(w; \varphi_i) = \int_{\partial K_i} \frac{\partial w}{\partial \nu} \varphi_i \, ds = \int_{\partial \hat{K}_k} \frac{\partial \hat{w}}{\partial \hat{\nu}} \hat{\varphi}_k \, d\hat{s} =: \beta_k(\hat{w}) \, ,$$

where $\beta_k : H^2(\hat{K}_k) \to \mathbf{R}$ becomes a continuous linear operator on $H^2(\hat{K}_k)$. Now, we prove the following constructive estimate for the boundary integral term.

Theorem 3. *For all $w \in H^2(\Omega) \cap H_0^1(\Omega)$, the following inequality holds.*

$$|b(w; \varphi_i)| \leq C_{\mathrm{BH}}(\hat{K}_k) \, h\sqrt{2} \|\hat{\varphi}_k\|_{H^1(\hat{K}_k)} \|\triangle w\|_{L^2(\Omega)} \, .$$

Proof. First, we evaluate the operator norm of β_k. Owing to Gauss-Green theorem, Cauchy-Schwarz inequality and Proposition 1, we have

$$\begin{aligned}
|\beta_k(\hat{w})| &= \left| \int_{\partial \hat{K}_k} \frac{\partial \hat{w}}{\partial \hat{\nu}} \hat{\varphi}_k \, d\hat{s} \right| = \left| \int_{\hat{K}_k} \nabla \hat{w} \cdot \nabla \hat{\varphi}_k \, d\hat{x} \, d\hat{y} + \int_{\hat{K}_k} (\triangle \hat{w}) \hat{\varphi}_k \, d\hat{x} \, d\hat{y} \right| \\
&\leq \|\nabla \hat{w}\|_{L^2} \|\nabla \hat{\varphi}_k\|_{L^2} + \|\triangle \hat{w}\|_{L^2} \|\hat{\varphi}_k\|_{L^2} \\
&\leq \left[(\|\nabla \hat{w}\|_{L^2}^2 + \|\triangle \hat{w}\|_{L^2}^2)(\|\nabla \hat{\varphi}_k\|_{L^2}^2 + \|\hat{\varphi}_k\|_{L^2}^2) \right]^{1/2} \\
&\leq \sqrt{2} \|\hat{w}\|_{H^2(\hat{K}_k)} \|\hat{\varphi}_k\|_{H^1(\hat{K}_k)} \, .
\end{aligned}$$

Hence, we get the following inequality for the operator norm of β_k.

$$\|\beta_k\| \leq \sqrt{2}\|\hat{\varphi}_k\|_{H^1(\hat{K}_k)} . \tag{12}$$

Second, we apply Bramble-Hilbert Lemma 1 to $\hat{w}$. Take a polynomial function $p \in P_1(\hat{K}_k)$ arbitrarily. Then, $\beta_k(p) = 0$ holds, because the normal derivative of p is constant and $\hat{\varphi}_k$ is linear on each side of the quadrilateral $\partial \hat{K}_k$. Thus, owing to Lemma 1, we have

$$|\beta_k(\hat{w})| = \inf_{p \in P_1(\hat{K}_k)} |\beta_k(\hat{w} - p)| \leq \inf_{p \in P_1(\hat{K}_k)} \|\beta_k\| \, \|\hat{w} - p\|_{H^2}$$

$$\leq C_{\mathrm{BH}}(\hat{K}_k) \, \|\beta_k\| \, |\hat{w}|_{H^2(\hat{K}_k)} \leq C_{\mathrm{BH}}(\hat{K}_k) \, \|\beta_k\| \, h|w|_{H^2(\Omega)} . \tag{13}$$

At last, applying Proposition 1 and combining inequalities (12) and (13), we complete the proof. □

4 Numerical Results

We show a numerical result to confirm our proposed verification method works well. Let $f : H_0^1 \to L^2$ be the following map.

$$f(u) := 0.001\pi^2 u + 1.999 \sin(\pi x)\sin(\pi y) . \tag{14}$$

Then, $u = \pi^{-2}\sin(\pi x)\sin(\pi y)$ is a solution to (1). The function f satisfies Assumption 1. Furthermore, we extend the domain of definition of f to $X_{h0} + H_0^1$ with the same definition (14). Thus, f also satisfies Assumption 2. Note that interval arithmetic allows us to deal with parametric uncertainty of a PDE, which is discussed by Nakao [7]. We omit the detail since our case is similar to [7].

Table 1. Numerical verification by Nakao's method with mesh size $h = 1/8$

# of iterations	Candidate set U		Another candidate set V		$V \subseteq U$
	$\max_i |v_i|$	α	$\max_i |d_i|$	β	
1	0.00525252	0.1	0.0132067	0.123234	
2	0.013867	0.173234	0.013209	0.123246	OK

Table 2. Numerical verification by proposed method with mesh size $h = 1/8$

# of iterations	Candidate set U		Another candidate set V		$V \subseteq U$
	$\max_i |v_i|$	α	$\max_i |d_i|$	β	
1	0.00101012	0.02	0.89045	0.0795599	
2	0.899354	0.0895599	0.91856	0.0820849	
3	0.927745	0.0920849	0.919445	0.0821645	OK

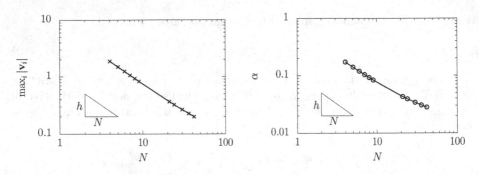

Fig. 2. Asymptotic behaviors of a-posteriori errors $\max_i|v_i|$ and α with respect to N

We show numerical results by Nakao's method and by our proposed method in Tables 1 and 2, respectively. The 2nd and 3rd columns are the size of the candidate set U. Here $|v_i|$ denotes $\max\{|x| \mid x \in v_i\}$. The 4th and 5th columns represent another candidate set V where d_i (resp. β) is computed by L.H.S. of (9) (resp. (6)). The 6th column verifies (6) and (10). We consider $\max_i|v_i|$ and α as a-posteriori errors for an FE approximate solution u_h, that is, these values in the last row represent the difference between the numerical solution and the exact solution. We see the a-posteriori error $\max_i|v_i|$ in Table 2 is larger than one in Table 1. This is because the boundary integral term $b(\,\cdot\,;\varphi_i)$ is enclosed by an interval with a relatively large width in proposed method. In Fig. 2, we also show asymptotic behaviors of the a-posteriori errors by proposed method as the mesh size h tends to 0. We observe both $\max_i|v_i|$ and α decrease in $O(h)$.

5   Summary

We successfully generalize Nakao's method for the nonconforming $\mathcal{P}_1$ FEM. The key idea is formulating the BVP as a fixed-point problem on $X_{h0} + H_0^1(\Omega)$. Owing to the discontinuity of the nonconforming $\mathcal{P}_1$ FE function φ_i, a boundary integral term $b(\,\cdot\,;\varphi_i)$ arises in our verification conditions, which makes it difficult to obtain a candidate set numerically satisfying the assumption of the Schauder fixed-point theorem. In order to resolve this difficulty, we get an $O(h)$ estimate for $b(\,\cdot\,;\varphi_i)$ and thus we replace the boundary integral term with an interval of $O(h)$ width. Finally, we show a numerical test to confirm that our verification method works well.

There are several future works related to our result. A numerical verification for the planar linear elasticity problem (or other elliptic BVPs from engineering), which may cause a locking effect, should be considered so that we compare a-posteriori errors by Nakao's method and by our method. In order for this, a possible extension of our result is, for example, using the FEs proposed by Lee et al. [5]. There is a disadvantage in our method which should be improved in future that the boundary integral term is enclosed by an interval with a relatively large width. Thus, other verification approaches than [7] also should be considered.

References

1. Babuška, I., Suri, M.: On locking and robustness in the finite element method. SIAM J. Numer. Anal. **29**(5), 1261–1293 (1992)
2. Ciarlet, P.G.: The Finite Element Method for Elliptic Problems. Society for Industrial and Applied Mathematics, Philadelphia (2002)
3. Dupont, T., Scott, L.: Polynomial approximation of functionals in sobolev spaces. Math. Comput. **34**, 441–463 (1979)
4. Grisvard, P.: Elliptic Problems in Nonsmooth Domains. Classics in Applied Mathematics. SIAM, Boston (1985)
5. Lee, C.O., Lee, J., Sheen, D.: A locking-free nonconforming finite element method for planar linear elasticity. Adv. Comput. Math. **19**, 277–291 (2003)
6. Liu, X.: Analysis of error constants for linear conforming and nonconforming finite elements. Ph.D. thesis, University of Tokyo, March 2009
7. Nakao, M.T.: A numerical approach to the proof of existence of solutions for elliptic problems. Jpn J. Appl. Math. **5**(2), 313–332 (1988)

Bibliography of Prof. Walter Krämer

Werner Hofschuster[1], Evgenija D. Popova[2], and Ralph Baker Kearfott[3]

[1] Faculty of Mathematics and Natural Sciences, Bergische Universität Wuppertal,
Gaußstr. 20, 42097 Wuppertal, Germany
hofschuster@math.uni-wuppertal.de
[2] Institute of Mathematics and Informatics, Bulgarian Academy of Sciences,
Acad. G. Bonchev str., block 8, 1113 Sofia, Bulgaria
epopova@bio.bas.bg
[3] Department of Mathematics, University of Louisiana,
Lafayette, LA 70504-1010, USA
rbk@louisiana.edu

Abstract. This is a bibliography of prof. Walter Krämer's (1952–2014) work.

Keywords: Bibliography · Enclosure methods · Mathematical software · Result verification · Standard and special functions · Multiple/arbitrary precision computations · High performance verified computing · Computer algebra systems · C-XSC

Walter Krämer (1952–2014) studied mathematics and computer science at the University of Karlsruhe. After graduating, he was a faculty assistant and research assistant at the Institute of Applied Mathematics, University of Karlsruhe. He received his PhD in 1987 and habilitated in 1993. During 1993–1999, Prof. Dr. Krämer was managing director and visiting professor at the Institute for Scientific Computing and Mathematical Modelling at the University of Karlsruhe. From the winter semester 1999/2000 until his death, he was a full professor of Scientific Computing/Software Engineering in the Department of Mathematics, University of Wuppertal.

Professor Krämer's research was on the development of numerical methods with result verification and software tools supporting them. A major focus of the activities of W. Krämer and his research group at the University of Wuppertal was on the maintenance and further development of the open-source programming language extension C-XSC.

Here we present a tentatively complete bibliography of professor Krämer's work. The electronic version of this bibliography contains fields not listed in the printed version, including cross-references, ISBN/ISSN, keywords, some abstracts, more hyperlinks (when electronic version of the article is available), etc.
The BibTeX source is available from
    http://www2.math.uni-wuppertal.de/wrswt/pubs-WKraemer.bib
and
    http://interval.louisiana.edu/Kraemer-bib/pubs-WKraemer.bib.

© Springer International Publishing Switzerland 2016
M. Nehmeier et al. (Eds.): SCAN 2014, LNCS 9553, pp. 281–290, 2016.
DOI: 10.1007/978-3-319-31769-4

References

1. Bohlender, G., Grüner, K., Kaucher, E., Klatte, R., Krämer, W., Kulisch, U., Miranker, W.L., Rump, S.M., Ullrich, C., Wolff von Gudenberg, J.: Pascal-SC: a Pascal for contemporary scientific computation. Research Report RC 9009, IBM Thomas J. Watson Research Center, Yorktown Heights, New York (1981)
2. Bohlender, G., Böhm, H., Grüner, K., Kaucher, E., Klatte, R., Krämer, W., Kulisch, U., Miranker, W.L., Rump, S.M., Ullrich, C., Wolff von Gudenberg, J.: Matrix Pascal. Res. Rep. RC 9577, IBM Thomas J. Watson Research Center, Yorktown Heights, New York (1982) Published in [5]
3. Bohlender, G., Böhm, H., Grüner, K., Kaucher, E., Klatte, R., Krämer, W., Kulisch, U., Rump, S.M., Ullrich, C., Wolff von Gudenberg, J.: Proposal for arithmetic specification in Fortran 8x. Interner Bericht Des Instituts F. Angew. Math., Universität Karlsruhe, Karlsruhe, Germany (1982)
4. Bohlender, G., Böhm, H., Braune, K., Grüner, K., Kaucher, E., Klatte, R., Krämer, W., Kulisch, U., Miranker, W.L., Ullrich, C., Wolff von Gudenberg, J.: Application module: scientific-computation for Fortran 8x. Modified proposal for arithmetic specification according to guidelines of the X3J3-meetings in Tulsa and Chapel Hill. Interner bericht, Instituts F. Angew. Math. Universität Karlsruhe, Karlsruhe, Germany (1983)
5. Bohlender, G., Böhm, H., Grüner, K., Kaucher, E., Klatte, R., Krämer, W., Kulisch, U., Miranker, W.L., Rump, S.M., Ullrich, C., Wolff von Gudenberg, J.: Matrix Pascal. In: Kulisch, U.W., Miranker, W.L. (eds.) A new approach to scientific computation. Notes and Reports in Computer Science and Applied Mathematics, vol. 7, pp. 311–384. Academic Press, New York (1983) see also [2]
6. Bohlender, G., Böhm, H., Grüner, K., Kaucher, E., Klatte, R., Krämer, W., Kulisch, U., Miranker, W.L., Rump, S.M., Ullrich, C., Wolff von Gudenberg, J.: Arithmetic specification in Fortran 8x. In: Ford, B., Rault, J.C., Thomasset, F. (eds.) Tools, Methods and Languages For Scientific and Engineering Computation, Amsterdam, The Netherlands, pp. 213–243. Elsevier Science Publishers, North Holland (1984)
7. Krämer, W.: Dokumentation und Fehlerabschätzungen der hochgenauen ACRITH (Intervall-) Standardfunktionen (inverse Funktionen), 220p. Booklet, IBM Deutschland (1984)
8. Braune, K., Krämer, W.: Standard functions for intervals with maximum accuracy. In: Proceedings of 11th IMACS Congress on System Simulation and Scientific Computation, Oslo, Norway, August 5–9, 1985, vol. I, pp. 167–170 (1985) see also [9]
9. Braune, K., Krämer, W.: High-accuracy standard functions for intervals. In: Ruschitzka, M. (ed.) Computer Systems: Performance and Simulation, pp. 341–347. Elsevier Science Publishers B.V. (North-Holland) (1986) see also [8]
10. Krämer, W.: Fehlerabschätzungen der hochgenauen komplexen ACRITH Intervallstandardfunktionen (inverse Funktionen). Booklet, IBM Deutschland (1986)
11. Braune, K., Krämer, W.: High-accuracy standard functions for real and complex intervals. In: Kaucher, E., Kulisch, U., Ullrich, C. (eds.) Computer Arithmetic: Scientific Computation and Programming Languages, pp. 81–114. Teubner, Stuttgart (1987)
12. Krämer, W.: Inverse Standardfunktionen für reelle und komplexe Intervallargumente mit a priori Fehlerabschätzungen für beliebige Datenformate. Dissertation, Universität Karlsruhe, Karlsruhe, Germany (1987)

13. Krämer, W.: Mehrfachgenaue reelle und intervallmäßige staggered-correction Arithmetik mit zugehörigen Standardfunktionen, pp. 1–80. Bericht, Inst. f. Angew. Mathematik, Universität Karlsruhe (1988)
14. Krämer, W.: Inverse standard functions for real and complex point and interval arguments with dynamic accuracy. In: Kulisch, U., Stetter, H.J. (eds.) Scientific Computation with Automatic Result Verification. Computing Supplementum, vol. 6, pp. 185–212. Springer, Vienna (1988)
15. Cordes, D., Krämer, W.: Vom Problem zum Einschließungsalgorithmus. In: Kulisch, U. (ed.) Wissenschaftliches Rechnen mit Ergebnisverifikation, pp. 167–181. Vieweg, Braunschweig (1989)
16. Krämer, W., Walter, W.V.: FORTRAN-SC: a FORTRAN extension for engineering/scientific computation with access to ACRITH: General information notes and sample programs, pp. 1–51. Booklet, IBM Deutschland GmbH, Stuttgart (1989)
17. Krämer, W.: Fehlerschranken für häufig auftretende Approximationsausdrücke. Z. Angew. Math. Mech. **69**(4), t44–t47 (1989)
18. Krämer, W.: Berechnung der Gammafunktion $\Gamma(x)$ für reelle Punkt- und Intervallargumente. Z. Angew. Math. Mech. **70**(6), T581–T584 (1990)
19. Krämer, W.: Highly accurate evaluations of program parts with applications. In: Ullrich, C. (ed.) Contributions to Computer Arithmetic and Self-validating Numerical Methods. IMACS Annals on Computing and Applied Mathematics, vol. 7, pp. 397–410. J.C. Baltzer AG, Basel (1990)
20. Krämer, W.: Einschluß eines Paares konjugiert komplexer Nullstellen eines reellen Polynoms. Z. Angew. Math. Mech. **71**(6), T820–T824 (1991)
21. Krämer, W.: Genaue Berechnung von komplexen Polynomen in mehreren Variablen. Wissenschaftliche Zeitschrift der Technischen Hochschule Leipzig Heft 9, pp. 401–407 (1991)
22. Krämer, W.: Die Berechnung von Standardfunktionen in Rechenanlagen. In: Chatterji, S., Kulisch, U., Laugwitz, D., Liedl, R., Purkert, W. (eds.) Jahrbuch Überblicke Mathematik, pp. 97–115. Vieweg, Braunschweig (1992)
23. Krämer, W.: Computation of interval bounds for elliptic integrals. In: Herzberger, J., Atanassova, L. (eds.) Computer Arithmetic and Enclosure Methods (Oldenburg, 1991), pp. 289–298. North-Holland, Amsterdam (1992)
24. Krämer, W.: Evaluation of polynomials in several variables with high accuracy. In: Kaucher, E., Markov, S., Mayer, G. (eds.) Computer Arithmetic, Scientific Computation and Mathematical Modelling. IMACS Annals on Computing and Applied Mathematics, vol. 12, pp. 239–249. Baltzer, Basel (1991)
25. Krämer, W.: Verified solution of eigenvalue problems with sparse matrices. In: Brezinski, C., Kulisch, U. (eds.) Computational and Applied Mathematics, I. Algorithms and Theory, pp. 277–287. North-Holland, Amsterdam (1992)
26. Krämer, W.: Die Berechnung von Funktionen und Konstanten in Rechenanlagen. Habilitationsschrift, University of Karlsruhe, Karlsruhe, Germany (1993)
27. Krämer, W., Barth, B.: Computation of interval bounds for Weierstrass' elliptic function $\wp(z)$. In: Albrecht, R., Alefeld, G., Stetter, H. (eds.) Validation Numerics: Theory and Applications. Computing Supplementum, vol. 9, pp. 147–159. Springer, Wien (1993)
28. Krämer, W.: Eine portable Langzahl- und Langzahlintervallarithmetik mit Anwendungen. Z. angew. Math. Mech. **73**(7–8), T849–T853 (1993)
29. Krämer, W.: Multiple-precision computations with result verification. In: Adams, E., Kulisch, U. (eds.) Scientific Computing with Automatic Result Verification. Mathematics in Science and Engineering, vol. 189, pp. 325–356. Academic Press Inc., Boston (1993)

30. Krämer, W.: Numerische Berechnung von Schranken für π. In: Chatterji, S., Fuchssteiner, B., Kulisch, U., Liedl, R. (eds.) Jahrbuch Überblicke Mathematik, pp. 57–72. Vieweg, Braunschweig (1993)

31. Bohlender, G., Krämer, W., Miranker, W.L.: Grading of basic arithmetical operations and functions. Technical Report RC 19593 (86059), IBM Research Division, T. J. Watson Research Center (1994)

32. Krämer, W.: Bericht über die Begutachtung des IWRMM im Dezember 1993. Preprint IWRMM 94/3, IWRMM, University of Karlsruhe, Karlsruhe, Germany (1994)

33. Krämer, W., Kulisch, U., Lohner, R.: Numerical Toolbox for Verified Computing II — Advanced Numerical Problems. unpublished draft, Karlsruhe (1994)

34. Krämer, W.: Schranken für Werte der Weierstraßschen $\wp$-Funktion. Z. angew. Math. Mech. **74**(6), t660–t662 (1994)

35. Krämer, W.: Grundlagen des verifizierten numerischen Rechnens (Bases of verified computing). Z. angew. Math. Mech. **75**(Suppl 2), s425–s428 (1995)

36. Krämer, W.: Validated function evaluation using polynomial approximation from truncated Chebyshev series. Reliab. Comput. Supplementum, 113–113 (1995)

37. Hofschuster, W., Krämer, W.: Ein rechnergestützter Fehlerkalkül mit Anwendung auf ein genaues Tabellenverfahren, 35p. Preprint IWRMM 96/5, IWRMM, University of Karlsruhe, Karlsruhe, Germany (1996)

38. Krämer, W., Wedner, S.: Computing narrow inclusions for Cauchy principal value integrals. In: Alefeld, G., Frommer, A., Lang, B. (eds.) Scientific Computing and Validated Numerics, pp. 45–51. Mathematical Research, Akademie Verlag, Berlin (1996)

39. Krämer, W.: Sichere und genaue Abschätzung des Approximationsfehlers bei rationalen Approximationen. Bericht 3/1996, Forschungsschwerpunkt Computerarithmetik, Intervallrechnung und Numerische Algorithmen mit Ergebnisverifikation (CAVN), Karlsruhe, Germany (1996)

40. Krämer, W., Wedner, S.: Two adaptive Gauss-Legendre type algorithms for the verified computation of definite integrals. Reliab. Comput. **2**(3), 241–253 (1996)

41. Krämer, W., Niethammer, W., (eds.): Tagungsband zum Workshop Wissenschaftliches Rechnen in den Ingenieurwissenschaften, Februar 1996, Karlsruhe. Preprint IWRMM 96/6, IWRMM, University of Karlsruhe, Karlsruhe, Germany (1996)

42. Niethammer, W., Krämer, W. (eds.): Wissenschaftliches Rechnen in der Universität Karlsruhe. Universität Karlsruhe (1996)

43. Blomquist, F., Krämer, W.: Algorithmen mit garantierten Fehlerschranken für die Fehler- und die komplementäre Fehlerfunktion. Preprint IWRMM 97/3, IWRMM, University of Karlsruhe, Karlsruhe, Germany (1997)

44. Hofschuster, W., Krämer, W.: A computer oriented approach to get sharp reliable error bounds. Reliab. Comput. **3**(3), 239–248 (1997)

45. Hofschuster, W., Krämer, W.: A fast public domain interval library in ANSI C. In: Sydow, A. (ed.) Proceedings of the 15th IMACS World Congress on Scientific Computation, Modelling and Applied Mathematics, vol 2, pp. 395–400 (1997)

46. Krämer, W.: A priori worst-case error bounds for floating-point computations. In: Lang, T., Muller, J.M., Takagi, N. (eds.) 13th IEEE Symposium on Computer Arithmetic: Proceedings, July 6–9, 1997, Asilomar, California, USA. Symposium on Computer Arithmetic, vol. 13, pp. 64–73. IEEE Computer Society Press (1997) See also [52]

47. Krämer, W.: Eine Fehlerfaktorarithmetik für zuverlässige a priori Fehlerabschätzungen. Bericht 5/1997, Forschungsschwerpunkt Computerarithmetik, Intervallrechnung und Numerische Algorithmen mit Ergebnisverifikation (CAVN), Karlsruhe, Germany (1997)
48. Krämer, W.: Effective range computations for rational functions. In: Sydow, A., (ed.) Proceedings of the 15th IMACS World Congress on Scientific Computation, Modelling and Applied Mathematics, vol. 2, pp. 421–426 (1997)
49. Bantle, A., Krämer, W.: Ein Kalkül für verläßliche absolute und relative Fehlerabschätzungen. Preprint IWRMM 98/5, IWRMM, University of Karlsruhe, Karlsruhe, Germany (1998)
50. Hofschuster, W., Krämer, W.: FI_LIB, eine schnelle und portable Funktionsbibliothek für reelle Argumente und reelle Intervalle im IEEE-double-Format, 227p. Preprint IWRMM 98/7, IWRMM, University of Karlsruhe, Karlsruhe, Germany (1998)
51. Krämer, W.: Automatisierte a-priori-Fehlerabschätzung bei gleitkommamäßiger Ausdrucksauswertung. Z. angew. Math. Mech. 78(Suppl 3), 979–980 (1998)
52. Krämer, W.: A priori worst case error bounds for floating-point computations. IEEE Trans. Comput. 47(7), 750–756 (1998) see also [46]
53. Krämer, W.: Constructive error analysis. J.UCS 4(2), 147–163 (1998)
54. Bantle, A., Krämer, W.: Implementierung, Handhabung und Beispielanwendungen eines verlässlichen Vorwärtsfehlerkalküls. Bericht 2/1999, Forschungsschwerpunkt Computerarithmetik, Intervallrechnung und Numerische Algorithmen mit Ergebnisverifikation, Karlsruhe, Germany (1999)
55. Bräuer, M., Krämer, W.: Rückwärtsmethode zur automatischen Berechnung von worst-case Fehlerschranken. Bericht 3/1999, Forschungsschwerpunkt Computerarithmetik, Intervallrechnung und Numerische Algorithmen mit Ergebnisverifikation (CAVN), Karlsruhe, Germany (1999)
56. Büdding, G., Fausten, D., Krämer, W., Möllers, T., Traczinski, H.: Verifikation von Lösungen ausgewählter Probleme aus der Modellierung von Manipulatoren. Schriftenreihe des Fachbereichs Mathematik, Universität Duisburg, Gesamthochschule, 463 (1999)
57. Geulig, I., Krämer, W.: Intervallrechnung in Maple — Die Erweiterung intpakX zum Paket intpak der Share-Library. Report, IWRMM, University of Karlsruhe, Karlsruhe, Germany (1999)
58. Krämer, W.: Gleichmäßige (Rundungs-)Fehlerschranken für Gleitkommaalgorithmen über Datenbereichen. Z. angew. Math. Mech. 79(Suppl 1), 245–246 (1999)
59. Krämer, W.: Modifikationen eines Satzes von Ehlich/Zeller zur Wertebereichsbestimmung bei rationalen Funktionen. Z. angew. Math. Mech. 79(Suppl 3), s865–s866 (1999)
60. Bantle, A., Krämer, W.: Automatische Bestimmung von relativen worst case Fehlerschranken. Z. angew. Math. Mech. 80, S819–S820 (2000)
61. Hofschuster, W., Krämer, W.: Mathematical function software on the web – are such codes useful for verification algorithms? Reliab. Comput. 6(2), 207–218 (2000)
62. Hofschuster, W., Krämer, W.: Rechnerisches Nachvollziehen und Analyse der praktischen Umsetzung des Finanzausgleichsgesetzes. Preprint BUW-WRSWT 2000/1, Bergische Universität Wuppertal, Wuppertal, Germany (2000)
63. Kaucher, E., Krämer, W.: Lösungseinschließung bei unendlichen linearen Gleichungssystemen. Z. angew. Math. Mech. 80(Suppl 3), S823–S824 (2000)

64. Krämer, W., Blomquist, F.: Algorithms with guaranteed error bounds for the error function and the complementary error function. Preprint BUW-WRSWT 2000/2, Bergische Universität Wuppertal, Wuppertal, Germany (2000)
65. Bräuer, M., Hofschuster, W., Krämer, W.: Steigungsarithmetiken in C-XSC. Preprint BUW-WRSWT 2001/3, Bergische Universität Wuppertal, Wuppertal, Germany (2001)
66. Hofschuster, W., Krämer, W., Wedner, S., Wiethoff, A.: C-XSC 2.0 — C++ class library for extended scientific computing. Preprint BUW-WRSWT 2001/1, Bergische Universität Wuppertal, Wuppertal, Germany (2001)
67. Krämer, W., Bantle, A.: Automatic forward error analysis for floating point algorithms. Reliab. Comput. **7**(4), 321–340 (2001)
68. Krämer, W., Geulig, I.: Interval calculus in Maple — the extension intpakX to the package intpak of the share-library. Preprint BUW-WRSWT 2001/2, Bergische Universität Wuppertal, Wuppertal, Germany (2001)
69. Krämer, W., Wolff von Gudenberg, J. (eds.): Scientific Computing, Validated Numerics, Interval Methods. Kluwer Academic/Plenum, Boston/Dordrecht/London (2001)
70. Lerch, M., Tischler, G., Wolff von Gudenberg, J., Hofschuster, W., Krämer, W.: The interval library filib++ 2.0 – design, features and sample programs. Preprint BUW-WRSWT 2001/4, Bergische Universität Wuppertal, Wuppertal, Germany (2001)
71. Hofschuster, W., Krämer, W.: C-XSC 2.0 — a C++ class library for extended scientific computing. Preprint BUW-WRSWT 2002/4, Bergische Universität Wuppertal, Wuppertal, Germany (2002)
72. Krämer, W.: Advanced software tools for validated computing. In: Proceedings of the 31st Spring Conference of the Union of Bulgarian Mathematicians, pp. 344–355 (2002)
73. Miehe, D., Krämer, W., Hofschuster, W.: Visualization of resulting sets coming from multiplication and division of complex intervals. Preprint BUW-WRSWT 2002/3, Bergische Universität Wuppertal, Wuppertal, Germany (2002)
74. Hölbig, C.A., Krämer, W.: Selfverifying solvers for dense systems of linear equations realized in C-XSC. Preprint BUW-WRSWT 2003/1, Bergische Universität Wuppertal, Wuppertal, Germany (2003)
75. Krämer, W., Wolff von Gudenberg, J.: Extended interval power function. Reliab. Comput. **9**(5), 339–347 (2003)
76. Popova, E.D., Krämer, W.: Parametric fixed-point iteration implemented in C-XSC. Preprint BUW-WRSWT 2003/3, Bergische Universität Wuppertal, Wuppertal, Germany (2003)
77. Blomquist, F., Hofschuster, W., Krämer, W.: Realisierung der hyperbolischen Cotangens-Funktion in einer Staggered Correction Intervallarithmetik in C-XSC. Preprint BUW-WRSWT 2004/3, Bergische Universität Wuppertal, Wuppertal, Germany (2004)
78. Blomquist, F., Hofschuster, W., Krämer, W.: Sichere a priori Abschätzungen und Realisierung der Funktion $\sqrt{x^2 - 1}$ in C-XSC. Preprint BUW-WRSWT 2004/5, Bergische Universität Wuppertal, Wuppertal, Germany (2004)
79. Blomquist, F., Hofschuster, W., Krämer, W.: Sichere a priori Fehlerabschätzung und Implementierung der Funktion zweier Variabler $\log(\sqrt{x^2 + y^2})$. Preprint BUW-WRSWT 2004/1, Bergische Universität Wuppertal, Wuppertal, Germany (2004)

80. Hofschuster, W., Krämer, W.: C-XSC 2.0 — a C++ library for extended scientific computing. In: Alt, R., Frommer, A., Kearfott, R., Luther, W. (eds.) Numerical Software with Result Verification. LNCS, vol. 2991, pp. 15–35. Springer, Berlin (2004)

81. Hölbig, C.A., Krämer, W., Diverio, T.A.: An accurate an efficient selfverifying solver for systems with banded coefficient matrix. In: Joubert, G.R., Nagel, W.E., Peters, F.J., Walter, W.V. (eds.) Parallel Computing: Software Technology, Algorithms, Architectures and Applications, PARCO 2003. Advances in Parallel Computing, vol. 13, pp. 283–290. Elsevier, Amsterdam (2004)

82. Krämer, W., Popova, E.D.: Zur Berechnung von verlässlichen Außen- und Inneneinschließungen bei parameterabhängigen linearen Gleichungssystemen. Proc. Appl. Math. Mech. (PAMM) 4(1), 670–671 (2004)

83. Blomquist, F., Krämer, W.: Computing a priori error bounds for floating-point evaluations of arithmetic expressions. Preprint BUW-WRSWT 2006/1, Bergische Universität Wuppertal, Wuppertal, Germany (2006)

84. Blomquist, F., Hofschuster, W., Krämer, W., Neher, M.: Complex interval functions in C-XSC. Preprint BUW-WRSWT 2005/2, Bergische Universität Wuppertal, Wuppertal, Germany (2005)

85. Blomquist, F., Hofschuster, W., Krämer, W.: Reliable computation of the complex interval function arcsin(z) in C-XSC. Preprint BUW-WRSWT 2005/1, Bergische Universität Wuppertal, Wuppertal, Germany (2005)

86. Blomquist, F., Hofschuster, W., Krämer, W.: Real and complex Taylor arithmetic in C-XSC. Preprint BUW-WRSWT 2005/4, Bergische Universität Wuppertal, Wuppertal, Germany (2005)

87. Krämer, W.: Generalized intervals and the dependency problem. Proc. Appl. Math. Mech. (PAMM) 6(1), 683–684 (2006)

88. Krämer, W.: Pitfalls in Maple. Preprint BUW-WRSWT 2006/7, Bergische Universität Wuppertal, Wuppertal, Germany (2006)

89. Lerch, M., Tischler, G., Wolff von Gudenberg, J., Hofschuster, W., Krämer, W.: FILIB++, a fast interval library supporting containment computations. ACM Trans. Math. Software 32(2), 299–324 (2006)

90. Blomquist, F., Hofschuster, W., Krämer, W.: Vermeidung von Über- und Unterlauf und Verbesserung der Genauigkeit bei reeller und komplexer staggered Intervall-Arithmetik. Preprint BUW-WRSWT 2007/8, Bergische Universität Wuppertal, Wuppertal, Germany (2007)

91. Grimmer, M., Krämer, W.: An MPI extension for verified numerical computations in parallel environments. In: Arabnia, H.R., Yang, J.Y., Yang, M.Q. (eds.) Proceedings of the 2007 International Conference on Scientific Computing, CSC 2007, June 25–28, 2007, Las Vegas, Nevada, USA, pp. 111–117. CSREA Press (2007)

92. Krämer, W.: Accurate computation of chaotic dynamical systems. In: Aggarwal, A. (ed.) Proceedings of the 8th WSEAS International Conference on Mathematics and Computers in Biology and Chemistry, MCBC 2007, pp. 74–79. WSEAS (2007)

93. Krämer, W.: Bugs, errors, and unexpected results in computer algebra packages. Proc. Appl. Math. Mech. (PAMM) 7(1), 2140009–2140010 (2007)

94. Krämer, W.: Computing and visualizing solution sets of interval linear systems. Serdica J. Comput. 1(4), 455–468 (2007)

95. Krämer, W.: intpakX — an interval arithmetic package for Maple. In: International Symposium on Scientific Computing, Computer Arithmetic and Validated Numerics, 2006. SCAN 2006. 12th GAMM - IMACS, IEEE Computer Society Press (2007) (Article number 4402417)

96. Krämer, W.: Introduction to the Maple power tool intpakX. Serdica J. Comput. **1**(4), 469–504 (2007)

97. Popova, E.D., Krämer, W.: Inner and outer bounds for the solution set of parametric linear systems. J. Comput. Appl. Math. **199**(2), 310–316 (2007)

98. Blomquist, F., Hofschuster, W., Krämer, W.: A modified staggered correction arithmetic with enhanced accuracy and very wide exponent range. In: Cuyt, A.A.M., Krämer, W., Luther, W., Markstein, P.W. (eds.) Numerical Validation in Current Hardware Architectures. LNCS, vol. 8021, pp. 41–67. Springer, Berlin (2008) see also [107]

99. Blomquist, F., Hofschuster, W., Krämer, W.: Real and complex staggered (interval) arithmetics with wide exponent range (in German). Preprint BUW-WRSWT 2008/1, Bergische Universität Wuppertal, Wuppertal, Germany (2008)

100. Cuyt, A.A.M., Krämer, W., Luther, W., Markstein, P.W.: Summary — numerical validation in current hardware architectures. In: Cuyt, A.A.M., Krämer, W., Luther, W., Markstein, P.W. (eds.) Numerical Validation in Current Hardware Architectures. LNCS, vol. 08021. Springer, Berlin (2008) see also [108]

101. Grimmer, M., Krämer, W.: An open source parallel interval solver for systems of linear Fredholm integral equations of the second kind. In: Arabnia, H.R. (ed.) Proceedings of the 2008 International Conference on Scientific Computing, CSC 2008, July 14–17, 2008, Las Vegas, Nevada, USA, pp. 204–210. CSREA Press (2008)

102. Hofschuster, W., Krämer, W., Neher, M.: C-XSC and closely related software packages. In: Cuyt, A., Krämer, W., Luther, W., Markstein, P. (eds.) Numerical Validation in Current Hardware Architectures. LNCS, vol. 8021. Springer, Berlin (2008) see also [109]

103. Kolberg, M., Krämer, W., Zimmer, M.: A note on solving problem 7 of the SIAM 100-digit challenge using C-XSC. In: Cuyt, A., Krämer, W., Luther, W., Markstein, P. (eds.) Numerical Validation in Current Hardware Architectures. LNCS, vol. 8021. Springer, Berlin (2008) see also [110]

104. Luther, W., Cuyt, A.A.M., Krämer, W., Markstein, P.W.: Abstracts collection - numerical validation in current hardware architectures. In: Cuyt, A.A.M., Krämer, W., Luther, W., Markstein, P.W. (eds.) Numerical Validation in Current Hardware Architectures. LNCS, vol. 8021. Springer, Berlin (2008) see also [108]

105. Popova, E.D., Krämer, W.: Visualizing parametric solution sets. BIT **48**(1), 95–115 (2008)

106. Zimmer, M., Krämer, W.: Fast (parallel) dense linear interval systems solvers in C-XSC using error free transformations and BLAS. In: Cuyt, A.A.M., Krämer, W., Luther, W., Markstein, P.W. (eds.) Numerical Validation in Current Hardware Architectures. LNCS, vol. 8021. Springer, Berlin (2008) see also [111]

107. Blomquist, F., Hofschuster, W., Krämer, W.: A modified staggered correction arithmetic with enhanced accuracy and very wide exponent range. In: Cuyt, A.A.M., Krämer, W., Luther, W., Markstein, P.W. (eds.) Numerical Validation in Current Hardware Architectures.. LNCS, vol. 5492, pp. 41–67. Springer, Berlin (2008) see also [98]

108. Cuyt, A., Krämer, W., Luther, W., Markstein, P. (eds.): Numerical Validation in Current Hardware Architectures. LNCS, vol. 5492. Springer, Berlin (2009) see [100, 104]

109. Hofschuster, W., Krämer, W., Neher, M.: C-XSC and closely related software packages. In: Cuyt, A.A.M., Krämer, W., Luther, W., Markstein, P.W. (eds.) Numerical Validation in Current Hardware Architectures. LNCS, vol. 5492, pp. 68–102. Springer, Berlin (2009) see also [102]

110. Kolberg, M.L., Krämer, W., Zimmer, M.: A note on solving problem 7 of the SIAM 100-digit challenge using C-XSC. In: Cuyt, A.A.M., Krämer, W., Luther, W., Markstein, P.W. (eds.) Numerical Validation in Current Hardware Architectures. LNCS, vol. 5492, pp. 250–261. Springer, Berlin (2009) see also [103]
111. Krämer, W., Zimmer, M.: Fast (parallel) dense linear system solvers in C-XSC using error free transformations and BLAS. In: Cuyt, A.A.M., Krämer, W., Luther, W., Markstein, P.W. (eds.) Numerical Validation in Current Hardware Architectures. LNCS, vol. 5492, pp. 230–249. Springer, Berlin (2009) see also [106]
112. Krämer, W.: Computer-assisted proofs and symbolic computations. Serdica J. Comput. 4(1), 73–84 (2010)
113. Krämer, W.: High performance verified computing using C-XSC. In: Reiser, R.H.S., Pilla, M.L. (eds.) IntMath-TSD: Interval Mathematics and Connections in Teaching and Scientific Development, pp. 3–14. Universidade Federal de Pelotas, Ed. Universitria (2010)
114. Krämer, W.: Verification methods and symbolic computations. Albanian J. Math. 4(4), 123–133 (2010)
115. Popova, E.D., Krämer, W.: Communicating functional expressions from *Mathematica* to C-XSC. In: Fukuda, K., Hoeven, J.v.d., Joswig, M., Takayama, N. (eds.) Mathematical Software, ICMS 2010. LNCS, vol. 6327, pp. 354–365. Springer, Berlin (2010)
116. Popova, E.D., Krämer, W., Russev, M.: Integration of C-XSC automatic differentiation in Mathematica. Preprint BUW-WRSWT 2010/1, Bergische Universität Wuppertal, Wuppertal, Germany (2010)
117. Popova, E.D., Kolev, L., Krämer, W.: A solver for complex-valued parametric linear systems. Serdica J. Comput. 4(1), 123–132 (2010)
118. Zimmer, M., Krämer, W., Bohlender, G., Hofschuster, W.: Extension of the C-XSC library with scalar products with selectable accuracy. Serdica J. Comput. 4(3), 349–370 (2010)
119. Zimmer, M., Krämer, W., Hofschuster, W.: Sparse matrices and vectors in C-XSC. Reliab. Comput. 14(1), 138–160 (2010)
120. Blomquist, F., Hofschuster, W., Krämer, W.: C-XSC-Langzahlarithmetiken für reelle und komplexe Intervalle basierend auf den Bibliotheken MPFR und MPFI. Preprint BUW-WRSWT 2011/1, Bergische Universität Wuppertal, Wuppertal, Germany (2011)
121. Kolberg, M., Krämer, W., Zimmer, M.: Efficient parallel solvers for large dense systems of linear interval equations. Reliab. Comput. 15(3), 193–206 (2011)
122. Krämer, W.: C-XSC: a powerful environment for reliable computations in the natural and engineering sciences. In: Ding, Y., et al. (ed.) 2011 4th International Conference on Biomedical Engineering and Informatics (BMEI), vol. 4, pp. 2130–2134. IEEE (2011)
123. Krämer, W.: C-XSC, a sophisticated environment for reliable computing. In: Ratschan, S. (ed.) Proceedings of the 4-th International Conference on Mathematical Aspects of Computer and Information Sciences (MACIS 2011), pp. 115–125 (2011)
124. Popova, E.D., Krämer, W.: Characterization of AE solution sets to a class of parametric linear systems. C. R. Acad. Bulgare Sci. 64(3), 325–332 (2011)
125. Popova, E.D., Krämer, W.: Embedding C-XSC nonlinear solvers in Mathematica. C. R. Acad. Bulgare Sci. 64(1), 11–20 (2011)
126. Blomquist, F., Hofschuster, W., Krämer, W.: Umfangreiche C-XSCLangzahlpakete für beliebig genaue reelle und komplexe Intervallrechnung. Preprint BUW-WRSWT 2012/2, Bergische Universität Wuppertal, Wuppertal, Germany (2012)

127. Krämer, W., Blomquist, F.: Arbitrary precision complex interval computations in C-XSC. In: Wyrzykowski, R., Dongarra, J., Karczewski, K., Wasniewski, J. (eds.) Parallel Processing and Applied Mathematics. LNCS, vol. 7204, pp. 457–466. Springer, Berlin (2012)
128. Krämer, W.: Multiple/arbitrary precision interval computations in C-XSC. Computing **94**(2–4), 229–241 (2012)
129. Krämer, W., Zimmer, M., Hofschuster, W.: Using C-XSC for high performance verified computing. In: Jónasson, K. (ed.) Applied Parallel and Scientific Computing. LNCS, vol. 7134, pp. 168–178. Springer, Berlin (2012)
130. Zimmer, M., Krämer, W., Popova, E.D.: Solvers for the verified solution of parametric linear systems. Computing **94**(2–4), 109–123 (2012)
131. Krämer, W.: High performance verified computing using C-XSC. Comput. Appl. Math. **32**(3), 385–400 (2013)
132. Zimmer, M., Rebner, G., Krämer, W.: An overview of C-XSC as a tool for interval arithmetic and its application in computing verified uncertain probabilistic models under Dempster-Shafer theory. Soft Comput. **17**(8), 1453–1465 (2013)

rence value as well as on the player's payoff function. For more examples with different preference values and payoff functions, please visit http://gametheory.cs. cinvestav.mx/Examples_AF_Analysis_of_Strategies.pdf.

5   Discussion

The analysis of strategies in American football gaming can be flexible done using both, the players' preference values on the plays and depending on specific circumstances on the match. Once the players' preference values in Table 3 – 4 are modified, we can obtain different payoff values, therefore other strategy profiles that satisfy NE conditions for the different valuations. This is a real behavior approach: the individual player's preferences and assessments regulate the actions decision making during a match, at least partially. The strategy profiles can be approached by computer simulations of American football gaming [10] and measure how good are these strategy profiles to improve the team performance.

6   Conclusion

In simulations of American football gaming, the analysis of strategies can use the normal form game setting, so the strategy profiles are valued using the every player's payoff functions to find out those that fit the Nash equilibrium conditions for the whole team. These Nash equilibrium strategy profiles can be used for actions decision making on the basis of a full strategic reasoning, essential for a team's success.

Acknowledgments. To the Mexican Council of Science and Technology (Conacyt) by the A. Yee's PhD degree grant 261089, and the R. Rodríguez's MSc grant 555159.

References

1. Baker, R.D., McHale, I.G.: Forecasting exact scores in National Football League games. Int. J. Forecasting 29, 122–130 (2013)
2. Gonzalez, A.J., Gross, D.L.: Learning tactics from a sports game-based simulation. Int. J. Comput. Simul. 5, 127–148 (1995)
3. Song, C., Boulier, B.L., Stekler, H.O.: The comparative accuracy of judgmental and model forecasts of American football games. Int. J. Forecasting 23, 405–413 (2007)
4. Alaways, L.W., Hubbard, M.: Experimental determination of baseball spin and lift. J. Sport Sci. 19, 349–358 (2001)
5. Jinji, T., Sakurai, S., Hirano, Y.: Factors determining the spin axis of a pitched fastball in baseball. J. Sport Sci. 29, 761–767 (2011)
6. MacMahon, C., Starkes, J.L.: Contextual influences on baseball ball-strike decisions in umpires, players, and controls. J. Sport Sci. 26, 751–760 (2008)
7. Alvarado, M., Rendón, A.Y.: Nash equilibrium for collective strategic reasoning. Expert Syst. Appl. 39, 12014–12025 (2012)

8. Deutsch, S.J., Bradburn, P.M.: A simulation model for American football plays. Appl. Math. Model. 5, 13–23 (1981)
9. McGrew, A.G., Wilson, M.J.: Decision making: approaches and analysis. Manchester University Press (1982)
10. Alvarado, M., Yee, A., Fernández, J.: Simulation of American football gaming. Advances in Sport Science and Computer Science 57, 227 (2014)
11. American Football Coaches Association: Offensive Football Strategies. Human Kinetics, Champaign, IL (2000)
12. Camp, W., Badgley, C.S.: American Football. Createspace Independent Pub. (2009)
13. Gifford, C.: American Football Tell me about Sport. Evans Publishing, London (2009)
14. Nash, J.: Non-Cooperative Games. The Annals of Mathematics 54, 286–295 (1951)
15. http://www.nfl.com/

Strategies for Reducing the Complexity of Symbolic Models for Activity Recognition

Kristina Yordanova, Martin Nyolt, and Thomas Kirste

University of Rostock,
18055 Rostock, Germany
{kristina.yordanova,martin.nyolt,thomas.kirste}@uni-rostock.de

Abstract. Recently, in the field of activity recognition a number of approaches that utilise probabilistic symbolic models have been proposed. Such approaches rely on the combination of symbolic state-space models and probabilistic inference techniques in order to recognise the user activities in situations with uncertainty. One problem with such approaches is the huge state space that can be generated just by a few rules. In this work we investigate the effects of a mechanism for reducing the model complexity on symbolic level. To illustrate the approach, we present one possible strategy and discuss its effects on the model size and the probability of selecting the correct action in an office scenario.

Keywords: symbolic human behaviour models, context awareness, activity recognition.

1 Introduction and Motivation

Recently, a number of emerging approaches for context-aware activity and intention recognition that utilise probabilistic symbolic human behaviour models have been proposed [3–6, 8]. Such approaches encode prior knowledge about the user behaviour in the form of rules that are later expanded to form the model state space. To perform activity or intention recognition the transitions between the states are assigned probabilities in order to cope with ambiguous data and missing or erroneous sensor readings. For example, Hiatt et al. use the ACT-R production system [1] to which probabilistic simulation analysis is performed to determine the different execution paths and their probabilities. Alternatively, Ramírez et al. use the Planning Domain Definition Language (PDDL) to encode the user behaviour in the form of precondition-effect action templates. Later, the model is mapped onto a partially observable Markov decision process and the transition probabilities are assigned based on the distance from the current state to the goal [6]. A similar approach is the one proposed by Krüger et al. where a PDDL-like notation is used but the model is mapped either onto a Hidden Markov Model or particle filter [4].

One disadvantage in such kind of approaches is that it is easy to generate huge models by defining several rules. For example, Ramírez et al. present the

G. Agre et al. (Eds.): AIMSA 2014, LNAI 8722, pp. 295–300, 2014.

recognition of kitchen tasks where 16 actions are considered, the corresponding rules for which generate almost 70 000 model states [6]. This increases the number of states that are reachable from the current state and the number of valid execution sequences thus requiring more options for the next action to be executed. This often causes problems in recognising the correct action especially in cases where the observations provide ambiguous information.

To solve this problem there are two options. The first one is on the probabilistic level – to introduce better heuristics for calculating the probabilities of the transitions between the different states. Theoretically, that will result in only the actions representing the actual user behaviour being selected. Such heuristics are the goal distance [2, 6], the ACT-R cognitive heuristics [1, p. 132–137], the landmarks [7] etc.

The second solution is on the modelling level. In it different modelling strategies can be used to reduce the model size. This in turn results in less reachable states and less possible plans which increase the probabilities assigned to the actions. In this work we investigate how modelling strategies influence the model size and dynamics and practically test whether they are able to increase the probabilities of the correct actions. To do that in Section 2 we present a more formal definition of symbolic state space models for activity recognition. In Section 3, the mechanism for reducing the model complexity on the symbolic level is discussed and later an example is given to to illustrate its effects on the model. Finally, the paper concludes with a short discussion and future work (Section 4).

2   Symbolic State Space Models

Symbolic state space models describe the underlying system behaviour in terms of states and transitions between these states. Representatives of this kind of models are the combination of PDDL with probabilistic reasoning [4, 6], or the combination of ACT-R with probabilistic reasoning [3]. Such models have an initial world state and a goal state that the system strives to reach by making transitions from state to state. More formally, a symbolic state space model M is a tuple (Pr, S, A, Pl), where Pr is the set of predicates which are boolean functions that provide statements about the model world state. The predicates in the model then build up the model states. Given a state $s \in S$ where S and if we have 3 predicates in the model each of which can be either true or false, then a possible state will be $s = (true, true, false)$. There is one special state in the model s_0 which is the initial model state. Then there is a special subset of states $g \subseteq S$ called the goal states which represent all the predicates that have to hold in order the goal to be reached. The model's state space S is the set of all valid combinations of the existing predicates that lead from the initial to the goal state. All states that do not lead to the goal state are not counted. To reach from one state to another the model has transitions between states which are denoted by A. In the context of activity and intention recognition we call them *actions*. For two states $s, s' \in S$ we say that s' is reachable from s

Author Index

by $a \in A$, if the result of applying a in s is s'. We denote that with $s' = a(s)$. $Pl = \{p_1, \ldots, p_n\}$ is the set of all possible plans from the initial to the goal state, where $p_i = (a_1, \ldots, a_m)$ for a goal $g \subseteq S$ and an initial state $s_0 \in S$ is a finite sequence of actions a_j such that $s_l \in g$ where $s_l = a_m(\cdots a_2(a_1(s_0))\cdots)$.

Assigning probabilistic structure of symbolic state space models is not discussed here as it is not the focus of this work.

3   Strategies for Reducing the Model Complexity

The basic mechanism behind reducing the model size is the lock predicates that are used as constraints. The locks used here allow or block the access to a certain resource in the shared environment when more than one agents act in the environment, or when the execution of different actions has influence on the shared environment. This results in only a small subset of all possible states of the model being reachable, and only a subset of the plans being valid. Locks are implemented in the action's preconditions and effects, where they are defined like any other predicate.

To illustrate how the locks reduce the model complexity below we present an example of a behaviour pattern, namely phases. Fig. 1 shows the idea behind

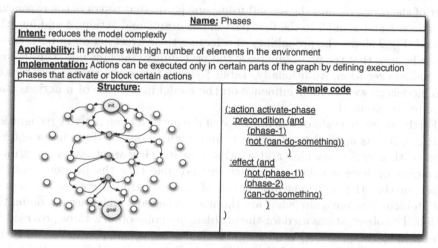

Name: Phases
Intent: reduces the model complexity
Applicability: in problems with high number of elements in the environment
Implementation: Actions can be executed only in certain parts of the graph by defining execution phases that activate or block certain actions

Structure:

Sample code

```
(:action activate-phase
   :precondition (and
      (phase-1)
      (not (can-do-something))
   )
   :effect (and
      (not (phase-1))
      (phase-2)
      (can-do-something)
   )
)
```

Fig. 1. A pattern describing the modelling of phases

them. They aim at reducing the model complexity by restricting the execution of certain actions only to parts of the state space graph by forcing the model to pass through the phases states. In that manner, the number of states that lead from the initial to the goal state is reduced, in the same time increasing the probability of selecting the correct action. It can be seen in the sample code, that the action representing the phase uses a predicated called *can-do-something*.

When the action is executed, it allows the execution of all the remaining actions that have this predicate in their preconditions, which was previously impossible. In that manner the actions can be restricted to certain phase of the behaviour execution which is essential for problems where many objects are involved and where only certain instantiations of the actions are applicable to the objects. To show the influence of the mechanism, a dummy example was built where

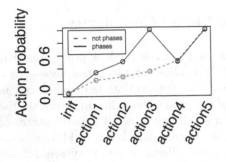

Fig. 2. Probability of selecting the correct action

Table 1. M_1 does not use phases, while M_2 is a model with phases

Param.	M_1	M_2	for m actions needed to reach goal	
			M_1	M_2
A	5	5	n	n
Pr	10	10	$2n$	$2n$
S	32	12	2^n	$\sum_{i\in I} 2^{|X_i|}$
Pl	120	4	$n!$	$p! \prod_{i\in I}(|X_i|-1)!$

five actions have to be executed in order to reach the goal state. To solve the problem, two models were built: the first allowing the actions execution in any part of the problem, and the second using one phase that restricts the execution of actions 1 and 2 and 3 to the first part of the graph, and actions 4 and 5 to the second. Fig 2 shows the probability of selecting the correct action in both cases. It can be seen that the phases increase the probability (the blue line) throughout the model execution. Additionally, Table 1 shows the model parameters for the two models as well as their influence on the model in the case of n actions that have to be executed.

Furthermore, to evaluate the influence of the pattern to an activity recognition problem, it was applied to an office scenario where one to three users enter a room with a coffee machine and a printer in it. They want either to print a document or have a coffee. It is however possible that the printer is out of paper, or that the coffee machine is out of water or ground coffee. The aim of the problem is to recognise what are the users doing and which user is doing the action. The observations used for the problem are noisy observations provided by sensors in the carpet. They send a signal whenever a person is standing on them. The formalism used to model the problem is Computational Causal Behaviour Models [4]. Fig. 3 compares the number of plans, number of states, and the achieved accuracy for intuitive model and one using phases. The results are for one to three users acting in the environment. It can be seen that the number of plans and states is seriously decreased especially for the case of 3 users acting in the environment, where the states were reduced from more than nine hundred million to under a million. This also increased the model accuracy in the case of applying phases, which stands to show that reducing the model complexity increases the model performance.

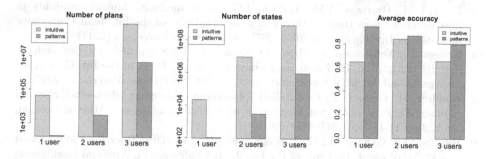

Fig. 3. Comparison between intuitive model and such using phases to reduce the model complexity

4 Discussion and Conclusion

In this work we presented a mechanism for reducing the model complexity of probabilistic symbolic models on the symbolic level. We showed that it is possible to decrease the model size and to increase the probability of selecting the correct actions by introducing lock mechanisms. Through them it is possible to decrease the size of the model. This in turn leads to increasing the probability of selecting the correct action as there are less available actions that can be executed from a given state.

Applying such mechanism to the model could be essential in improving its ability to recognise the user behaviour. It also allows dealing with the problem on symbolic level as opposed to probabilistic level, where in order to select the correct heuristics one should have a solid background knowledge on the user cognition and behaviour patterns.

One problem that could arise with reducing the model complexity, is that the behaviour variability the model is able to explain is also reduced. This indicates that models utilising locks have less predictive power than general models. Restricting the model too much could lead to model overfitting where it is able to explain just a small fraction of the behaviour variations. On the other hand, designing a more general model leads to the problem of reducing the probability of recognising the correct action. In that sense, unless good action selection heuristics are introduced, one should find the middle ground between the model's ability to explain behaviour variability, and the model's performance. In the future, the effect of different strategies on real activity recognition problems is to be tested.

References

1. Anderson, J.R.: The Architecture of Cognition. Harvard University Press, Cambridge (1983)
2. Baker, C.L., Saxe, R., Tenenbaum, J.B.: Action understanding as inverse planning. Cognition 113(3), 329–349 (2009)

3. Hiatt, L.M., Harrison, A.M., Trafton, J.G.: Accommodating human variability in human-robot teams through theory of mind. In: Proc. of the Int. Joint Conference on Artificial Intelligence, pp. 2066–2071. AAAI Press, Barcelona (2011)
4. Krüger, F., Yordanova, K., Hein, A., Kirste, T.: Plan synthesis for probabilistic activity recognition. In: Filipe, J., Fred, A.L.N. (eds.) Proc. of the Int. Conference on Agents and Artificial Intelligence, pp. 283–288. SciTePress, Barcelona (2013)
5. Ramírez, M., Geffner, H.: Probabilistic plan recognition using off-the-shelf classical planners. In: Proc. of the Nat. Conference of Artificial Intelligence, Atlanta, Georgia, USA, July 11-15, pp. 1211–1217 (2010)
6. Ramírez, M., Geffner, H.: Goal recognition over POMDPs: Inferring the intention of a POMDP agent. In: Proc. of the Int. Joint Conference on Artificial Intelligence, pp. 2009–2014. AAAI Press, Barcelona (2011)
7. Richter, S., Westphal, M.: The LAMA planner: Guiding cost-based anytime planning with landmarks. Journal of Artificial Intelligence Research 39(1), 127–177 (2010)
8. Trafton, J.G., Hiatt, L.M., Harrison, A.M., Tamborello, F.P., Khemlani, S.S., Schultz, A.C.: ACT-R/E: An embodied cognitive architecture for human-robot interaction. Journal of Human-Robot Interaction 2(1), 30–55 (2013)

Author Index

Printed in the United States
by Baker & Taylor Publisher Services

Printed in the United States
by Baker & Taylor Publisher Services